AVANT-PROPOS

Les travaux de recherche et de développement en vue d'évaluer les performances à long terme des systèmes de stockage des déchets radioactifs constituent à l'heure actuelle l'un des aspects les plus importants de la gestion des déchets radioactifs. Les activités liées à l'évaluation des performances occupent une place importante dans le programme en cours du Comité AEN de la gestion des déchets radioactifs (RWMC). On a reconnu qu'il s'agissait là d'un domaine prioritaire en rapide évolution dans lequel se fait fortement sentir le besoin d'une coopération et d'une coordination internationales. Afin de concentrer les travaux sur les sujets les plus appropriés dans ce domaine, un "Groupe consultatif sur l'évaluation des performances des systèmes d'évacuation" (PAAG) a été constitué en vue de formuler des orientations et des avis à l'intention du RWMC concernant ces questions. Le PAAG se compose de représentants des pays Membres occupant des positions clefs dans les programmes nationaux d'évaluation des performances, notamment au sein des organismes publics et privés chargés de ces programmes ainsi que des organes réglementaires. Ces travaux ont pour finalité de promouvoir la qualité et la crédibilité des techniques d'évaluation de la sûreté applicables au stockage des déchets radioactifs.

Parmi les activités menées actuellement par l'AEN dans ce domaine figurent l'établissement de bases de données géochimiques, l'action en faveur de la mise au point de modèles et de programmes de calcul ainsi que de l'exécution de travaux de vérification et de validation, le patronage de projets internationaux de recherche et de développement et l'organisation de réunions de travail, groupes de travail et cours sur des questions d'actualité choisies.

Le PAAG a recommandé à l'AEN d'organiser une réunion de travail sur les études et la modélisation du comportement des phénomènes au voisinage immédiat (champ proche) des dépôts de déchets de faible et de moyenne activité susceptibles d'influer sur le rejet de radionucléides dans l'environnement. De nombreux pays ont actuellement engagé un processus de planification ou de construction de dépôts pour des déchets de faible et de moyenne activité et l'évaluation des performances du champ plus proche retient considérablement l'attention.

On trouvera dans le présent compte rendu, le texte des communications soumises pour présentation lors de la réunion de travail, ainsi qu'un résumé et des conclusions établis par le Secrétariat de l'AEN et les présidents de séances sur la base des débats qui ont eu lieu lors de la réunion. Les opinions exprimées sont celles des auteurs et n'engagent en aucune façon les pays Membres de l'OCDE.

SOME OTHER NEA PUBLICATIONS

QUELQUES AUTRES PUBLICATIONS DE L'AEN

Shallow Land Disposal of Radioactive Waste *Reference levels for the Acceptance of long-lived Radionuclides,* **1987** (A Report by an NEA Expert Group

Enfouissement des déchets radioactifs à faible profondeur *Niveaux de référence pour l'admission des radionucléides à vie longue,* **1987**
(Rapport d'un groupe d'experts de l'AEN)

Free on request – Gratuit sur demande

Uncertainty Analysis for Performance Assessments of Radioactive Waste Disposal Systems
(Proceedings of the Seattle Workshop, 1987)

Analyse des incertitudes dans l'évaluation des performances des systèmes d'évacuation des déchets radioactifs
(Compte rendu d'une réunion de travail de Seattle-1987)

£12.00 US$25.00 F120.00 DM52.00

System Performance Assessments for Radioactive Waste Disposal, 1986
(Proceedings of an NEA Workshop, Paris, 1985)

Evaluation des performances des systèmes d'évacuation des déchets radioactifs, 1986
(compte rendu d'une réunion de travail de l'AEN, Paris 1985)

ISBN 92-64-02831-5

£12.00 US$24.00 F120.00 DM53.00

Radioactive Waste Disposal – *In Situ Experiments in Granite*
(Proceedings of the 2nd NEA/Stripa Project Symposium, Stockholm 1985)

Évacuation des déchets radioactifs – *Expériences in situ dans du granite*
(Compte rendu du 2e Symposium AEN/Projet de Stripa, Stockholm 1985)

ISBN 92-64-02728-9

£16.80 US$34.00 F168,00 DM75.00

The Radiological Impact of the Chernobyl Accident in OECD Countries
(Report by an NEA Group of Experts, 1987)

Les incidences radiologiques de l'accident de Tchernobyl dans les pays de l'OCDE
(Rapport établi par un Groupe d'experts de l'AEN, 1987)

£16.40 US$31.00 F140,00 DM60.00

Chernobyl and the Safety of Nuclear Reactors in OECD Countries (1987)
ISBN 92-64-12975-8

Tchernobyl et la sûreté des réacteurs nucléaires dans les pays de l'OCDE (1987)
ISBN 92-64-22975-2

£11.00 US$23.00 F110,00 DM47.00

ADDENDUM

PROCEEDINGS OF AN NEA WORKSHOP ON NEAR-FIELD ASSESSMENT OF REPOSITORIES FOR LOW AND MEDIUM LEVEL RADIOACTIVE WASTE

COMPTE RENDU D'UNE RÉUNION DE TRAVAIL DE L'AEN SUR L'ÉVALUATION DU CHAMP PROCHE DES DÉPÔTS DE DÉCHETS RADIOACTIFS DE FAIBLE ET MOYENNE ACTIVITÉ

BADEN
SWITZERLAND / SUISSE
23-25 Nov. 1987

Organised by the
OECD NUCLEAR ENERGY AGENCY
in co-operation with the
National Cooperative for the Storage
of Radioactive Waste, NAGRA, Switzerland

Organisée par
L'AGENCE DE L'OCDE POUR L'ÉNERGIE NUCLÉAIRE
en coopération avec
La Société Coopérative Nationale pour
l'Entreposage de Déchets Radioactifs, CEDRA, Suisse

and increase solubilities. More generally, it is desirable to use well understood materials in order to reduce uncertainties in risk estimates.

The remainder of this paper deals exclusively with the disposal of LLW and ILW in deep underground repositories. Section 2 presents an overview of the near-field safety case, section 3 describes the status of near-field assessment modelling and section 4 gives some typical results. Finally, the outlook for the future is summarised in section 5.

2. OVERVIEW OF NEAR-FIELD SAFETY CASE

The normal evolution scenario for the near-field of a deep cementitious repository is as follows. A schematic view of the near-field evolution is shown in fig. 1. The steel containment is expected to remain effective for some hundreds of years in the alkaline environment induced by the concrete, assuming that the repository is saturated soon after closure[11]. The corrosion products from steel reinforcement and canisters are expected to ensure chemically reducing conditions in the concrete pore water in advance of canister failure[12]. Also, during the first few hundred years the concrete acts as a physical and chemical barrier to incoming corrodants and as an extra independent barrier to radionuclide. During this physical containment period short-lived radionuclides, with half-lives less than about thirty years, will decay substantially.

Eventually, it will no longer be prudent to rely on the physical integrity of concrete and steel barriers. However, they will still perform an extremely important role by chemically conditioning the pore water to be alkaline and reducing for very long periods of time[12,13]. Under these conditions, the aqueous concentrations of many long-lived radionuclides are severely restricted. In addition, sorption of radionuclides on near-field materials plays an equally important part in reducing pore water concentrations.

The slow water flow through a repository, which is guaranteed by a suitable choice of host geology, allows chemical equilibrium to be established in a repository. it is felt that arguments based on chemical equilibrium are far more robust and reliable than those which make use of transfer kinetics such as leaching rates.

Thus the rate of release of radionuclides in groundwater is zero for at least several hundred years, and is then determined by equilibrium concentrations and diffusion and/or advection in slowly moving water.

There are a number of phenomena which could modify the above picture. For example, gas generation from the anaerobic corrosion of metals and the degradation of organic materials could lead to an enhanced fracture permeability of the concrete as discussed elsewhere in this Workshop[7]. The outflow of gas will clearly effect the time to completely resaturate the repository. Also, the degradation of organic materials in the waste produces complexing agents which could enhance radionuclide solubilities and reduce sorption[8]. Natural humic and fulvic acids entering the repository from the surrounding geology, and other colloids, could also lead to an increase in the mobility of some radionuclides.

All of these topics are being addressed with the Nirex Safety Assessment Research Programme[3-8,14].

TABLE OF CONTENTS
TABLE DES MATIERES

Session I - Séance I

KEY ISSUES IN DESIGN AND ASSESSMENT OF THE NEAR-FIELD IN L/ILW REPOSITORIES

PRINCIPAUX PROBLEMES DANS LA CONCEPTION ET L'EVALUATION DU CHAMP PROCHE DANS LES DEPOTS DE DECHETS DE FAIBLE ET MOYENNE ACTIVITE

Chairman - Président : Dr. C. McCOMBIE (Switzerland)

Session II - Séance II

GAS PRODUCTION AND GAS RELEASE

PRODUCTION ET LIBERATION DES GAZ

Chairman - Président : Dr. P. ZUIDEMA (Switzerland)

Session III - Séance III

CHEMICAL EFFECTS IN THE NEAR-FIELD

EFFETS CHIMIQUES DANS LE CHAMP PROCHE

Chairman - Président : Dr. I. NERETNIEKS (Sweden)

Session IV - Séance IV

NEAR-FIELD TRANSPORT MODELLING

MODELISATION DU TRANSPORT DANS LE CHAMP PROCHE

Chairman - Président : Dr. D. HODGKINSON (United Kingdom)

Session V - Séance V

DISCUSSION, CONCLUSION

DISCUSSIONS, CONCLUSIONS

Chairman - Président : Dr. C. McCOMBIE (Switzerland)

Based upon the discussions, a summary and conclusions were prepared by the NEA Secretariat and the session chairmen and it is reported on page 10 of these proceedings. (Version française page 14)

SUMMARY AND CONCLUSIONS

Background

The problem of safely disposing of High Level Radioactive Wastes (HLW) from nuclear power generation is receiving a great deal of attention in OECD Member countries and has been the main subject of the NEA activities on waste management. However, many countries view the disposal of low and medium level waste (L/MLW) as of more immediate concern. This should meet the same safety criteria as proposed for HLW but, because of very much larger volumes, the lower activity contents and the larger variety of waste forms, many different conditioning methods and disposal concepts have been used or are being considered. Sea dumping has been practiced by some countries up to 1983; shallow land burial in different ways has been performed during many years, for example in the UK, the United States and France; and disposal in deep abandoned mines or in specially constructed rock caverns is now being initiated, for instance, in the Federal Republic of Germany and Sweden, respectively.

All steps of the nuclear fuel cycle and also industrial and medical use of radioelements generate low and medium level wastes. These arise in a variety of physical and chemical forms, ranging from potentially contaminated trash to well characterised waste streams from nuclear fuel reprocessing. In general, these wastes can contain both short-lived and long-lived radionuclides depending on the origin. If the waste streams mainly contain short-lived nuclides (eg. Co-60, Cs-137, Sr-90), then the time period of concern in the safety assessments is limited to some hundreds of years. If, however, significant quantities of long-lived elements are contained in the wastes time periods of the same order of magnitude as for HLW disposal have to be considered in the safety assessments.

There is no universal definition of low and medium level waste. In the context of this workshop, all waste types, except heat-generating HLW and spent fuel, were considered as being L/MLW. This means that effects in the near-field caused by heat generation could be disregarded.

Purpose of the near-field and main topics of the workshop

In general the major purposes of the near-field are to provide:

- an engineered framework for waste packages (structures for emplacement, backfilling of voids, protection against subsidence or direct intrusion)
- physical containment for short-lived radionuclides
- chemical conditions to minimize the release of radionuclides.

Thus, the near-field is the first major barrier to releases of radionuclides. In contrast to the far-field (geosphere) and the biosphere it is man-made. The design and construction can, within certain technical/economical limits, be chosen to provide the above mentioned features. The impact upon concept/design development and near-field performance assessment is therefore particularly important. The assessment should be based on a sound scientific understanding of all parts of the system. The future behaviour need not to be predicted in minute detail but enough needs to be understood to give confidence that no harmful releases of radioactivity can occur. To gain this level of understanding and to demonstrate it to responsible authorities and the public is one of the major tasks in any nuclear waste disposal programme.

The research related to the near-field covers a very broad area, ranging from the detailed study and modelling of individual phenomena and processes to the modelling of the release from the near-field to the geosphere. Much of the R&D is repository concept specific because of the differency in materials chosen for the repository system and their interaction with the surrounding geology. The basic phenomena and processes, however, are similar for most types of repositories.

While a number of different disposal concepts (such as disposal in salt domes, abandoned mines, rock caverns and near-surface structures above the water table) were presented at the Workshop, the main theme of the Workshop was disposal of conditioned waste in engineered repositories constructed largely from concrete. Over the last few years there has been a significant advance in the approaches and knowledge needed for the safety assessments of such repositories. This workshop gave a timely opportunity for Member countries of OECD to summarize the state of knowledge and to discuss matters of common interest and focus on future research and development requirements.

General overviews of disposal concepts and integrated safety assessments were presented in the first session of the Workshop. The remainder of the Workshop focused on phenomena of particular importance for the long-term safety and their possible implications on repository concepts and design. A major emphasis was on the chemical behaviour of repository near-fields and potential release of radionuclides to the far-field.

It has been realised in the last few years that gas generation by corrosion of steel and organic degradation of waste could be of such a magnitude that it has, in some cases, to be taken into account in repository concepts. To address these issues a special session on gas generation and gas release was included in the Workshop.

Conclusions and recommendations

Inventories

In contrast to HLW, where radionuclide inventories and waste form compositions are well characterised, the diversity of L/MLW means that special attention has to be given to determination of nuclide inventories and major waste form constituents. Some of the long-lived radionuclides which can be of importance for the long term safety (for example, I-129, Cl-36, Ca-41) are difficult to measure and the inventories of these have to be estimated. There are several procedures for this. It was felt that further information exchange between Member countries on the approaches and methods being used would be beneficial. It was also recommended that published performance assessments of L/ILW repository systems should include a description of the waste inventory which is detailed enough to explain the sources of the dominant radionuclides.

Chemical effects

The choice of cement dominated near-fields has many advantages in addition to that from practical engineering considerations it is a natural choice. Most importantly, it provides for very long times a stable and predictable alkaline and reducing chemical environment. Under such conditions some nuclides are highly insoluble.

A major factor in the safety assessments that have been performed to date is the retention by sorption in the near-field. A number of measurement programmes have been initiated on this topic and some were reported at the Workshop. However, because of the wide range of elements and environmental conditions of concern due to the diversity of materials in the wastes, further work is required in this area. In addition it was felt that more fundamental work is needed to understand the basic sorption processes. One particular concern was the influence of natural organic substances and complexants in the wastes and produced by degradation processes in the repository. Thus a major recommendation is to address this topic in a group limited to specialists directly involved in experimental studies on sorption together with a few performance assessors representing the users of sorption data. Such a group should meet to come up with specific proposals and priorities for future research. This could give a valuable input to national programmes and could also provide a basis for a co-ordination between the Member countries and their specialist groups performing the experimental work in this area.

Modelling of near-field phenomena

Significant advances were reported at the Workshop on the use of coupled chemistry and migration models. An example of this is the modelling of alteration of a bentonite liner due to the ingress of cementitious water. In general, this approach is providing an improved understanding of the evolving chemical environments which determine the migration behaviour of radionuclides. Even with modern computing facilities there are, however, significant practical limitations on the complexity of the problems which can be numerically solved. In future the implications of degradation for the physical integrity also need to be studied in more detail, especially its evolution with time.

Data needs

It was recognized that there is a need for better data on the transport properties in concrete and in possible backfills such as bentonite. Hydraulic conductivities and diffusivities in these materials are not well known in the long-term perspectives.

Gas generation

There is agreement that gas will be generated in repositories by corrosion of steel and degradation of organic materials. Gas generation by radiolysis is expected to be negligible by comparison. While the rate of production is expected to be very low under anaerobic conditions it cannot be excluded that a gas phase would develop within a repository. A crucial parameter is steel corrosion rates in the waste, in canisters and in the reinforcement of engineered stuctures. This has led to some designs which provide engineered venting of the near-field to assure that no excessive overpressures will be built up within the repository. The whole topic of gas generation and of the influence of the gas phase on flow and transport in and around the repository is subject to considerable uncertainty; these topics will be addressed continuously in research programmes and design studies presented at the meeting.

Modelling of releases from the near-field

Recent progress in predictive modelling of potential radionuclide release from the near-field to the far-field was reported at the Workshop. An approach used in many cases is to include sorption and solubility constraints (if any) in the near-field, while the possible kinetic limitations on leaching of radionuclides from the waste form is often not taken into account. These calculations indicate that an appropriately designed near-field constitutes an important barrier to radionuclide releases and that it can make a significant contribution to the overall safety of the disposal systems. Future work will aim at improving further the reliability of predictions of the near-field barrier.

A general observation from this Workshop is that assessments of the near-field performance of repositories for low- and medium level waste have progressed a lot over the last 5 to 10 years. Major phenomena and processes of significance for potential releases have been identified and extensively studied. State-of-the-art modelling and computer calculations are being used to make predictions over long time periods. Repository projects now under implementation in several Member countries give evidence of a close interaction between near-field performance assessments and repository design and construction. Research and development now under way will add further to the quality and credibility of the assessment techniques which are being used to evaluate the long-term safety of these radioactive waste repositories.

RESUME ET CONCLUSIONS

Généralités

Le problème de l'évacuation, dans des conditions de sûreté, des déchets de haute activité issus de la production d'électricité d'origine nucléaire retient considérablement l'attention dans les pays Membres de l'OCDE et constitue le principal sujet des activités que l'AEN consacre à la gestion des déchets. Toutefois, de nombreux pays considèrent que l'évacuation des déchets de faible et de moyenne activité constitue une préoccupation plus immédiate. Cette évacuation doit satisfaire aux mêmes critères de sûreté que ceux proposés pour les déchets de haute activité, mais en raison des volumes beaucoup plus importants, des teneurs plus faibles en radioactivité et de la plus grande variété des formes de déchets, de nombreux modes d'évacuation et méthodes de conditionnement différents ont été utilisés ou sont envisagés. L'immersion en mer a été pratiquée par certains pays jusqu'en 1983 ; il a été procédé pendant de nombreuses années à un enfouissement à faible profondeur dans le sol de différentes manières, notamment au Royaume-Uni, aux Etats-Unis et en France, et le stockage dans des mines désaffectées à grande profondeur ou dans des excavations spécialement aménagées dans des roches est maintenant entrepris respectivement en République fédérale d'Allemagne et en Suède par exemple.

Toutes les étapes du cycle du combustible nucléaire et, également les utilisations industrielles et médicales des radioéléments engendrent des déchets de faible et de moyenne activité. Ces derniers se présentent sous toute une gamme de formes physiques et chimiques allant de détritus susceptibles d'être contaminés à des effluents bien caractérisés issus du retraitement du combustible nucléaire. D'une façon générale ces déchets peuvent contenir des radionucléides tant à vie courte qu'à vie longue selon leur origine. Si ces effluents contiennent principalement des nucléides à vie courte (Co-60, Cs-137, Sr-90, par exemple), la période prise en considération dans les évaluations de sûreté se limite alors à quelques centaines d'années. Si au contraire des quantités significatives d'éléments à vie longue sont renfermées dans ces déchets, il faut prendre en considération dans les évaluations de sûreté des durées du même ordre de grandeur que dans le cas de l'évacuation des déchets de haute activité.

Il n'existe pas de définition universelle des déchets de faible et de moyenne activité. Dans le contexte de la présente réunion de travail, tous les types de déchets, à l'exception des déchets calogènes de haute activité et du combustible irradié, sont considérés comme étant des déchets de faible et de moyenne activité. Il s'ensuit que l'on a pu faire abstraction des effets causés dans le champ proche par la production de chaleur.

Fonction du champ proche et principaux thèmes de la réunion

D'une façon générale, le champ proche a principalement pour fonction de fournir :

- un cadre ouvragé pour les colis de déchets (structures de mise en place, remblayage des espaces vides, protection contre l'affaissement ou l'intrusion directe) ;

- un confinement physique pour les radionucléides à vie courte ;

- des conditions chimiques permettant de réduire au minimum le rejet de radionucléides.

Ainsi, le champ proche constitue la première grande barrière s'opposant aux rejets de radionucléides. Par opposition au champ lointain (géosphère) et à la biosphère, il est artificiel. La conception et la construction peuvent, dans certaines limites techniques et économiques, être choisies de façon à lui conférer les caractéristiques susmentionnées. L'incidence sur l'élaboration du concept et des plans et sur l'évaluation des performances du champ proche est donc particulièrement importante. Cette évaluation doit se fonder sur des connaissances scientifiques solides de toutes les parties du système. Il n'est pas nécessaire de prévoir dans tous ses détails le comportement futur mais il faut en savoir suffisamment pour être assuré qu'aucun rejet nocif de radioactivité ne pourra se produire. Parvenir à ce niveau de connaissance et en fournir la preuve aux autorités responsables et au public constitue l'une des principales tâches de tout programme de stockage des déchets nucléaires.

Les travaux de recherche relatifs au champ proche couvrent un très vaste domaine, allant de l'étude et de la modélisation détaillées des divers phénomènes et processus à la modélisation du rejet à partir du champ proche vers la géosphère. Les travaux de R-D sont pour une large part spécifiques à chaque concept de dépôt, en raison de la différence dans les matériaux choisis pour le système de dépôt et de l'interaction de ceux-ci avec le milieu géologique alentour. Les phénomènes et processus fondamentaux sont toutefois analogues dans le cas de la plupart des types de dépôts.

Bien qu'un certain nombre de modes de stockage différents (tels que le dépôt dans des dômes de sel, des mines désaffectées, des excavations rocheuses et des structures proches de la surface et situées au-dessus de la nappe phréatique) ont été présentés au cours de la réunion, celle-ci avait pour thème principal le stockage de déchets conditionnés dans des dépôts en structures ouvragées, construits principalement en béton. Ces dernières années, des progrès notables ont été réalisés dans les méthodes et connaissances requises pour les évaluations de la sûreté de tels dépôts. La présente réunion de travail a offert aux pays Membres de l'OCDE une occasion opportune de dresser un bilan des connaissances et de débattre de questions d'intérêt commun, tout en s'axant sur les besoins futurs en matière de recherche et de développement.

Au cours de la première séance de la réunion de travail, des aperçus généraux ont été présentés sur les modes de stockage et les évaluations de sûreté intégrées. Le reste de la réunion a été axé sur les phénomènes qui revêtent une importance particulière pour la sûreté à long terme, et sur ce qu'ils peuvent impliquer pour les concepts et les plans de dépôt. L'accent a tout particulièrement été mis sur le comportement chimique des champs proches des dépôts et sur le rejet potentiel de radionucléides vers le champ lointain.

On s'est rendu compte ces dernières années que la production de gaz par corrosion de l'acier et dégradation organique des déchets, pourrait prendre une ampleur telle que, dans certains cas, il faille en tenir compte dans les concepts des dépôts. Afin de traiter ces questions, il a été prévu que la réunion comporterait une séance spéciale sur la production et le rejet de gaz.

Conclusions et recommandations

Inventaires

Contrairement aux déchets de haute activité, dont les inventaires de radionucléides et les compositions sont bien définis, la diversité des déchets de faible et de moyenne activité signifie qu'une attention particulière doit être accordée à la détermination des inventaires de radionucléides et des principaux constituants de ces déchets. Certains radionucléides à vie longue qui peuvent revêtir de l'importance pour la sûreté à long terme (I-129, Cl-36, Ca-41, par exemple) sont difficiles à mesurer et les inventaires de ces éléments doivent être faits. Il existe plusieurs procédures à cet effet. Il a été considéré qu'un échange continu d'informations sur les approches et les méthodes utilisées serait bénéfique. Il a également été recommandé de faire figurer dans les évaluations publiées des performances des dépôts de déchets de faible et de moyenne activité une description de l'inventaire des déchets qui soit suffisamment détaillée pour permettre de connaître les sources des radionucléides prédominants.

Effets chimiques

Le choix de champs proches dans lesquels le ciment est le matériau prédominant offre de nombreux avantages outre le fait que, pour des considérations techniques pratiques, c'est là un choix naturel. Avant tout, ce matériau permet de disposer pendant de très longues durées d'un milieu chimique alcalin et réducteur qui est stable et dont le comportement est prévisible. Placés dans de telles conditions, certains nucléides sont hautement insolubles.

Dans les évaluations de sûreté qui ont été exécutées à ce jour, la rétention par sorption dans le champ proche constitue un important facteur. On a entrepris à ce sujet, un certain nombre de programmes de mesure dont certains ont fait l'objet de rapports à la réunion de travail. Toutefois, en raison du large éventail d'éléments et de conditions liées au milieu, qui présentent de l'intérêt par suite de la diversité des matériaux présents dans les déchets, des travaux complémentaires sont nécessaires dans ce domaine. En outre, on a estimé qu'il faut entreprendre des études plus fondamentales pour comprendre les aspects à la base des phénomènes de sorption. Un sujet particulier de préoccupation concerne l'influence des substances organiques naturelles et des agents complexants présents dans les déchets et produits par des processus de dégradation dans le dépôt. Ainsi, il est notamment recommandé d'aborder ce sujet au sein d'un groupe limité à des spécialistes intervenant directement dans des études expérimentales sur la sorption, auxquels se joindraient quelques évaluateurs des performances représentant les utilisateurs de données sur la sorption. Un tel groupe se réunirait en vue de formuler des propositions concrètes et de définir des priorités pour les travaux futurs de recherche. Cela pourrait constituer un apport intéressant aux programmes nationaux et également jeter les fondements d'une coordination entre les pays Membres et leurs groupes de spécialistes menant des travaux expérimentaux dans ce domaine.

Modélisation des phénomènes en champ proche

D'importants progrès ont été signalés lors de la réunion en ce qui concerne l'utilisation de modèles couplés de chimie et de migration. A titre d'exemple, on peut mentionner la modélisation de l'altération d'un revêtement de bentonite due à l'entrée d'eau ayant interagi avec du ciment. D'une façon générale, cette méthode permet de mieux comprendre l'évolution des milieux chimiques qui déterminent le comportement des radionucléides du point de vue de la migration. Même avec des moyens de calcul modernes, il existe toutefois des limites pratiques considérables visant la complexité des problèmes qui peuvent être résolus numériquement. A l'avenir, il faudra également étudier plus en détail ce que la dégradation implique pour l'intégrité physique, en particulier son évolution dans le temps.

Besoins en matière de données

On a reconnu qu'il est nécessaire d'améliorer les données relatives aux propriétés de transport dans le béton et dans d'éventuels matériaux de remblayage, tels que la bentonite. Les conductivités et diffusivités hydrauliques dans ces matériaux ne sont pas bien connues dans des perspectives à long terme.

Production de gaz

On s'accorde à penser que la corrosion de l'acier et la dégradation des matières organiques entraîneront la production de gaz dans les dépôts. En comparaison, la production de gaz par radiolyse devrait être négligeable. Alors que le taux de production sera vraissemblablement très faible dans des conditions anaérobies, on ne peut exclure qu'une phase gazeuse se développe à l'intérieur d'un dépôt. Un paramètre crucial est constitué par les taux de corrosion de l'acier dans les déchets, dans les conteneurs et dans les armatures des structures ouvragées. Cela a conduit à adopter certains modèles qui prévoient des dispositifs d'éventage du champ proche pour faire en sorte qu'aucune surpression excessive ne s'accumule à l'intérieur du dépôt. L'ensemble de la question de la production de gaz et de l'influence de la phase gazeuse sur les écoulements et le transport dans le dépôt et alentour est empreint d'importantes incertitudes ; ces questions continueront à être traitées dans les programmes de recherche et les études de conception qui ont été présentés lors de la réunion.

Modélisation du rejet à partir du champ proche

Il a été rendu compte au cours de la Réunion de travail des progrès récemment réalisés dans la modélisation prévisionnelle du rejet potentiel de radionucléides à partir du champ proche vers le champ lointain. Une méthode utilisée dans de nombreux cas consiste à inclure les contraintes (éventuelles) de sorption et de solubilité dans le champ proche, alors que les limitations cinétiques possibles visant la lixiviation de radionucléides à partir de la forme de déchets ne sont souvent pas prises en compte. Ces calculs montrent qu'un champ proche convenablement conçu représente une importante barrière s'opposant aux rejets de radionucléides et qu'il peut apporter une contribution notable à la sûreté globale des systèmes de stockage. Les travaux futurs auront pour but d'améliorer encore la fiabilité des prévisions afférentes à la barrière que constitue le champ proche.

Au cours de cette réunion de travail, on a observé d'une façon générale que les évaluations relatives aux performances du champ proche des dépôts de déchets de faible et de moyenne activité ont considérablement progressé au cours des cinq à dix dernières années. Les principaux phénomènes et processus susceptibles de revêtir de l'importance pour les rejets potentiels ont été cernés et amplement étudiés. On a recours à une modélisation et à des calculs informatiques de pointe pour formuler des prévisions portant sur des périodes prolongées. Dans plusieurs pays Membres, des projets de dépôt présentement mis en oeuvre témoignent d'une interaction étroite entre les évaluations des performances du champ proche et la conception ainsi que la construction de dépôts. Les travaux de recherche et de développement actuellement menés confèreront davantage de qualité et de crédibilité aux techniques d'évaluation qui sont utilisées pour apprécier la sûreté à long terme de ces dépôts de déchets radioactifs.

SESSION I

KEY ISSUES IN DESIGN AND ASSESSMENT OF THE NEAR-FIELD IN L/ILW REPOSITORIES

SEANCE I

PRINCIPAUX PROBLEMES DANS LA CONCEPTION ET L'EVALUATION DU CHAMP PROCHE DANS LES DEPOTS DE DECHETS DE FAIBLE ET MOYENNE ACTIVITE

Chairman - Président

C. McCOMBIE

(Switzerland)

INTEGRATED NEAR-FIELD ASSESSMENTS FOR NIREX REPOSITORY CONCEPTS

D P Hodgkinson*, D A Lever+, P C Robinson*, and P W Tasker+

*Environmental Sciences Group, INTERA/Exploration Consultants Limited, Highlands Farm, Greys Road, Henley-on-Thames, Oxfordshire RG9 4PS, UK
+Theoretical Physics Division, Harwell Laboratory, Oxfordshire OX11 ORA, UK

ABSTRACT

In the United Kingdom, UK Nirex Ltd. are responsible for the disposal of low- and intermediate-level radioactive waste. The current programme is considering various options for a deep repository for these wastes. Shallow disposal of low level waste, which has been considered previously, is not now UK policy. In all of these options a near field consisting largely of concrete or cement has been envisaged. The Nirex research programme investigates all aspects of such a near field and its consequences for radiological safety after closure of the repository.

For the purposes of safety assessment, the results of the research are combined into a simplified source term model. The model STRAW takes data such as container lifetimes, waste inventory, solubility and sorption of the radio elements and calculates near-field solution concentrations. The code fulfills two functions. First, it predicts the source of radionuclides to the far field, which can then be used as part of a total safety assessment. Secondly, it can be used to examine the role of the near-field chemistry, waste inventory etc. on the overall near-field performance. Results from the STRAW model will be presented to illustrate these points in the context of a safety assessment for the disposal of low- and intermediate-level waste.

EVALUATIONS INTEGREES DU CHAMP PROCHE POUR LES CONCEPTS DE DEPOTS DE LA NIREX

RESUME

Au Royaume-Uni, la société britannique Nirex Limited est responsable du stockage des déchets de faible et de moyenne activité. Le programme actuel envisage diverses options pour un dépôt à grande profondeur destiné à ces déchets. L'évacuation à faible profondeur de déchets de faible activité, qui était auparavant étudiée, n'est plus l'option actuellement retenue au Royaume-Uni. Dans toutes ces solutions, on a envisagé un champ proche constitué en grande partie par du béton ou du ciment. Le programme de recherche de la Nirex porte sur tous les aspects d'un tel champ proche et ses conséquences pour la sûreté radiologique après la fermeture du dépôt.

A des fins d'évaluation de la sûreté, les résultats des recherches sont intégrés dans un modèle simple du terme source. Le modèle STRAW utilise des données telles que les durées de vie des conteneurs, l'inventaire de déchets, la solubilité et la sorption des radioéléments, et calcule les concentrations des solutions se trouvant dans le champ proche. Le programme de calcul remplit deux fonctions. Premièrement, il établit des prévisions relatives à la source de radionucléides pour le champ lointain; celles-ci peuvent alors être utilisées en tant qu'élément de l'évaluation globale de sûreté. Deuxièmement, il peut servir à examiner l'influence de la chimie du champ proche, de l'inventaire de déchets, etc. sur l'ensemble des performances du champ proche. Les résultats obtenus grâce au modèle STRAW seront présentés afin d'illustrer ces aspects dans le contexte d'une évaluation de la sûreté afférente au stockage de déchets de faible activité.

1. INTRODUCTION

The organisation now known as United Kingdom Nirex Limited was established in July 1982 to develop and implement the national strategy for the safe disposal of low- and intermediate-level wastes (LLW and ILW) arising in the UK[1]. Until May 1987 Nirex were planning to develop two repositories, namely a near-surface engineered disposal facility in clay for LLW and a deep repository for ILW. However, following a change in national policy, Nirex now intend to develop a single deep repository for LLW and ILW.

In the course of obtaining approvals for the construction of a repository, Nirex must satisfy themselves of the safety and acceptability of their plans and will have to present a safety case which demonstrates compliance with stringent radiological safety criteria laid down by the regulatory authorities[2]. The safety target requires that at any time the risk to an individual of a serious health effect is less than one in a million in a year, which is equivalent to an annual dose of 0.1 mSv. Moreover, doses should be As Low as Reasonably Achievable (ALARA) below this figure, social and economic factors being taken into account.

Nirex are adopting a scientific approach to meeting these criteria. The scientific basis for the safety case is being studied in the Nirex Safety Assessment Research Programme (NSARP)[3-6]. The scope of this research programme is discussed in more detail in related papers at this Workshop[7,8].

The general safety philosophy being adopted by Nirex is that the waste should be packaged and the repository should be constructed and located such that radioactivity remains contained until it has decayed substantially[9]. Moreover, any subsequent migration should be inherently very slow.

Post-emplacement radiological assessments of Nirex repository concepts are carried out by the Disposal Safety Assessment Team (DSAT). This team has recently performed a preliminary assessment of LLW disposal in near-surface clay in which intrusion pathways, gaseous pathways and natural disruptive events were considered in addition to the more familiar groundwater transport scenario[10].

One of the major conclusions of the near-surface disposal assessment was that human intrusion appears to be the dominant transfer mechanism although the risk estimates are very uncertain. Thus near-surface repositories should be designed to minimise intrusion risks. This could be done by increasing the depth of burial beyond the usual depth of building foundations. Uncertainties in these risk estimates could be reduced by improved understanding of the physical form of repository materials at the time of intrusion.

Nirex design concepts emphasise the use of steel containment together with cementitious materials for grouting, backfill and the repository structure itself. These are manageable and well understood engineering materials which also provide physical and chemical constraints on radionuclide transfer. An iterative approach is required for the specification of concrete formulations in order to balance engineering and radionuclide migration considerations. For example, a particular issue is whether certain types of organic additives should be used to improve the flow properties of concrete during repository construction since they, or their degradation products, could reduce sorption

3. NEAR-FIELD ASSESSMENT MODELLING

This section describes the status of the source-term modelling for a repository containing LLW and ILW inside steel containment and backfilled with concrete. The model is only concerned with the behaviour of radionuclides once physical containment can no longer be relied on. It is assumed that all of the waste containers lose their integrity simultaneously. However, in practice they would fail over a period of time which would result in lower doses[15].

The source-term model reviewed here, STRAW[16,17], has been designed to include the near-field phenomena which are currently thought to be the most important. Detailed geometrical and other non-essential factors have been omitted in order to increase the reliability and robustness of the predictions[18]. Also, such simplifications facilitate the interpretation and presentation of the results. A further guiding principle in constructing the model was that it should only use parameters which could be readily measured or estimated.

The primary purpose of the model is to provide the flux of radionuclides entering the geosphere. However, it has also been found useful to define a performance measure which reflects the behaviour of the near field alone. Radiological toxicity has been used for this purpose. This is defined as the factor by which the repository water must be diluted so that an individual using it as the sole source of drinking water (two litres per day) receives an annual dose of 1 mSv. Equivalently it is the annual dose, in units of mSv, which an individual receives if the undiluted repository water is used as the sole source of drinking water. By this means radionuclides can be compared on an equal footing. It is of course recognised that undiluted repository water would be undrinkable for others reasons, e.g. high alkalinity.

The repository region is considered to be a porous medium which, since it contains degraded material, has a permeability greater than that of the surrounding geosphere. The water flow through the repository is therefore determined by that in the surrounding geosphere. This is assumed to be sufficiently low that chemical equilibrium exists within the repository. Moreover, flow and diffusion are implicitly assumed to have equalised concentrations throughout the repository region.

The chemical constraints in the model are linear reversible sorption onto a fraction of the solid surfaces and solubility limits for the sum of the concentrations of all stable and active isotopes of each elements. It is assumed that leaching kinetics of radionuclides from the waste are rapid, so that the total inventory is always available for dissolution in the water and sorption on the solids. With these assumptions the concentration of an element is solubility limited only if the amount of that element in the repository exceeds the sum of the amount needed to saturate the repository pore water and the corresponding amount sorbed onto the repository materials. Otherwise, the concentration of each isotope of that element is determined by the inventory and by sorption.

The time dependence of the model is determined by radioactive decay and ingrowth and by the radionuclide fluxes to the geosphere. These are assumed to be proportional to the repository pore water concentration, with a constant of proportionality representing advection and/or diffusion.

The above source-term model is incorporated into the STRAW computer code. The non-linear first order differential equations are solved numerically using Gear's method[19]. However in certain special cases the equations have analytical solutions which have been used to verify the coding and numerical methods.

4. TYPICAL RESULTS

In order to illustrate the use of the model we consider a typical example[16]. This involves the disposal of 10^5 m^3 of cemented ILW[20] in a repository with a total volume of 10^6 m^3 and a cross-sectional area of 2×10^5 m^2, perpendicular to the flow direction. Solubility limits were calculated with the chemical equilibrium code PHREEQE[21] using a recently revised thermodynamic database[22]. Sorption values were based on available experimental data and inferences from expected similarities in chemical behaviour. To be conservative only 10% of the repository materials were assumed to be available for sorption. In view of the many uncertainties an extensive sensitivity study has been performed[16].

The toxicity of the repository pore water, in the base case, as a function of time following containment failure at 300 years is shown in figure 2. This shows the seven nuclides with highest toxicity (Pb-210, Nb-94, Sr-90, Cs-137, Nb-93m, Ra-226, Am-241). The dominant nuclide, Pb-210, arises from two sources, namely from Ra-226 at early times and from U-238 at later times.

The toxicity gives an indication of the reduction factors required from other processes, such as dilution, dispersion and decay in both the geosphere and in the various biosphere pathways. It is a method of scoping the problem; it is not a method of calculating likely doses to any particular individual. It is emphasised that the doses from the near-field pore water will not be incurred in practice because: i) it would not be easy to extract two litres of pore water from the repository; ii) the pore water would be so alkaline that it could not be drunk; and iii) the short-lived daughters, particularly Pb-210, would decay away if the pore water moved any distance from the repository. This last point arises because the uranium is solubility limited and in the repository Pb-210 is in secular equailbrium with the total amount of U-238, whereas in the geosphere it would only be in equilibrium with the dissolved and sorbed portion. For this reason, the nuclides having the highest near-field toxicity should not necessarily be investigated in the greatest detail.

Finally we note that if the chemical constraints of solubility limits and sorption are omitted from the calculation then the overall shape of the toxicity curves is not significantly changed but the toxicity is about one thousand times higher. This quantifies the effectiveness of the chemical constraints.

As noted earlier the primary purpose of this model is in calculating fluxes to the geosphere. A useful way of presenting these is in terms of effective leach rates. These are simply the fraction of each radionuclide release per year. These are the same for all isotopes of a particular element. Since these are time dependent we present the initial leach rate and the maximum value up to 10^6 years in table 1. It is noted that many of these are considerably less than the value of 10^2/year used in some early assessments of ILW disposal[23].

5. OUTLOOK

There are a number of ways in which the present model could be extended. These include: the time evolution of pH and Eh, and their effect on solubilities and sorption coefficients; the leach resistance of the waste; a distribution of canister failure times; diffusion in the near-field; variability within the repository; coupling between gas production, saturation and corrosion.

ACKNOWLEDGEMENT

This work was funded by UK Nirex Ltd. as part of their Safety Studies programme.

REFERENCES

1. Ginniff, M.E. : "Management of Radioactive Wastes by UK Nirex, Nuclear Europe, 3, 10, 1986.
2. Department of the Environment, Disposal Facilities on Land for Low and Intermediate-Level Radioactive Wastes: Principles for the Protection of the Human Environment (1984).
3. Hodgkinson, D.P. et al., The Nirex Safety Assessment Research Programme : Annual Report for 1985/86, UKAEA Report AERE-R.12331 (1987).
4. Cooper, M.J. and Hodgkinson, D.P. : 1986/87 Annual Report., NSS/R101, (1987).
5. Cooper, M.J. and Hodgkinson, D.P. : Nirex Safety Assessment Research Programme Bibliography, UKAEA Report AERE Bib 206, (1987).
6. Hodgkinson, D.P. : The Nirex Safety Assessment Research Programme on Near-Field Effects in Cementitious Repositories, UKAEA Report AERE R.12514. (1987) and Radioactive Waste Management and the Nuclear Fuel Cycle (n press).
7. Lever, D.A. and Rees, J.H. : Gas Generation and Migration for Nirex Repository Concepts, Proceedings of the NEA Workshop on Near-Field Assessment of Repositories for Low and Medium Level Waste, Baden, November 1987 (to be published).
8. Ewart, F.T. et al. : Experimental and Modelling Studies of the Near-Field Chemistry for Nirex Repository Concepts, Proceedings of the NEA Workshop on Near-Field Assessment of Repositories for Low and Medium Level Waste, Baden, November 1987 (to be published).
9. Flowers, R.H., Keen, N.J. and Rae, J. : Performance of Near-Field Barriers in the Development of Waste Packaging, Proceedings of the International Seminar on Radioactive Waste Products : Suitability for Final Disposal, Julich, June 1985.
10. Hodgkinson, D.P. and Robinson, P.C. : Nirex Near-Surface Repository Project, Preliminary Radiological Assessment: Summary, UKAEA Report NSS/A100 (1987).
11. Marsh, G.P., Taylor, K.J., Sharland, S.M. and Tasker, P.W. : An Approach for Evaluating the Generalised Localised Corrosion of Carbon Steel Containers for Nuclear Waste Disposal, Proceedings of a Symposium on the Scientific Basis for Nuclear Waste Management, Materials Research Society, Boston, 1986.

12. Sharland, S.M., Tasker, P.W. and Tweed, C.J., The Evolution of Eh in the Pore Water of a Model Nuclear Waste Repository, UKAEA Report AERE-R.12442 (1986).
13. Atkinson, A. : The Time Dependence of pH within a Repository for Radioactive Waste Disposal, UKAEA Report AERE-R.11777 (1985).
14. Ewart, F.T. and Tasker, P.W. : Chemical Effects in the Near-Field, Proceedings of Waste Management 1987, Tuscon, Arizona, March 1987 (to be published).
15. Swedish Nuclear Fuel Supply Co/Division KBS, Final Storage of Spent Nuclear Fuel - KBS-3, Vol. IV - Safety, 1983.
16. Robinson, P.C., Hodgkinson, D.P., Tasker, P.W., Lever, D.A., Windsor, M.E., Grime, P.W. and Herbert, A.W., A Solubility-Limited Source-Term Model for the Geological Disposal of Cemented Intermediate-Level Waste, UKAEA Report AERE R.11854, 1986.
17. Windsor, M.E. : STRAW: A Source-Term Code for Buried Radioactive Waste, UKAEA Report AERE-R.12368, 1986.
18. Pigford, T.H. : Reliable Predictions of Waste Performance in a Geologic Repository, Paper presented at the ANS/CNS/AESJ/ENS International Topical Meeting on High-Level Nuclear Waste Disposal - Technology and Engineering, Pasco, Washington, USA, 1985.
19. Gear, C.W. : The Automatic Integration of Stiff Ordinary Differential Equations, Proceedings of the IFIP Congress, 1968.
20. Fletcher, A.M., Wear, F.J., Haselden, H., Shepherd, J. and Tymons, B.J., The 1985 United Kingdom Radioactive Waste Inventory: January 1986, UK Department of the Environment Report DoE/RW386-087, 1986.
21. Parkhurst, D.L., Thorstenson, D.C. and Plummer, L.N., Report USGS/WRI 80-96, NTIS Tech.Rep. PB81-167801, 1980, Revised 1985.
22. Cross, J.E., Ewart, F.T. and Tweed, C.J. : Thermochemical Modelling with Application to Nuclear Waste Processing and Disposal, UKAEA Report, AERE R.12324, 1987.
23. Hill, M.D., Mobbs, S.F. and White, I.F. : An Assessment of the Radiological Consequences of Disposal of Intermediate Level Wastes in Argillaceous Rock Formations, National Radiological Protection Board Report NRPB-R126, 1981.

Table 1 Effective leach rate for each element for the base case

Element	Initial leach rate (per year)	Maximum leach rate up to 10^6 years (per year)
C	8.1×10^{-4}	7.7×10^{-3}
Ni	8.2×10^{-9}	8.3×10^{-9}
Se	7.7×10^{-3}	7.7×10^{-3}
Sr	1.6×10^{-3}	1.6×10^{-3}
Zr	1.3×10^{-7}	1.5×10^{-7}
Nb	7.7×10^{-3}	7.7×10^{-3}
Tc	1.5×10^{-5}	2.0×10^{-4}
Sn	7.7×10^{-3}	7.7×10^{-3}
I	7.7×10^{-3}	7.7×10^{-3}
Cs	1.6×10^{-3}	1.6×10^{-3}
Sm	2.7×10^{-9}	2.7×10^{-9}
Pb	7.7×10^{-3}	7.7×10^{-3}
Ra	1.6×10^{-3}	1.6×10^{-3}
Ac	2.0×10^{-4}	2.0×10^{-4}
Th	2.0×10^{-6}	2.0×10^{-6}
Pa	2.0×10^{-4}	2.0×10^{-4}
U	2.3×10^{-13}	2.3×10^{-13}
Np	2.0×10^{-6}	2.0×10^{-6}
Pu	7.3×10^{-9}	2.0×10^{-6}
Am	4.5×10^{-7}	2.0×10^{-6}
Cm	2.0×10^{-6}	2.0×10^{-6}

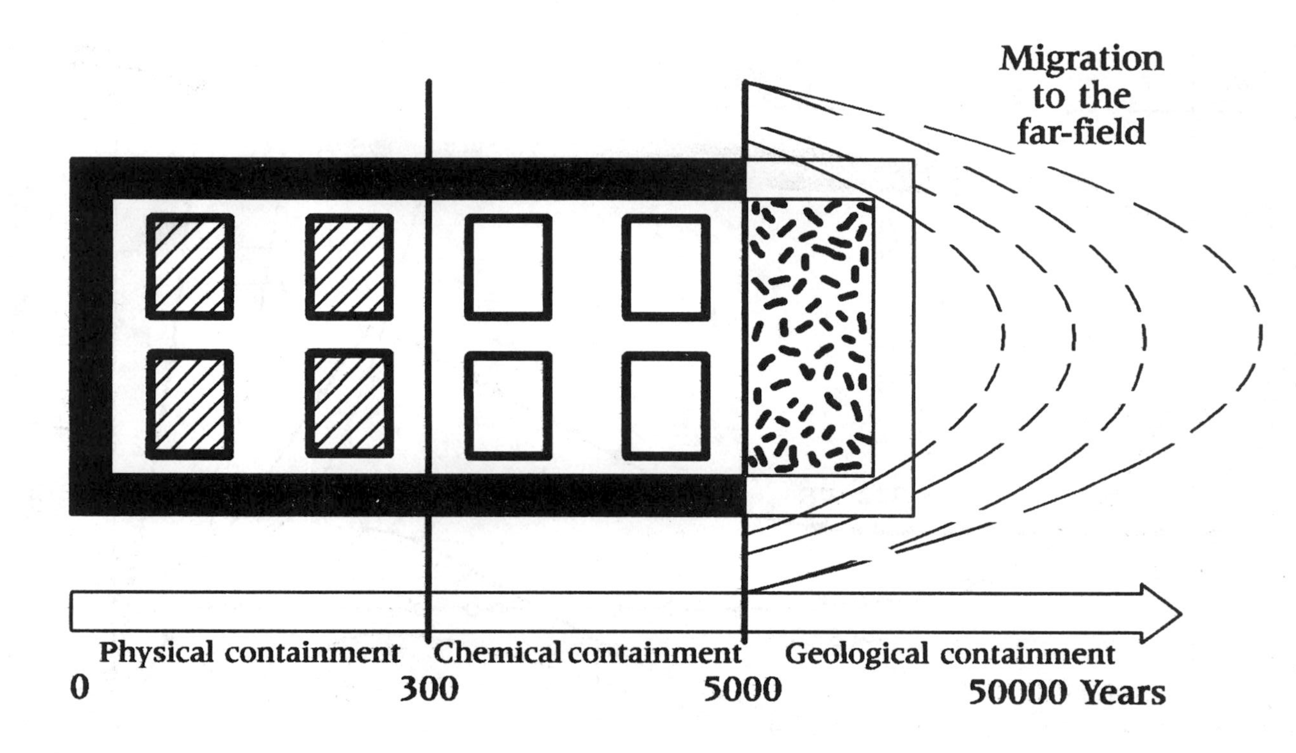

FIGURE 1 A SCHEMATIC VIEW OF THE EVOLUTION OF THE NEAR-FIELD OF A CEMENTITIOUS REPOSITORY

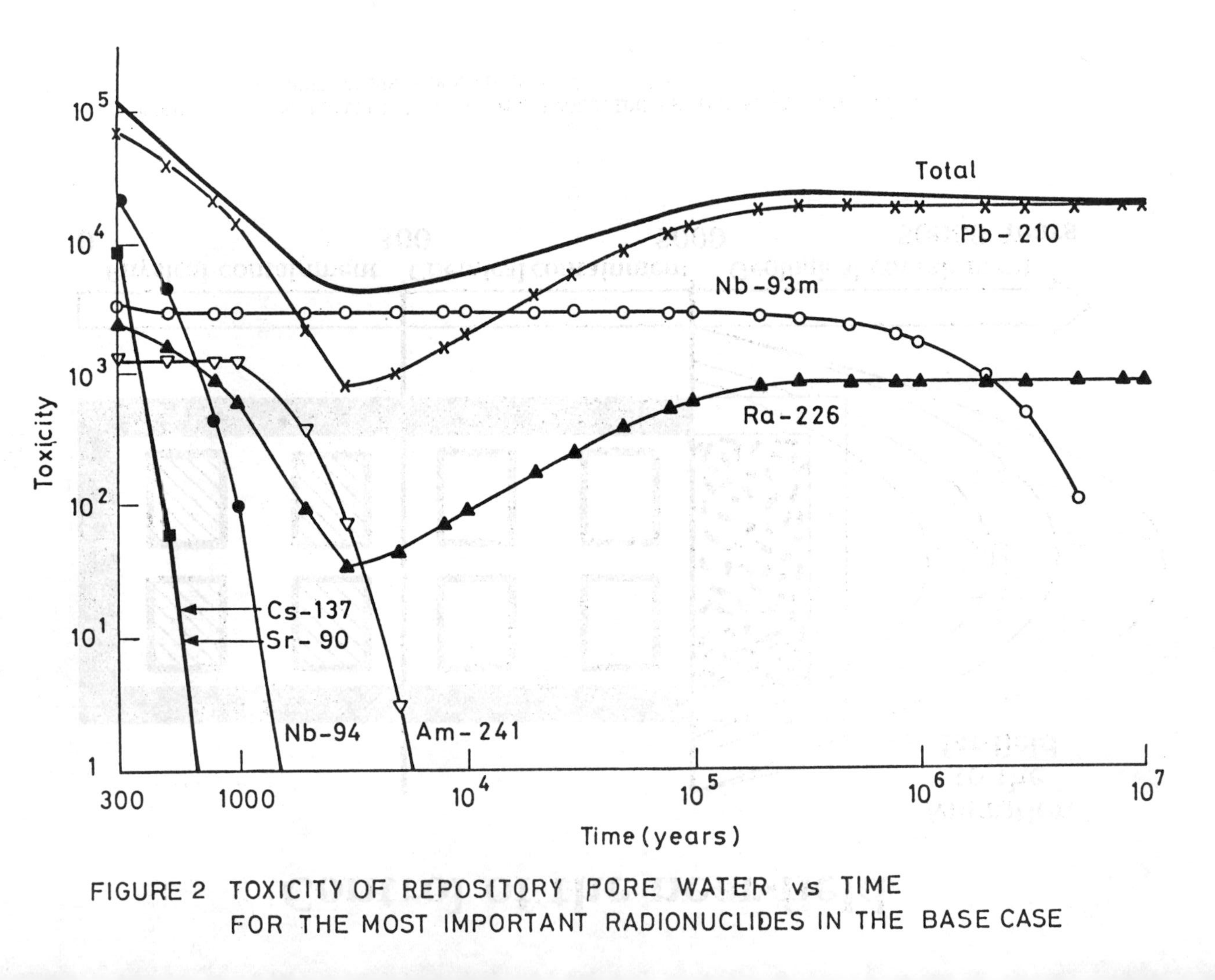

FIGURE 2 TOXICITY OF REPOSITORY PORE WATER vs TIME FOR THE MOST IMPORTANT RADIONUCLIDES IN THE BASE CASE

NEAR-FIELD ANALYSIS IN THE SAFETY ASSESSMENTS FOR LICENSING OF TWO ROCK CAVERN REPOSITORIES FOR LOW AND MEDIUM LEVEL WASTE

Timo Vieno and Henrik Nordman
Technical Research Centre of Finland
Helsinki, Finland

ABSTRACT

Repositories for low and medium level reactor waste from the Finnish nuclear power plants will be constructed in the bedrock of the power plant sites, at the depths of 50-125 meters. Preliminary Safety Analysis Reports (PSARs) have been submitted to the regulatory authorities in 1986 and excavation of the repositories is planned to be started in 1988. In the safety analyses several scenarios were considered in analysing release of radionuclides from the repositories into the bedrock. The results of the near-field analyses show that the multibarrier repository concepts consisting of conditioned waste, concrete barriers and crushed rock backfilling constitute compatible systems which effectively retard most radionuclides within the repository.

ANALYSE DU CHAMP PROCHE DANS LES EVALUATIONS DE SURETE DESTINEES A L'AUTORISATION DE DEUX DEPOTS DE DECHETS DE FAIBLE ET DE MOYENNE ACTIVITE DANS DES EXCAVATIONS ROCHEUSES

RESUME

Les dépôts destinés aux déchets de faible et de moyenne activité provenant des centrales nucléaires finlandaises seront construits dans le soubassement rocheux des sites de ces centrales à des profondeurs comprises entre 50 et 125 mètres. Des rapports préliminaires d'analyse de sûreté ont été soumis aux autorités chargées de la réglementation en 1986, et il est prévu de démarrer l'excavation de ces dépôts en 1988. Dans les analyses de sûreté, plusieurs scénarios ont été envisagés pour analyser le rejet de radionucléides à partir de ces dépôts dans le soubassement rocheux. Les résultats des analyses du champ proche montrent que les concepts de dépôt à barrières multiples constitués par des déchets conditionnés, des barrières de béton et un remblayage par de la roche broyée, représentent des systèmes compatibles qui retiennent de façon efficace la plupart des radionucléides à l'intérieur du dépôt.

1. INTRODUCTION

Repositories for low and medium level waste from the two Finnish nuclear power plants, Loviisa and Olkiluoto, will be excavated at depths of 50-125 m in the bedrock of the power plant sites. The repository designs are based on multiple engineered and natural barriers taking into consideration site- and waste-specific characteristics [1]. Comprehensive safety analyses were performed for the Preliminary Safety Analysis Reports (PSARs) of the repositories. The PSARs were submitted to the regulatory authorities in 1986 and their assessment is expected to be completed by the end of 1987. Construction of the repositories is planned to be started in 1988, with commissioning for operation possible in 1992. According to the present plans the repositories will be later expanded to provide disposal space also for decommissioning wastes.

2. REPOSITORY DESIGNS

2.1 The Loviisa repository

The bedrock of the Loviisa site on the island of Hästholmen consists of homogeneous low permeable rapakivi granite and two major fractured zones with higher permeability (Fig. 1). The groundwater on the island contains two zones of different salinity. The boundary between the upper, lens-like zone of fresh, flowing groundwater and the lower zone of saline, stagnant groundwater lies in the upper fracture zone varying between -60 and -140 m. The repository will be constructed between the two gently dipping fracture zones at the level of approximately -120 m. Accordingly, the repository will be situated in the zone of almost stagnant groundwater [2].

The repository will consist of a cavern for solidified wet waste and tunnels for dry maintenance waste (Fig. 2). Dry maintenance waste is contained in ordinary steel drums and no backfilling will be used in the disposal tunnels. For wet waste from the Loviisa power plant, a cementation process has been designed and licensed [1]. In the disposal cavern for solidified waste the most important release barriers consist of the conditioned waste itself, the reinforced concrete containers and the concrete walls together with a layer of crushed rock surrounding the pile of the waste packages (Fig. 3).

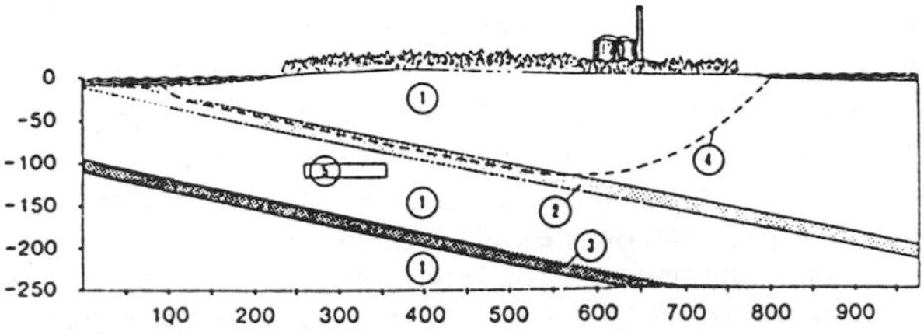

(1) Homogeneous intact rock (median $3 \cdot 10^{-9}$ m/s)
(2) Upper broken zone (median $1 \cdot 10^{-6}$ m/s)
(3) Lower broken zone (median $4 \cdot 10^{-7}$ m/s)
(4) Boundary between fresh and saline groundwater
(5) Location of the repository

Fig. 1. Hydrogeological model of the Loviisa site.

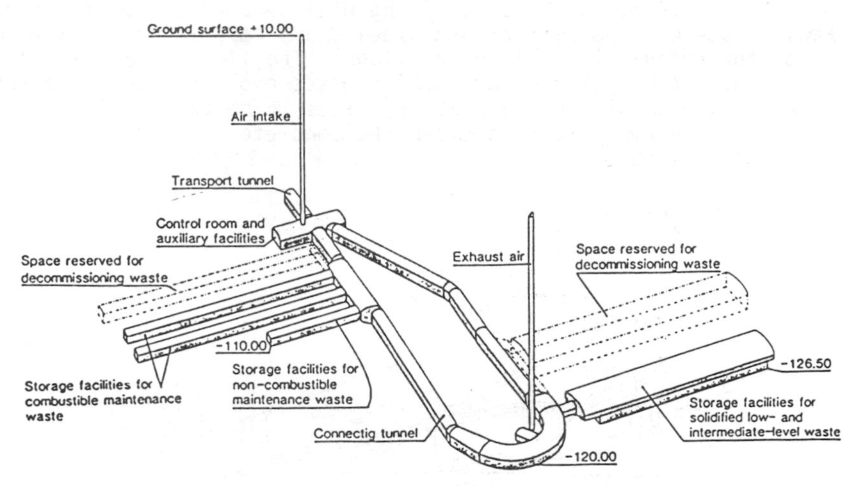

Fig. 2. The Loviisa repository.

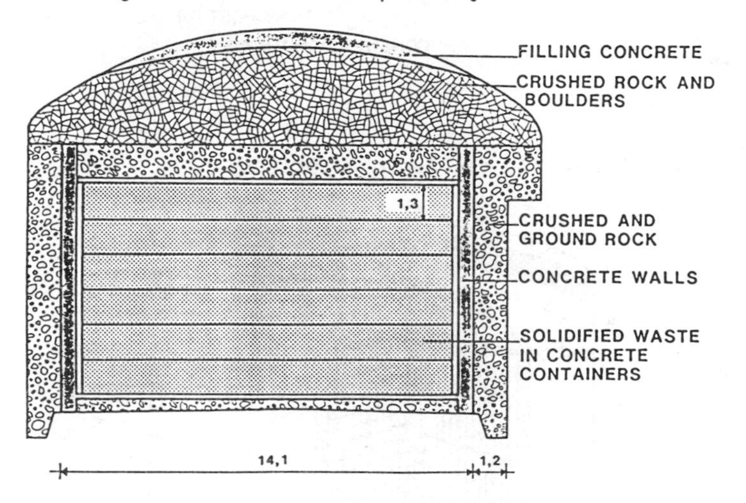

Fig. 3. Cavern for solidified waste in the Loviisa repository.

2.2 The Olkiluoto repository

At the Olkiluoto site the bedrock consists of an intact tonalite massive surrounded by more fractured micagneiss. Groundwater of the site is of fresh or brackish type with no great variations in salinity.

In the Olkiluoto repository two separate silos will be constructed, one for bituminized waste, the other for dry maintenance waste (Fig. 4). The diameter of the silos is 22 m and the height about 30 m. The silo for maintenance waste is a rock silo with internal structure of concrete walls.

The silo for bituminized waste consists of a thick-walled concrete silo inside the rock silo (Fig. 5). No backfilling will be used inside the concrete silo. Empty space will be left around steel drums containing bituminized waste to allow for potential swelling of bitumen. The lid of the silo will be provided with a gas lock system which enables gases evolving due to corrosion of steel drums or due to microbiological processes to be transported out of the concrete silo. The empty space between the concrete silo and the rock will be filled with crushed rock.

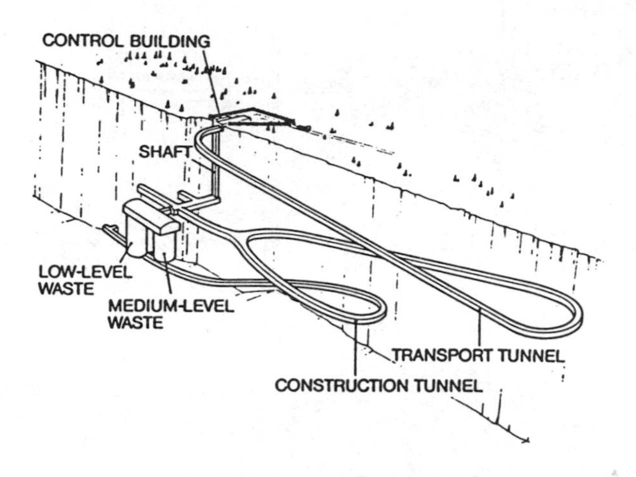

Fig. 4. The Olkiluoto repository.

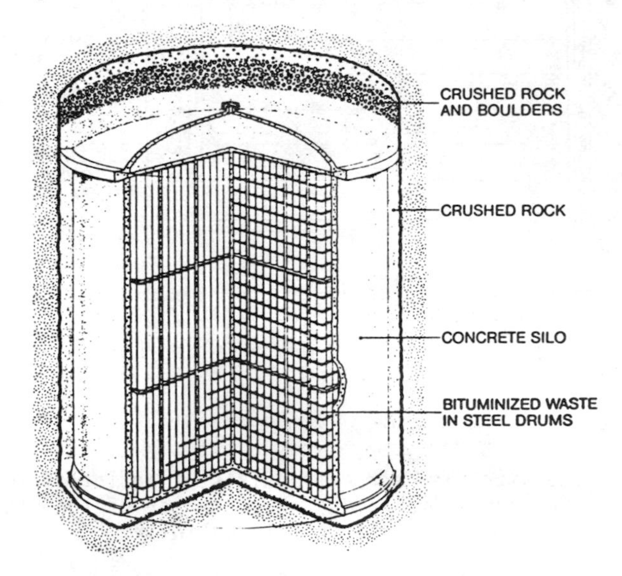

Fig. 5. Silo for medium level waste in the Olkiluoto repository.

3. GENERAL PRINCIPLES AND SCENARIOS FOR NEAR-FIELD ANALYSIS

General principles and selection of scenarios for the safety analyses were determined by licencing aspects and requirements. Several scenarios were considered in analysing release of radionuclides from the repositories into the bedrock [3].

Gradual degradation of engineered barriers was taken into consideration in all scenarios. In the case of the concrete containers to be used in the Loviisa repository, the already initially rather conservative diffusion coefficients in concrete were increased by a factor of 10 at 500 years and total degradation of all concrete structures in the repository was assumed to take place after 5000 years. After the latter time point concrete was assumed to be no longer functioning as a diffusion barrier. Consequently, only the sorption capacity of crumbled concrete is retarding radionuclides. In the case of the thick-wall reinforced concrete silo in the Olkiluoto repository, an increase of diffusion coefficients by a factor of 10 was assumed to take place at 1000 years and once again at 10000 years. Degradation of engineered barriers results also in an increase of groundwater flow through the repositories.

At the Loviisa site the repository will be situated in the zone of stagnant groundwater. Due to the uncertainty considering the long-term persistence of these rather favourable groundwater conditions around the repository, all release scenarios were splitted into two parallel scenarios. The two parallel groundwater scenarios are based a) on the present groundwater conditions and b) on an alternative assumption according to which the repository is situated in a region of fresh, flowing groundwater.

The safety margins of the disposal systems were explored by disturbed evolution and disruptive event scenarios. In each disturbed evolution scenario, one of the main barriers was assumed to be severely impaired. The consequences of a hypothetical total loss of performance of engineered barriers after 500 years were scrutinized due to a (potential) licencing requirement specifying that in the long-term natural barriers should alone be sufficient to restrict the flux of radioactive substances into the biosphere below a safe limit. In the other disturbed evolution scenarios the consequences of serious malfunctioning or damaging of the concrete silo for bituminized waste in the Olkiluoto repository were analysed. For instance, the gas lock of the concrete silo was assumed to become filled with water creating an additional diffusion pathway out of the concrete silo for the radionuclides. The consequences of cracking of the concrete silo were scrutinized, too.

In the disruptive event scenario chosen for the Loviisa repository, it was assumed that the two fractured zones lying above and below of the repository would be connected to each other by a water transporting cleft caused by a strong earthquake. It was further assumed that the cleft would intersect the cavern for solidified waste.

Selection of scenarios is discussed in more detail in ref. [3]. A summary of the scenarios employed in the near-field analyses is presented in Table I.

Table I Summary of near-field scenarios.

The Loviisa repository

- The repository is assumed to be situated a) in the zone of saline, stagnant groundwater and b) in a zone of fresh, flowing groundwater. In both cases the following scenarios have been considered:

  - o Basic scenario
  - o Disturbed evolution scenario: Hypothetical total loss of performance of engineered barriers after 500 years
  - o Distruptive event scenario: A cleft caused by a strong earthquake is intersecting the cavern for solidified wet waste at 500 years

The Olkiluoto repository

- The repository is situated in the zone of fresh, flowing groundwater. The following scenarios have been analysed:

  - o Basic scenario
  - o Realistic scenario
  - o Disturbed evolution scenario 1: The gas lock system of the lid of the concrete silo for bituminized waste becomes filled with water
  - o Disturbed evolution scenario 2: Cracking of the concrete silo for bituminized waste at 100 years
  - o Disturbed evolution scenario 3: Hypothetical total loss of performance of engineered barriers after 500 years

---

4. MODELLING

The release of radionuclides from the waste packages and their subsequent transport through engineered barriers was analysed by the numerical compartment model REPCOM [4]. Each barrier is devided into several compartments to simulate diffusive and convective transport of radionuclides and their retardation due to sorption. Mass transfer across the interfaces of different materials is considered by infinitesimal boundary compartments consisting of two subcompartments. Concentrations of nuclides in the subcompartments are determined by the properties of the interfacing materials.

For a repository model consisting of n compartments a system of n linear differential equations is obtained. The equation system is solved by the matrix exponent method applying the eigenvalues and eigenvectors of the coefficient matrix. The obtained solutions for the activity inventories of the radionuclides in the compartments are hence continuous in time.

The REPCOM model is flexible and does not require extensive computing resources. The basic model is one-dimensional but parallel pathways can be taken into account by connecting compartments to each others. The model can hence be applied to almost any kind of repository geometry and it is fairly easy to take into consideration different release barriers and boundary conditions. Solubility limits can be taken into account by the REPCOM model; they were, however, not considered in the analyses of reactor waste repositories. Gradual degradation of engineered barriers can be accounted for by altering parameters at fixed time points.

The main output from the near-field model is the flux of radionuclides from the repository into the bedrock. More profound performance analysis of the near-field system can be based on the distributions of radionuclides within the repository as a function of time and on the infinite time integrals of the radionuclide concentrations in the different parts of the repository system. The latter quantities indicate where the radionuclides decay and what fraction is released from the repository into the surrounding bedrock.

5. DATA

The data for the performance analyses were derived from site investigations and laboratory experiments. The final choice of parameter values was made in co-operation with the pertinent specialists in accordance with the general principle of conservatism of the analysis.

The groundwater flow through the repositories was estimated in the comprehensive groundwater flow analyses of the sites [2,5]. The assumptions concerning the distribution of the flow within in the repositories (Table II) were based on pessimistic judgements.

For dry maintenance waste it was conservatively assumed that their caverns will be filled with groundwater and the radionuclides will be dissolved immediately after sealing of the repository. The choice of the parameter values characterising the most import properties of the engineered barriers for solidified wet waste were based on several site and material specific studies [6,7,8,9]. The data presented in Tables II and III are applicable for the Loviisa repository. The corresponding data for the Olkiluoto repository are presented in ref. [4].

Table II Groundwater flow through the Loviisa repository in the scenarios, where the repository is assumed to be situated in a zone of fresh, flowing groundwater.

Repository room	Groundwater turnover fraction (a^{-1}) Basic scenario	Disturbed evolution scenario	Disruptive event scenario
Tunnels for dry maintenance waste	$8 \cdot 10^{-4}$	n.a.	n.a.
Cavern for solidified wet waste			
o cavern			
- 0... 500 a	} $4 \cdot 10^{-4}$	} $4 \cdot 10^{-4}$	$4 \cdot 10^{-4}$
- > 5000 a			$1 \cdot 10^{-2}$
o water volume inside the pile of waste packages			
- 0... 500 a	$2 \cdot 10^{-4}$	$2 \cdot 10^{-4}$	$2 \cdot 10^{-4}$
- 500...5000 a	$2 \cdot 10^{-4}$	$7 \cdot 10^{-4}$	$5 \cdot 10^{-3}$
> 5000 a	$7 \cdot 10^{-4}$	$7 \cdot 10^{-4}$	$2 \cdot 10^{-2}$

Table III Apparent diffusivities (D_a) and distribution coefficients (K_d) in materials in the Loviisa repository.

Element	D_a (m^2/s) cemented [1]) waste	D_a (m^2/s) concrete [1])	K_d (m^3/kg) cemented waste	K_d (m^3/kg) concrete	K_d (m^3/kg) crushed rapakivi granite saline gw	fresh gw
C	$1 \cdot 10^{-13}$	$1 \cdot 10^{-14}$	$1 \cdot 10^{0}$	$1 \cdot 10^{0}$	$2 \cdot 10^{-3}$	$2 \cdot 10^{-3}$
Co	$1 \cdot 10^{-13}$	$1 \cdot 10^{-14}$	$1 \cdot 10^{-1}$	$2 \cdot 10^{-1}$	$1 \cdot 10^{-1}$	$1 \cdot 10^{-1}$
Ni	$1 \cdot 10^{-13}$	$1 \cdot 10^{-14}$	$4 \cdot 10^{-1}$	$1 \cdot 10^{0}$	$5 \cdot 10^{-1}$	$6 \cdot 10^{-1}$
Sr	$1 \cdot 10^{-13}$	$1 \cdot 10^{-13}$	$1 \cdot 10^{-3}$	$1 \cdot 10^{-3}$	$6 \cdot 10^{-4}$	$1 \cdot 10^{-2}$
Tc	$2 \cdot 10^{-10}$	$2 \cdot 10^{-11}$	$1 \cdot 10^{-4}$	$1 \cdot 10^{-4}$	$2 \cdot 10^{-4}$	$5 \cdot 10^{-4}$
I	$2 \cdot 10^{-11}$	$2 \cdot 10^{-12}$	$1 \cdot 10^{-3}$	$1 \cdot 10^{-3}$	$2 \cdot 10^{-4}$	$5 \cdot 10^{-4}$
Cs	$4 \cdot 10^{-12}$	$1 \cdot 10^{-13}$	$5 \cdot 10^{-2}$	$1 \cdot 10^{-1}$	$7 \cdot 10^{-1}$	$1 \cdot 10^{0}$
Pu	$1 \cdot 10^{-13}$	$1 \cdot 10^{-13}$	$1 \cdot 10^{-2}$	$2 \cdot 10^{-2}$	$1 \cdot 10^{-2}$	$1 \cdot 10^{-1}$

[1]) Apparent diffusivities in concrete were increased by a factor of 10 at 500 years.

6. RELEASE OF RADIONUCLIDES FROM THE REPOSITORIES INTO THE BEDROCK

6.1 The Loviisa repository

In the scenarios where the Loviisa repository is assumed to be situated in the zone of saline, stagnant groundwater, near-field analysis was expanded to cover diffusion of radionuclides in the rock layer separating the repository from the zone of fresh, flowing groundwater. The thickness of the rock layer between the top of the repository and the upper fractured zone in the bedrock of the island is at least 10 meters. The diffusion through the rock has a clear reducing effect on the releases rates of the radionuclides from the repository into the zone of fresh, flowing groundwater.

For low level waste it was conservatively assumed that the tunnels will be filled with groundwater and the radionuclides will be dissolved immediately after sealing of the repository. In the scenario of fresh, flowing groundwater, release of radionuclides from these tunnels is hence determined by the groundwater flow through the tunnels and maximum release rates appear immediately after sealing of the repository. Low level waste contains, however, only insignificant amounts of long-lived radionuclides and the activities of short-lived radionuclides released from the repository decay effectively during their migration in the geosphere.

The effect of groundwater conditions can be seen by comparing Figures 6 and 7 where the release rates from the cavern for solidified wet waste are presented. In the scenario where the repository is assumed to be situated in the zone of stagnant groundwater, radionuclides have to diffuse through the engineered barriers in the repository and the 10 m thick rock layer to escape from the near-field. Consequently, only long-lived radionuclides are released and the release pulses become flat (Fig. 6). In the other groundwater scenario radionuclides are transported from the repository into surrounding bedrock by groundwater flowing through the repository. The resulting release rates are higher and the effects of the assumed increase of diffusion

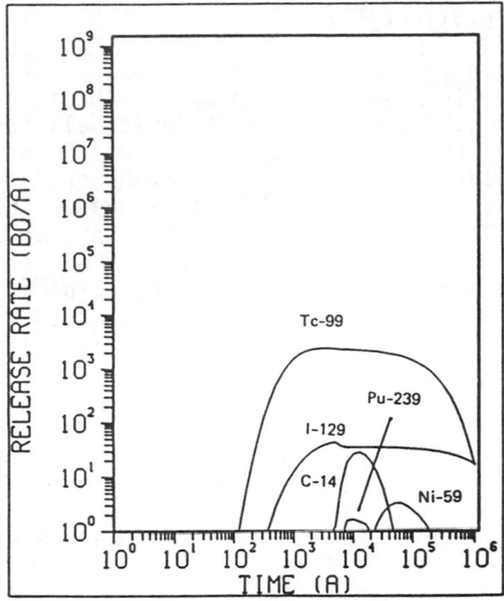

Fig. 6. Radionuclide release rates from the cavern for solidified wet waste in the Loviisa repository; basic scenario; stagnant groundwater.

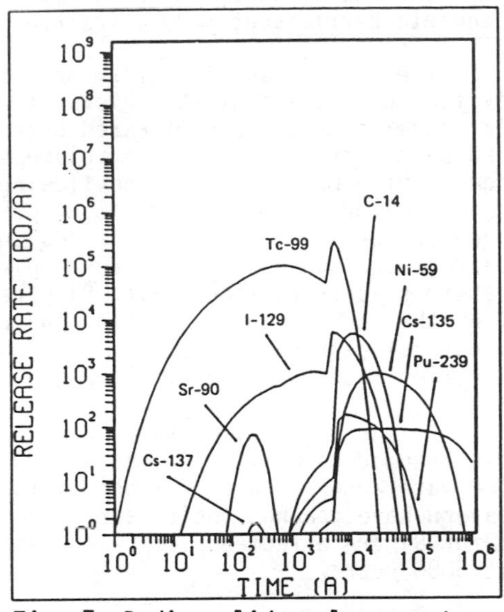

Fig. 7. Radionuclide release rates from the cavern for solidified wet waste in the Loviisa repository; basic scenario; flowing groundwater.

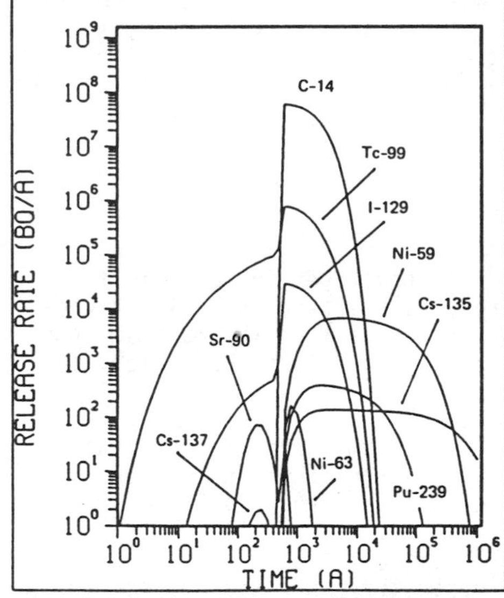

Fig. 8. Radionuclide release rates from the cavern for solidified wet waste in the Loviisa repository; disturbed evolution scenario; flowing groundwater.

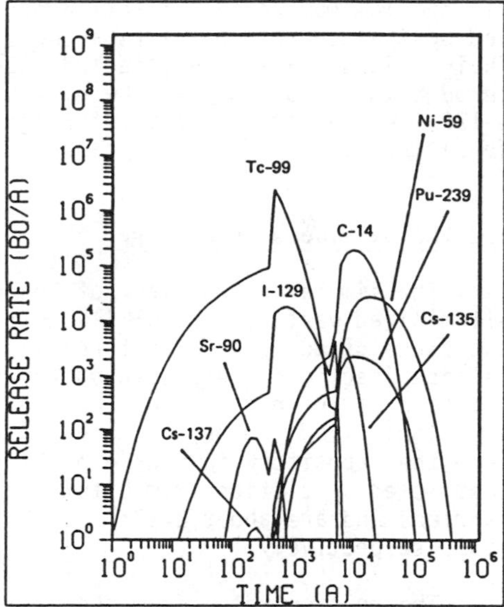

Fig. 9. Radionuclide release rates from the cavern for solidified wet waste in the Loviisa repository; disruptive event scenario; flowing groundwater.

coefficients in concrete at 500 years and the later total degradation of the concrete barriers at 5000 years are more distinct (Fig. 7).

The total loss of performance of engineered barriers after 500 years - evaluated in the disturbed evolution scenario - was modelled by hypothetically assuming that the engineered barriers would instantaneously "emit" all the radioactivity into the water volume of the repository and "be converted" into non-sorbing materials. In the flowing groundwater scenario, the transport of radionuclides from the repository cavern is hence primarily determined by the groundwater flow through the repository. Radionuclide release rates into the bedrock are suddenly and very sharply increased as a consequence of the pessimistic assumptions (Fig. 8). The crushed rock backfilling manufactured from the rock excavated during construction of the repository was regarded as a natural barrier and has a retarding effect on the radionuclides having high sorption factors (Cs-135, Ni-59 and Pu-239).

In the disruptive event scenario the cavern for solidified wet waste was assumed to be intersected by a water transporting cleft at 500 years when the water phase concentrations of the most significant radionuclides in the cavern have reached their peak values. The release rates of radionuclides (Fig. 9) are effected also by the assumed degradation of engineered barriers at 5000 years.

The maximum annual release fractions (maximum annual release divided by the initial inventory) of some long-lived radionuclides from the cavern for solidified wet waste are presented in Fig. 10. It can be observed that the hypothetical total loss of performance of engineered barriers increases above all the maximum release fraction of C-14. Ni-59, Cs-135 and Pu-239 are retarded by the concrete barriers as well as by the crushed rock backfilling. Their release rates are effected most by the increased groundwater flow through the repository in the disruptive event scenario. Release of Tc-99 and I-129, which migrate easily through all the engineered barriers, are not sensitive to scenario assumptions.

6.2 The Olkiluoto repository

The simulated behaviour of long-lived radionuclides in the silo for bituminized waste in the Olkiluoto repository is relieved in Table IV where the place of decay and total release fractions as well as the maximum annual release fractions are presented for the realistic and basic scenarios. In addition of the nuclides presented in Table IV, bituminized waste contain also long-lived iodine and technetium isotopes which will be totally released from the repository in course of time. Their maximum annual release rates are restricted by elution from bitumen and delays due to other engineered barriers and are about $2 \cdot 10^{-4}$ and $2 \cdot 10^{-3}$ in the basic and realistic scenarios, respectively.

The release rates from the Olkiluoto repository into the bedrock are generally somewhat higher than in the case of the Loviisa repository. The main reason is that the groundwater conditions at the Olkiluoto site are not as favourable as at the Loviisa site. The results of the near-field analysis of the Olkiluoto repository are presented and discussed in more detail in ref. [4].

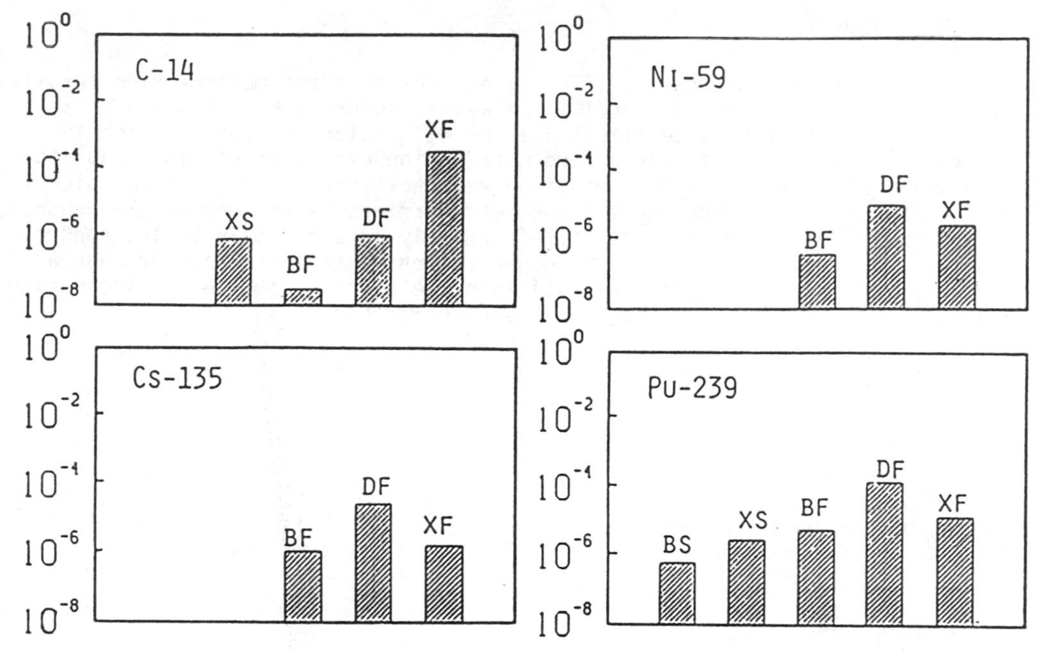

Fig. 10. Maximum annual release fractions of initial inventories of some radionuclides from the cavern for solidified wet waste in the Loviisa repository.

BS: Basic scenario; stagnant groundwater
XS: Total loss of performance of engineered barriers; stagnant groundwater
BF: Basic scenario; flowing groundwater
DF: A cleft intersecting the cavern; flowing groundwater
XF: Total loss of performance of engineered barriers; flowing groundwater

Table IV Place of decay and maximum annual release fractions of some long-lived radionuclides from the silo for bituminized waste in the Olkiluoto repository.

	C-14		Ni-59		Cs-135		Pu-239	
	real.	cons.	real.	cons.	real.	cons.	real.	cons.
Place of decay								
Bitumen	0.11	0.01	0.009	0.001	<0.001	<0.001	0.03	0.003
Water in the silo	0.007	0.002	<0.001	<0.001	<0.001	<0.001	0.07	0.02
Concrete walls	0.88	0.94	0.83	0.33	0.03	0.003	0.75	0.18
Backfilling around the concrete silo	0.001	<0.001	0.11	0.03	0.20	0.007	0.09	0.06
Repository together	0.999	0.96	0.96	0.37	0.23	0.01	0.94	0.27
Released into the geosphere	0.001	0.04	0.04	0.63	0.77	0.99	0.06	0.73
Max. annual release fraction of initial inventory	$5 \cdot 10^{-8}$	$4 \cdot 10^{-6}$	$3 \cdot 10^{-7}$	$2 \cdot 10^{-5}$	$1 \cdot 10^{-6}$	$3 \cdot 10^{-5}$	$8 \cdot 10^{-7}$	$8 \cdot 10^{-5}$

7. DISCUSSION

The results of the near-field analyses show that the multibarrier repository concepts consisting of conditioned waste, concrete barriers and crushed rock backfilling constitute compatible and well balanced systems which in normal conditions effectively retard most radionuclides within the repository. Among the most important long-lived radioactive elements, cesium, nickel and plutonium are retarded by the concrete barriers as well as by the crushed rock backfilling, whereas carbon is effectively retarded only by the concrete barriers. Consequently the assumed degradiation of concrete barriers has a clear effect on release rates of C-14 into the bedrock. Iodine and technetium migrate relatively easily through all engineered barriers.

Several extreme scenarios were considered in the analyses. In the overall performance assessments of the repositories the radiological impacts have been estimated for each scenario. The results show that the location of the repositories at the depths of 50...125 meters in the bedrock ensure safety even if the engineered barriers are assumed to be severely impaired [3]. An addiotional safety feature for at least some thousand years is provided by siting of the repositories near the present coastline so that groundwater flowing through the repositories will be discharged into the sea.

REFERENCES

1. Tusa, E. et al., "Final disposal of low and intermediate level waste from Finnish nuclear power plants", Proc. of Int. Symp. on Siting, Design and Construction of Underground Repositories for Radioactive Wastes. IAEA-SM-289, p. 231-240, 1986.
2. Simola, K., "Loviisa nuclear power station - Final disposal of radioactive waste - Groundwater simulation of the Hästholmen island." Nuclear Waste Commission of Finnish Power Companies, Report YJT-85-32, 1985.
3. Vieno, T., "Safety analysis of disposal of low and intermediate level waste", Proc. of Finnish-German Symposium on Nuclear Waste Management, Espoo, September 1986. (To be published).
4. Vieno, T. et al., "Performance analysis of a repository for low and inter-intermediate level reactor waste", Scientific Basis for Nuclear Waste Management X, p. 381-392, 1986.
5. Winberg, M., "Safety analysis of the disposal of low and intermediate level reactor waste from the Olkiluoto nuclear power plant. Part II: Groundwater flow in the bedrock", Nuclear Waste Comission of Finnish Power Companies, Report YJT-86-21, 1986. (In Finnish).
6. Hietanen, R. et al.,"Sorption of strontium, cesium, nickel, iodine and carbon in concrete", Nuclear Waste Commission of Finnish Power Companies, Report YJT-84-04, 1984.
7. Hietanen, R. et al.,"Sorption of cesium, strontium, iodine, nickel, and carbon in sand in groundwater/concrete environments", Nuclear Waste Commission of Finnish Power Companies, Report YJT-84-04, 1984.
8. Pinnioja, S. et al., "Summary of sorption into Finnish bedrock of the most significant reactor waste nuclides", Nuclear Waste Commission of Finnish Power Companies Report YJT-86-02, 1986. (In Finnish).
9. Muurinen, A. et al., "Diffusion measurements in concrete and compacted bentonite", Scientific Basis for Nuclear Waste Management VI, p. 777-784, 1983.

PERFORMANCE ASSESSMENTS FOR THE DISPOSAL OF LOW-LEVEL WASTES INTO AN IRON-ORE FORMATION

Richard Storck
Gesellschaft für Strahlen- und Umweltforschung mbH München
Institut für Tieflagerung
Theodor-Heuss-Straße 4
D-3300 Braunschweig, Germany F.R.

Abstract

Radioactive wastes producing negligible heat production are planned to be disposed of into a deep iron ore formation in Lower Saxony. For this repository at the Konrad site a performance assessment has been carried out as part of the licensing procedure.

As a first part of the exercise the flow of groundwater has been modelled and scenarios for nuclide migration have been established. The assumptions for the groundwater modelling and the results from the calculations are explained and the resulting scenarios are described.

In the second part the release of radionuclides from the near field has been investigated. The release from near field is influenced by the degradation of the waste forms, the retention of radionuclides by sorption and precipitation and the groundwater flow through the repository system.

Finally in part three of the exercise the migration of radionuclides through the geosphere is considered. The simplified model used for migration calculations has been extracted from the groundwater modelling. For migration modelling, sorption and dispersion phenomena, and dilution by fresh water are taken into account.

The results obtained from the groundwater modelling, the release calculations, and the migration modelling are presented. The relevance of the considered phenomena are discussed.

STOCKAGE DES DECHETS DE FAIBLE ACTIVITE DANS UNE FORMATION DE MINERAI DE FER : EVALUATIONS DES PERFORMANCES

RESUME

Il est projeté de stocker des déchets radioactifs dégageant des quantités négligeables de chaleur, dans une formation profonde de minerai de fer en Basse Saxe. Ce dépôt se trouvant sur le site de Konrad, a fait l'objet d'une évaluation des performances dans le cadre de la procédure d'autorisation.

Dans une première phase des travaux, on a modélisé l'écoulement de l'eau souterraine et établi des scénarios relatifs à la migration des nucléides. La présente communication expose les hypothèses retenues pour la modélisation de l'eau souterraine et les résultats découlant des calculs, et décrit les scénarios ainsi obtenus.

Au cours d'une deuxième phase, on a étudié le rejet de radionucléides à partir du champ proche. Ce rejet est influencé par la dégradation des formes de déchets, par la rétention de radionucléides par sorption et précipitation, et par l'écoulement de l'eau souterraine à travers le dépôt.

Enfin, au cours d'une troisième phase des travaux, la migration des radionucléides à travers la géosphère est examinée. Le modèle simplifié utilisé pour les calculs de migration, a été tiré de la modélisation relative à l'eau souterraine. Pour modéliser cette migration, on tient compte des phénomènes de sorption et de dispersion, ainsi que de la dilution par l'eau douce.

Les résultats obtenus à partir de la modélisation de l'eau souterraine, les calculs concernant les rejets et la modélisation de la migration sont présentés. La pertinence des phénomènes étudiés est examinée.

Introduction

A potential site for the disposal of radioactive waste producing negligible heat in the Federal Republic of Germany is the abandoned iron-ore formation at the Konrad site. This site is located in the state of Lower Saxony close to the city of Braunschweig.

The Konrad mine was opened in 1960 and operated over 16 years mining some seven million tons of iron ore. After the closure of operation, the mine was object to feasibility studies for the disposal of radioactive waste. At the end of an extensive reseach and development work the general feasibility of the site was concluded. Then the licensing procedure was started in 1982 which is still under way. As part of this procedure the long term performance assessment was carried out for the normal evolution of the site.

Repository Planning

The actual planning of the repository is for 650 000 m^3 of waste volume. The waste will be disposed of in drifts of 40 m^2 cross section. The total volume of drifts necessary for disposal is 1 128 000 m^3. These drifts are especially excavated in a depth between 800 m and 1300 m. The existing mined out rooms with a total volume of 2 111 000 m^3 will not be used for disposal.

For disposal of the waste containers, a stacking technique will be applied. Remaining void volumes between the containers will be backfilled with crushed host rock. The diposal drifts will be sealed and all other drifts will be backfilled too. The two shafts will be sealed to a permeability which should be as low as the surrounding formation.

At the end of the operational phase of the mine the remaining void volume in the whole disposal area will be 739 000 m^3. This includes the porosity of the backfill and the void volume within the containers.

The Scenario

The mine workings at the Konrad site are located within the iron ore formation called Oxford. This is a low permeable formation which causes the dry conditions in the mine today. Over the long term during the post-operational phase the residual void volume in the backfilled repository will get resaturated from surrounding aquifer layers through the low permeable host formation and the neighboring clay layers. After resaturation and subsequent increase of the hydrostatic pressure, a groundwater flow through the repository area is expected. This water will carry the mobilized radionuclides from the repository through the geosphere to the subsurface area.

Water Intrusion

Although the conditions in the mine today are rather dry, some amount of water infiltration from the surroundings can be observed. By balancing the humidity of the air in the mine a flow rate of about 50 l/min out of the mine can be calculated.

Assuming a confined horizontal aquifer for the host formation, analytical models can be derived to calculate the infiltration rate as a function of time. Extrapolating this approach into the future, the fill up period for the total void volume in the repository can be calculated to about 100 years.

After resaturation of the void volume the flow of water towards the repository area will continue to increase the groundwater head in the repository area. The undisturbed conditions in the groundwater head will be reached asymptotically. Even after 10 000 years there is still a difference to the undisturbed conditions of a few percent.

In the water intrusion modelling, changes in void volume in the repository from backfill consolidation due to the rock pressure have not been taken into account. Because this has to be expected to some degree the above mentioned effects are neglected for performance assessments. An instantaneous resaturation and pressure increase are assumed at the beginning of the post-operational phase.

Groundwater Movements

To identify potential pathways from the repository to the biosphere and to establish models for nuclide migration, the goundwater flow pattern at the site has to be identified.

For groundwater modelling purposes a model area has to be fixed. The boundaries of the model area approximately follow the course of water divides and effluent streams or the course of impermeable formations such as salt domes. Simplified, they are prescribed as closed boundaries.

The model covers an area of 48.7 km x 13.5 km down to a depth of 2 400 m. The model volume is devided into 10 350 blocks for finite difference modelling with the code SWIFT.

The groundwater flow at the site is influenced mainly by groundwater recharge in the elevated region at the southern boundary and groundwater discharge in the low lands at the northern boundary. As shown in Figure 1, the maximum difference in groundwater head is about 150 m over a horizontal distance of 40 000 m.

The groundwater flow is calculated for two sets of permeabilities. The difference between the two sets is only in the permeability of the Unterkreide, which is a clay layer overlying the host formation. From the calculated velocity fields, pathways are obtained by performing particle tracking. The resulting pathways for the two sets of permeabilities are characterized in Table I and also shown in Figures 1 and 2.

For the higher permeability of the Unterkreide, there is a shorter pathway to the biosphere through the Unterkreide itself. For the lower permeability, the pathway stays within the host formation until it reaches the surface. In both cases the total flow rate through the repository area is calculated which gives the conditions for the near field modelling.

Table I: Groundwater modelling results: pathways through the geosphere

UNTERKREIDE PERM. in m/s	MIGRATION THROUGH:	PATHLENGTH in m	TRAVEL TIME in yr	FLOW RATE in m^3/yr
10^{-10}	Unterkreide	3 800	370 000	3 200
10^{-12}	Oxford	33 300	338 000	1 620

Modelling

The modelling is carried out seperately for the repository area, the geosphere, and the biosphere.

Near Field Modelling

In order to determine the release of radionuclides from the backfilled repository area, the following phenomena have to be considered:

- release of radionuclides from the waste package
- precipitation of radionuclides
- sorption of radionuclides on the backfill material
- water exchange within the repository
- flow of water through the repository area

A change in the total void volume in the backfilled repository area is neglected because this is expected to be a very slow process.

The release of radionuclides from the waste package only accounts for the waste matrix. Due to the possible mechanical load of the rock pressure on the containers, the barrier effect of containers are not taken into account. For modelling the release from the waste matrix, element-specific mobilization rates are taken from experiments which result in a total release after at least 600 years. The data used are given in Table II.

Precipitation of radionuclides is considered, if solubility limits are reached. Element-specific solubility limits are derived from laboratory experiments and experts opinion. For sorption of radionuclides the backfill material as well as the surface of the host rock is considered. The sorption process is modelled by constant Kd-values which are obtained from laboratory experiments using original rock samples. Solubility and sorption data are also given in Table II.

Water exchange within the repository is forced by the water flow through the repository area and also by small temperature and density effects as well as by gas production from corrosion and radiolysis. Because there are no distinct flow resistances in the drift system which prevent large scale convection of water an instantanous mixing up of mobilized radionuclides in the whole repository area to a homogeneous concentration is assumed.

Table II: Element-specific data sets for near field modelling

ELEMENT	MOBILIZATION PERIOD FOR CONCRETE WASTE PACKAGES in year	SOLUBILITY LIMITS in mol/l	SORPTION COEFFICIENT in cm^3/g
Cl	10	$1\ 10^{-2}$	0
Ca	15	$1\ 10^{-2}$	0
Se	40	$1\ 10^{-2}$	0.1
Tc	40	$7\ 10^{-4}$	0.4
I	10	$1\ 10^{-2}$	0
Pu	600	$2\ 10^{-7}$	500
Np	600	$3\ 10^{-5}$	70
Th	600	$1\ 10^{-7}$	200
U	600	$1\ 10^{-4}$	8

The groundwater flow rate through the repository area is derived from the groundwater modelling and is given in Table I. This flow rate is assumed to intersect the whole void volume in the repository. Taking a flow rate of 1620 m^3/a and a total volume of 739 000 m^3, an exchange of water will occur every 456 years.

For near field modelling two submodels have been developed and integrated into the code EMOS which calculates the release of the radionuclides from the near field. One submodel is for the release of radionuclides out of the waste package. The other submodel covers all the other phenomena mentioned for the repository area. The near field modelling is schematically represented in Figure 3.

Far Field Modelling

In order to determine the migration of radionuclides through the geosphere the following phenomena have to be considered:

- advection of radionuclides by groundwater movement
- dispersion and diffusion of radionuclides
- sorption of radionuclides
- dilution of nuclide concentration

For handling the migration of radionuclides along the two pathways, one-dimensional transport models are established. The numerical calculations for these models are performed with the one-dimensional version of the code SWIFT. The parameters characterizing the pathways through the geosphere and the movement through the different geological formations along the pathways are taken from particle tracking procedures which are the basis for the data given in Table I.

Dispersion of radionuclides is considered for the longitudinal direction only. Dispersion lengths of 30 m for the short pathway and of 200 m for the longer pathway are applied.

The sorption process is modelled in the same way as for the near field by assuming constant Kd-values. Data are taken from laboratory experiments for the different geological formations and chemical environments along the pathways.

During transport of radionuclides along a pathway, addition of uncontaminated water will reduce the radionuclide concentration. For the pathway through the host formation, fresh water will be added from underlying aquifer systems, which at the end results in a dilution by a factor of seven. When entering the subsurface area, additional fresh water in the quartenary aquifer will cause a further dilution by a factor of ten. For the pathway through the Unterkreide, the dilution is restricted to the effect in the quartenary formation only.

Biosphere Modelling

In order to determine the radiation exposure to an individual the following pathways to men have been considered:

- drinking water
- watering
- irrigation of grass, vegetable and corn
- fishing in surface lakes

Model approaches and data for the biosphere pathways considered are taken from national regulations. The consumption rates for the food chains are based on present conditions of life. Other conditions such as irrigation of agricultural areas and the plant species in use are taken as they are today. Individual doses are calculated for a maximum exposed individual as effective dose equivalent per year of intake. Dose factors required are taken from ICRP-30.

Results

Release rates from the near field for selected radionuclides are shown in Figure 4. The results are related to a flow rate of 1 620 m^3/a. The main results from the near field modelling are:

- Nuclide concentrations and release rates are effected by the mobilization only over a relatively short period of time. Hence, the mobilization from the waste package is of minor importance.

- The release time is determined by the sorption process. Non-sorbing elements such as Iodine are released according to the water exchange period in the repository of 456 years. The release of sorbing elements such as Uranium and Plutonium are delayed in accordance to their retardation factors.

- Precipitation of radionuclides is limited to Thorium due to its low solubility limit and large amount in terms of mass.

Contamination of subsurface groundwater is obtained by applying the release rates as input to the geosphere transport modelling. The main results are:

- Several radionuclides such as Ni-59, Cs-135, Np-237, and Pu-239 will never reach the subsurface area due to travel times which are much higher than their lifetimes.

- Contaminitation of the subsurface water will occur mainly at two points in time: non-sorbing radionuclides such as I-129 and Cl-36 appear after groundwater travel time of about 300 000 years for the pathway through the Oxford formation. The transport of Uranium isotopes causes contamination after more than 10 million years from the Uranium isotopes themselfes and their daughter products. The delay to the groundwater transport results from the sorption of Uranium.

From the contamination of the subsurface water, the radiation exposure is calculated at the time of its maximum. Comparing the different pathways through the biosphere, the irrigation pathway contributes mostly to the individual exposure.

As the final result of the exercise the individual doses are given. Figure 5 shows the doses as function of time for the pathway through the Oxford formation. Table III gives the maximum doses for both pathways. The content of Table III is limited to radionuclides with contributions of more than 10^{-10}Sv/a. The results can be summarized as follows:

- The pathway through the Oxford formation will result in subsurface contamination mainly after 300 000 years and after 10 000 000 years.

- The maximum dose at the first peak is caused by I-129 and the maximum dose at the second peak is caused by Uranium isotopes and their daughter products.

- The pathway through the Unterkreide formation will result in similar doses as for the Oxford pathway. The contamination from I-129 is delayed due to sorption effects in subsurface formations.

- The near field phenomena are of minor importance due to the relation between release times and travel times.

- The retention and dilution effects in the geosphere are of major importance.

The calculated radiation exposures to a maximum exposed individual for the normal evolution of the Konrad site are in the order of the variations in natural radiation exposure.

Table III: Inventories of the repository and calculated individual doses

Nuclide	Half-Life-Time in a	Inventory in kg	Pathway Unterkreide Time in a	Pathway Unterkreide Dose in Sv/a	Pathway Oxford Time in a	Pathway Oxford Dose in Sv/a
Cl- 36	3.0E05	8.9E-2	3.03E5	6.0E-08	3.22E5	3.0E-08
Ca- 41	8.1E04	3.8E-2	7.26E5	1.7E-11	3.06E5	1.2E-09
Se- 79	6.5E04	4.7E-2	1.05E6	1.0E-13	6.00E5	1.9E-10
Tc- 99	2.1E05	4.2E 1	1.12E6	4.2E-08	2.22E6	5.1E-10
I -129	1.6E07	2.4E 1	3.69E6	1.4E-06	3.28E5	6.8E-06
U -236	2.3E07	1.8E 2	4.80E7	4.5E-08	1.10E7	3.4E-07
Th-232	1.4E10	5.7E 4	5.00E8	1.6E-08	4.00E8	8.8E-09
U -238	4.5E09	5.4E 5	5.67E7	1.3E-06	1.10E7	3.0E-06
U -234	2.4E05	3.7E 0	5.67E7	1.4E-06	1.10E7	3.4E-06
Th-230	7.7E04	8.1E-2	5.67E7	1.8E-08	1.10E7	1.5E-08
Ra-226	1.6E03	3.8E-3	5.67E7	4.0E-07	1.10E7	1.1E-05
Pb-210	2.2E01	2.5E-5	5.67E7	1.4E-05	1.10E7	4.2E-06
U -235	7.0E08	2.5E 3	5.67E7	4.7E-08	1.10E7	1.1E-07
Pa-231	3.3E04	9.7E-3	5.67E7	6.9E-09	1.10E7	7.2E-09
Ac-227	2.2E01	6.3E-6	5.67E7	5.8E-08	1.10E7	3.0E-07

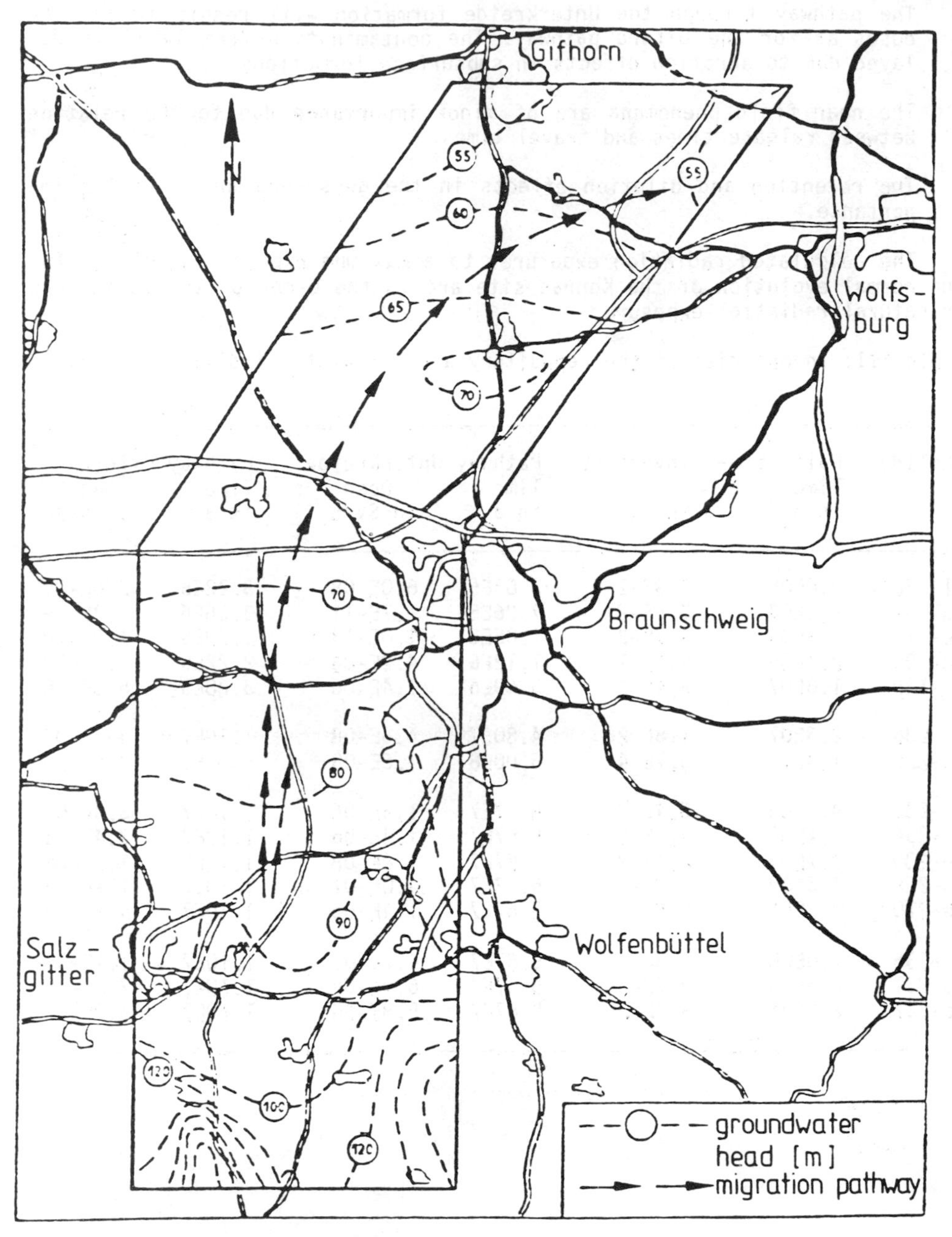

Figure 1: Model Area: boundaries, groundwater heads and pathways

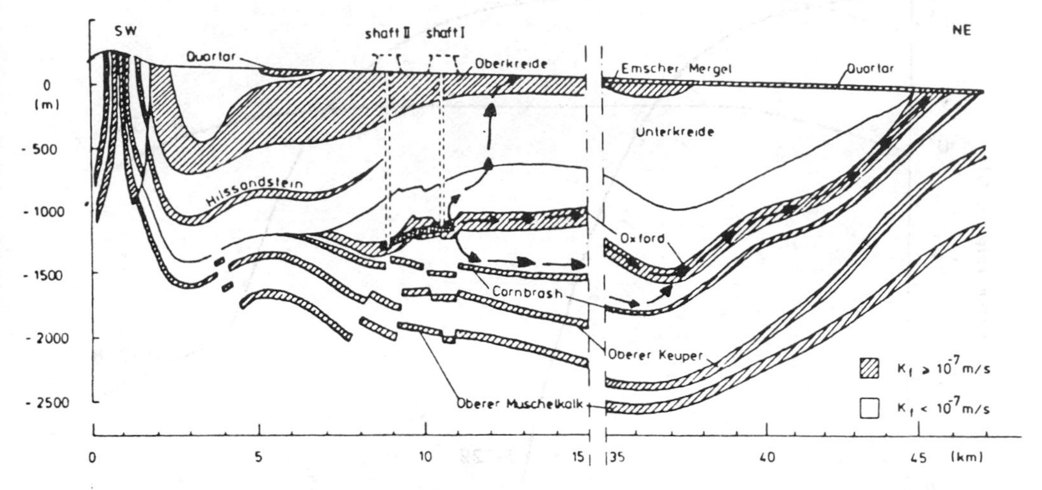

Figure 2: Vertical cross section along the pathway

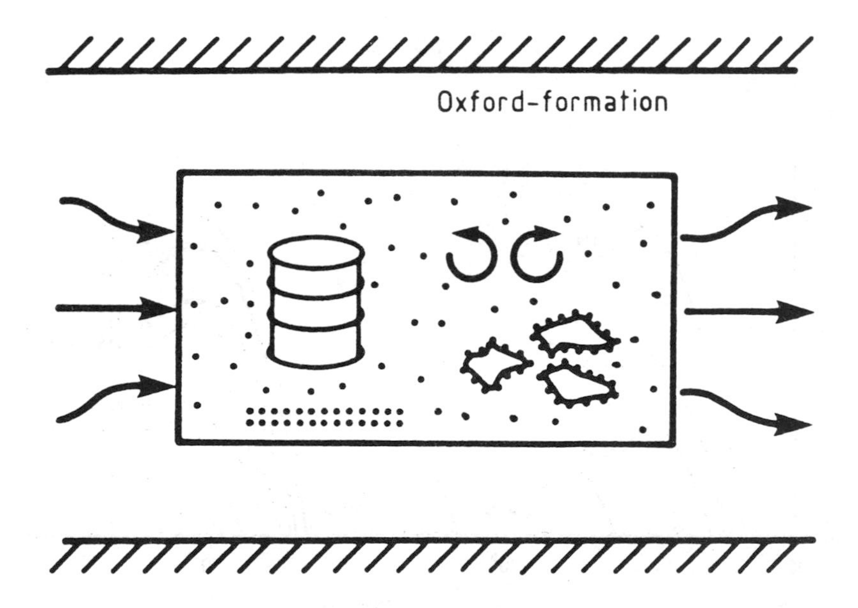

Figure 3: Schematic representation of the near field modelling

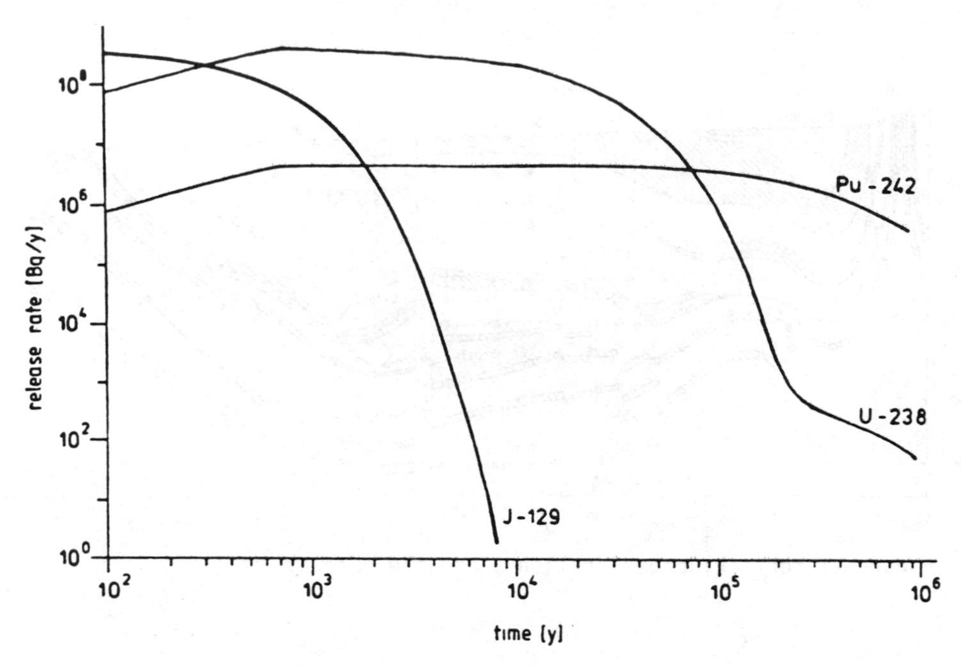

Figure 4: Release rates from near field for selected radionuclides and for a flow rate through the near field of 1620 m^3/a

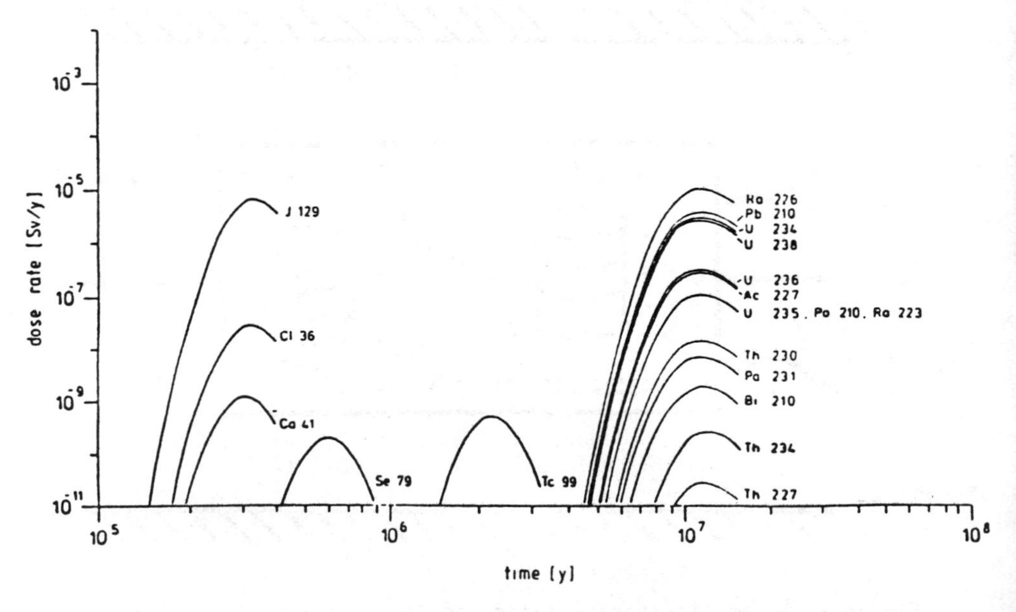

Figure 5: Individual doses for the pathway through the Oxford formation

IMPACT OF PERFORMANCE ASSESSMENT ON SWISS PROJECTS FOR A L/ILW REPOSITORY

C. McCombie, F. van Dorp, P. Zuidema
Nationale Genossenschaft für die Lagerung radioaktiver Abfälle
Nagra, 5401 Baden, Switzerland.

ABSTRACT

This paper gives an overview of performance assessments for the near field of a Swiss L/ILW repository. The first major applications in Switzerland were for the feasibility study "Project Gewähr 1985". Currently, increased efforts are being made for studies aimed at design and implementation of actual repositories; particular emphasis is placed upon optimization of site investigations and repository design, the selection of materials for engineered barriers and on detailed study of the impact on safety of waste specifications. For the near-field assessment mechanical, chemical, microbial, hydrological and nuclide transport processes are taken into account; current modelling and data requirements for these different topics vary considerably.

INCIDENCES DE L'EVALUATION DES PERFORMANCES SUR LES PROJETS SUISSES RELATIFS A UN DEPOT DE DECHETS DE FAIBLE ET DE MOYENNE ACTIVITE

RESUME

La présente communication donne un aperçu général des évaluations des performances relatives au champ proche d'un dépôt suisse de déchets de faible et de moyenne activité. Les premières applications importantes en Suisse étaient destinées à l'étude de faisabilité intitulée "Project Gewähr 1985". A l'heure actuelle, on consacre des efforts accrus à des études visant la conception et la réalisation de dépôts réels ; en particulier, l'accent est mis sur l'optimisation des recherches relatives au site et de la conception du dépôt, la sélection de matériaux destinés aux barrières ouvragées et sur l'étude détaillée des incidences sur la sûreté qu'ont les spécifications des déchets. Pour l'évaluation du champ proche, il est tenu compte des phénomènes mécaniques, chimiques, microbiens et hydrologiques, ainsi que des processus de transport des nucléides ; les besoins actuels en matière de modélisation et de données afférents à ces différents sujets varient considérablement.

1 INTRODUCTION

In principle, it is recognized that performance assessment should play a major rôle in the choice of concept, site and engineering design for radioactive waste repositories. Application of performance assessment in an iterative manner has been recommended in various published guidelines /1/,/2/,/3/. In practice, however, complete system assessments tend to be carried through when disposal projects are well advanced and technical options largely settled by other, less formal decision mechanisms. A basic problem with the application of quantitative assessments at an early stage is that the database can be so uncertain that the results predicted by the models do not allow one to discriminate between alternative options being compared.

Nevertheless, in successive phases of Swiss projects for the development of repositories for L/ILW, support for decision making has been provided by performance assessments of differing levels of sophistication and detail. Early work on selection of sites was guided by only qualitative input from safety assessors. The formulation of site characterization plans, however, has been based on increasingly explicit results from system modelling. Furthermore, formal requirements for a rigorous feasibility study (Project Gewähr /4/), led to rather detailed assessments at an early phase; the results of these are being used to guide further work. Finally, at the current stage in repository project development, interest is focussed upon the impact of performance assessment on choice of specific repository layout and on engineered barrier designs.

In this paper, we begin with a brief overview of the concepts for a Swiss L/ILW repository and of our modelling approaches. Emphasis is placed thereafter on the specific lessons learned from assessments to date and on the current modelling applications accompanying repository system design and implementation work.

2 SWISS CONCEPTS FOR L/ILW REPOSITORIES

Current Swiss planning allocates all wastes arising from our relatively small nuclear power programme to one of two repository types. A deep repository (approx. 1000 m) is considered for HLW and possibly limited volumes of longer-lived ILW. All other wastes are intended for disposal in rock caverns with some hundreds of metres of overburden; no near surface disposal is planned. The topography of those potential L/ILW sites which have already been selected makes access via a horizontal tunnel the obvious practical alternative. Combined tunnel and shaft designs are also feasible should this prove to be justified by safety considerations.

For the 240 GW(e)-y nuclear programme assumed in Project Gewähr, Table I gives an overview of the quantities and types of wastes involved and Table II summarizes the radionuclide contents of the important long-lived radioisotopes in the wastes. The current planning basis has been reduced to 160 GW(e)-y. It is assumed that L/ILW for disposal may originate from plant operation, from decommissioning or from reprocessing abroad. All wastes are solidified, the main matrix being cement although some bitumen and smaller quantities of polymers are included. The wastes will be stacked in an orderly fashion and remaining void spaces completely backfilled. The disposal caverns will be lined but the detailed design of the liner is not yet defined.

The first three potential sites selected by narrowing in (using criteria ranking by expert opinion) from an original 100 possibilities are in different host rocks; Piz Pian Grand in crystalline, Bois de la Glaive in anhydrite and Oberbauenstock in marl. A fourth site at Wellenberg, also in marl, has been added to the original list. The locations of all these sites are given on the map in Figure 1.

To date, first field investigations involving drilling, hydrologic testing and geophysical measurements have been carried out at Piz Pian Grand and Oberbauenstock. It should be noted that existing underground workings (auxiliary facilities for a road tunnel) at Oberbauenstock had previously yielded geotechnical data which led to this site being selected as a model for the Project Gewähr feasibility study. Local opposition has hampered intensive work at Bois de la Glaive. At Wellenberg a major field campaign is planned from the middle of 1988, assuming that all legal permits which are required in Switzerland for site studies have been obtained by then.

3 NEAR-FIELD PERFORMANCE ASSESSMENT MODELLING

No attempt is made in this paper to describe in detail the suite of calculational models used for performance assessments of Swiss L/ILW repository projects. Instead we briefly note the scientific/technical issues studied and indicate the modelling approaches which have been used to date. In Table III we give a concise list of the main physical and chemical processes considered in assessing the performance of the near-field of our L/ILW repository in the post-closure phase. Swiss efforts to date have been concentrated upon the long-term behaviour of the repository but significant impacts on repository planning are expected also from analyses of the operational phase. Most modelling has been at the detailed level treating single processes; no extensive use has been made of integrated models which take into account the coupling of various individual processes. Also, all modelling to date has been deterministic in nature. The near-field modelling has, however, been performed in the scope of total system analyses of a complete disposal system and information has thus been gained on the sensitivity of system behaviour to different near-field processes. The linkage of different near-field data sets and model calculations is illustrated schematically in Figure 2.

4 PERFORMANCE ASSESSMENTS OF L/ILW DISPOSAL SYSTEMS

4.1 Project Gewähr studies

The model system studied within the scope of the feasibility "Project Gewähr" was a horizontally accessed repository in a marl site at Oberbauenstock in central Switzerland. A schematic overview of the system is shown in Figure 3. The waste inventory was assumed to be 200'000 m^3 of cement-solidified L/ILW consisting of drums packed into large containers (2 x 2 x 4 m) and backfilled with cement grout. The entire repository caverns were also assumed backfilled with a special cement with high plasticity.

Analyses of near-field nuclide releases were performed for different scenarios and parameter values. Geosphere transport takes place in a network of localized disturbed zones (Ruschelzone) in the marl and different biosphere scenarios are treated. In addition to the groundwater release paths, direct release at long times following removal of overburden by extreme erosion was considered. Direct exposures due to human intrusion (construction of further tunnels) were also estimated.

The overall results of the assessments performed indicated that potential radiation doses are below the Swiss regulatory guidelines (0.1 mSv/y). At a more detailed level, relevant information for guiding further work was obtained on a range of near-field issues which are discussed in more detail later

4.2 Assessments accompanying site characterization work

The Project Gewähr studies of Oberbauenstock gave a starting basis for defining a site characterization programme. The original proposal was to drive a pilot tunnel into the potential siting zone at a relatively early stage of the site work. Government permits for geologic investigations, however, restricted first studies to drilling and geophysics. This means that characterization in depth has not yet been possible, especially since the topography gives rise to access problems which rule out extensive exploration from the surface. From existing tunnels near the disposal area, however, three bore-holes of varying depths have been drilled with appropriate down-hole hydrologic and geophysical testing; in addition seismic investigations have been performed along the tunnels.

The field programme was chosen to provide information for safety assessment modellers concerning the host rock geometry, the hydraulic conductivities of host rock and underlying formations, the gas permeability of marl, the detailed characteristics of water bearing fractures and the groundwater chemistry. All of the results from Oberbauenstock Phase 1 influencing further repository planning have not yet been fully evaluated. A novel result influencing the study of the repository near-field is that anomalously low hydraulic heads have been measured; these are most likely to be due to transients caused by the existence of de-gassed zones within the marl host rock.

Results of analyses of measurements from Oberbauenstock are providing direct input to formulation of the testing programme at the second marl site at Wellenberg. The location of boreholes and the corresponding test programmes at Wellenberg have been based upon extensive hydrogeologic modelling of the siting area. An important additional feature of the work at Wellenberg is that more specific testing of the host rock is planned from an initial test cavern at the edge of the formation. This should provide more data on near-field hydrology, rock stresses, decompression, excavation damage, discrete flow systems and sealability. However, relevant rock stresses and the effects of excavation can only be measured under appropriate overburden once one has tunnelled into the actual potential disposal area.

4.3 Assessments in support of disposal facility design

In Project Gewähr, performance assessments influenced system design primarily in that strong emphasis was placed upon designs of maximum simplicity which allow transparent analyses, easily checked in a formal review procedure. Relatively little attention was paid to aspects of practicability and economics. Now, projects are more advanced and analyses must be more realistic. Designs are therefore being developed on a site-specific basis; in particular technical concepts are being developed for the marl sites. Specific decisions with direct input from near-field performance assessment are the design of the engineered barriers (which is influenced by the hydraulic and gas permeabilities) and the layout of the tunnels and caverns (which is influenced by site hydrology and by the characteristics of the potential damaged zone around the excavations). These topics are discussed later.

5 KEY NEAR-FIELD ISSUES IDENTIFIED IN ASSESSMENTS TO DATE

In this section we summarize those results and conclusions of performance analyses to date which impact most upon investigation programmes accompanying repository development and in the following section 6 we attempt to narrow in to specific design issues.

5.1 Waste specific issues

For different release scenarios, maximum potential exposures can result from particular, unexpected waste types and specific radionuclides. One example is Cl-36 which is present in wastes due to activation processes. Although quantities are small, the nuclide is mobile and pessimistic scenarios involving release at a spring used by relatively few persons can lead to dose predictions for a small population not far below the regulated 0.1 mSv/y limit. A different example concerns exposure by extreme erosion in the far future; here the actinides (primarily from reprocessing wastes) predominate and doses are predicted which, although below natural levels, are around the defined limit.

In general, we can note the following, waste-specific points which are highlighted by the results of performance assessment studies:

- radionuclides of concern for groundwater release are not normally the actinides, but rather the high-solubility, low-sorption nuclides (e.g. Cl-36, I-129)

- such nuclides can be distributed over waste streams from plant operation, reprocessing and decommissioning

- dependent upon the scenario considered, either the total inventory of radionuclides or the specific activity can be more important

- not only the nuclide inventory, but also the chemical form of wastes can be important since this determines, for example, complexation potential, corrosion gas production etc.

The performance assessments therefore directly give guidance for investigations on waste inventories, on relevant sorption measurements and on experimental programmes on degradation of matrices.

5.2 Backfill requirements

The physical and chemical evolution of the backfill strongly affects the time-dependent release of radionuclides from the near field. If the hydraulic conductivity of the backfill is lower than that of the surrounding media, nuclide releases can take place primarily by slow diffusion rather than advection even in host rocks with significant groundwater flows. This is discussed in detail in a further paper in the workshop /4/. On the other hand, generation of significant amounts of gases cannot be excluded and these should be able to penetrate the backfill without inducing large pressures, so that a minimum gas permeability is required. The behaviour of gas is also covered in more detail in another paper /5/. The chemical composition of backfill (and waste matrices) should be as well defined as possible in order that chemical thermodynamic models can be applied to determine the evolution of pore water chemistry and hence elemental solubilities and speciation /6/. The amount of soluble radionuclides in the pore waters will be reduced if significant quantities can sorb on the backfill. For mechanical reasons and to avoid fluid pockets, filling should be as complete as possible. Plasticity in the backfill would be beneficial in softer host rocks with significant creep potential.

5.3 Liner for disposal caverns

For those wastes requiring increased levels of isolation a hydraulic barrier between backfilled waste and host rock is desirable. The design of such liners must allow for the presence of gas in the repository and for the slow application of lithostatic pressures in host rocks with significant creep. An external "crushable" zone may delay mechanical damage which leads to increased conductivity of the liner. Figure 4 gives an example of a potential near-field barrier system allowing for the factors mentioned. A significant problem is to predict the evolution of the hydraulic parameters determining the efficiency of the liner. Increased cracking may be partly compensated by blocking of fissures due to reprecipitation of materials dissolved by the groundwater.

5.4 Engineered barrier / host rock interactions

The requirements on liner (and on backfill) depend, of course, on the state of the host rock immediately around the caverns. If a decompressed zone leads to a much increased conductivity around the excavations, then this can provide a useful drainage or hydraulic cage effect - although there are problems in demonstrating that gradual blockage would not occur. Around the access tunnel as such, a decompressed conductive zone tends to provide potential short circuits to the biosphere and this must be countered by tunnel layout or by good sealing measures. The chemistry changes caused by the large amounts of concrete in the repository can have effects which penetrate out into the host rock. In particular, attention should be paid to the possible occurrence of sharp pH gradients since these can encourage precipitation processes or colloid formation.

6 DESIGN ISSUES WITH INPUT FROM NEAR-FIELD PERFORMANCE ASSESSSMENT

A major decision dependent upon the results of performance assessment is, of course, the inventory of wastes which can be stored in a given repository. Using safety analyses, acceptance critera for waste streams and even for individual waste packages can be derived. In parallel with the development of such criteria, appropriate quality assurance techniques and procedures must be defined; the challenge here is to determine a minimum suite of practicable measures which provide the information needed for assessment.

Once the wastes are within the repository, it is possible that they might need to be segregated at different locations. In proposed Swiss disposal facilities, the local hydrology can lead to significant differences in flow paths and transport times to the biosphere from different parts of the repository. In any case, separation of chemically different wastes will provide a simpler basis for assesssment and the varied handling requirements for differing activity levels will obviously influence the disposition of wastes in the facility. Since wastes are produced in independent nuclear facilities, where form and content have sometimes long-since been settled, there is limited opportunity for major design input from performance assessment to the specification of the actual wastes.

For the choice of engineered barriers within the repository more freedom exists in principle. Backfill can be non-existent, composed of material previously excavated, or specially tailored to the repository. In the Swiss case, the last option is favoured. Preferred material compositions have been proposed and programmes are being initiated on measurement of relevant properties and on engineering feasibility. Performance assessments help quantify the requirements on relevant parameters (e.g. those determining hydraulic conductivity, gas permeability, pore water chemistry and their long-term evolution). An important point to note in material selection is that potential advantages from tailoring the near field (by, for example, selective use of different additives) can be outweighed by the problems caused for the chemical modelling by the increased complexity of the system.

The liner must also be designed using input from assessments. Conventional engineering practice would lead to liners (if any) which provide support for a limited time allowing for continued drainage and for recurring inspection. In repository engineering, greater emphasis must be placed upon designs of maximum simplicity which allow for gas escape whilst providing a passive hydaulic barrier. There are differing attitudes towards the issue of engineered venting. Conceptually it is unattractive to deliberately introduce perforations into a liner system intended to isolate the wastes from groundwaters; on the other hand, if rupturing due to gas pressure build-up cannot be excluded, it is better to design the leakage positions into the system. It can then be easier to predict convincingly the performance of the repository near field over long times. The mechanical behaviour of the liner must also be predicted over long timescales within which lithostatic pressures will come to bear upon the filled repository.

The design of liners, the choice of excavation methods, the layout of caverns and the concepts for repository sealing can all depend upon the influence of any damaged zone caused by mining. It is relatively straightforward to model the effects of damaged zones but the necessary data is almost non-existent. Some work has been done on hard rocks, but in the marls of current interest in Switzerland little is known. Furthermore, relevant measurements will become possible only when one has access to test facilities in the proper host rock under the correct overburden. This means that preliminary design work must be based on conservative assumptions and must be flexible enough to allow for subsequent adaption as more and better data become available.

One example is the approach being used to plan access tunnel layout at the Wellenberg site. Extensive hydrologic modelling of the site gives indications of flow directions in the marl; the initial field measurements from the boreholes are designed to validate these predictions. Finally, a preliminary layout in which a long access tunnel, running in part against the undisturbed hydraulic gradient, is designed as a conservative starting point. Figure 5 shows a typical potential layout of caverns and tunnel at Wellenberg. During the actual site investigations, a more economic, direct tunnel layout can replace the first version if the gradients are different or if the potential flow paths along a sealed tunnel can be neglected because the sealing (including that of any damaged zone) is of sufficiently low hydraulic conductivity. Safety assessor, engineer and geologist must cooperate in designing field programmes from the surface or from the test cavern which will provide the necessary information for the required near-field hydrologic modelling.

7 SUMMARY AND CONCLUSIONS

Performance assessment has proceeded in Switzerland in parallel with development of concepts and designs for L/ILW repositories. The results of the assessments have already provided direct input for planning of investigation programmes in the laboratory and in the field; current work is aimed at contributing to the optimized planning of the technical barriers within the facility and of the operational procedures. In this paper we have identified and discussed specific near-field issues in which performance assessments play a crucial rôle. Future work will become increasingly detailed as site data accumulate, results from in-situ experiments become available and engineering designs become firmer. It is worthwhile to integrate performance asssessment at the earliest stages possible in order to minimise problems at the final stages of repository planning - namely the preparation of defensible documentation quantifying the achievable safety of the final disposal system to be implemented.

REFERENCES

/1/ IAEA: "Underground Disposal of Radioactive Wastes - Basic Guidance"; IAEA Safety Series No. 54, IAEA, Vienna/Austria, 1981

/2/ IAEA: "Safety Assessments for the Underground Disposal of Radioactive Wastes"; IAEA Safety Series No. 56, IAEA, Vienna/Austria, 1981

/3/ IAEA: "Control of Radioactive Waste Disposal into the Marine Environment - Procedures and Data"; IAEA Safety Series No. 61, IAEA, Vienna/Austria, 1983

/4/ Nagra: "Feasibility Study Project Gewähr 1985"; Nagra NGB 85-01 to NGB 8508 (in German) and NGB 85-09 (in English), Nagra, Baden/Switzerland, 1985

/5/ Wiborgh M., Pers K., Höglund L.O., van Dorp F.: "Modelling of radionuclide transport in the near field of Swiss L/ILW repositories"; Paper presented at the NEA-Workshop on Near-field performance assessments, Baden/Switzerland, 1987

/6/ Zuidema P., Höglund L.O.: "Impact of production and release of gas in a L/ILW repository - A summary of the work performed within the Nagra programme"; Paper presented at the NEA-Workshop on Near-field performance assessments, Baden/Switzerland, 1987

/7/ Berner U., Jacobsen J., McKinley I.: "The near-field chemistry of a Swiss L/ILW repository"; Paper presented at the NEA-Workshop on Near-field performance assessments, Baden/Switzerland, 1987

FIGURES

TABLES

Table I:
Total quantities (in t) of materials in the waste and its solidification matrix in the L/ILW repository

(Material quantities in t)	Operational waste	Reprocessing waste	Decommissioning waste	Waste from research, medicine + industry	Total
Solidification matrix:					
Cement	54'730	9'120	118'280	8'140	190'500
Bitumen	300	3'100	-	-	3'400
Polymers	100	-	-	-	100
Waste material:					
Steel	2'800	23'600	39'700	-	66'100
Al/Zn	-	600	6	-	600
Salt concentrates	300	1'800	8'500	-	10'600
Ash	200	-	100	-	300
Glass	400	-	-	-	400
Ion exchange resin	12'200	100	-	-	12'300
Concrete	20	-	4'300	-	4'300
Cellulose	-	5'390	-	-	5'300
Plastic	5	600	3'300	-	3'600
Other organic material	0	1	-	-	1
Other solid material	3'000	-	2'400	3'300	8'700

Table II:
Activity of important radionuclides in a L/ILW repository

Radio-nuclide	Reprocessing waste (Ci)	Operational waste (Ci)	Decommissioning waste (Ci)	Waste from research, medicine + industry (Ci)	Total (Ci)
C-14	-	6.1E+2	4.6E+3	2.5E+2	5.4E+3
Cl-36*	3.8E-1	7.2E-2	1.6E-1	-	6.0E-1
Ni-59	-	5.4E+2	4.6E+4	-	4.7E+4
Se-79	4.1E-1	-	-	-	4.1E-1
Tc-99	8.5E+0	8.0E+0	5.5E+0	3.1E-1	2.1E+1
Sn-126	1.1E+1	-	-	-	1.1E-1
I-129	1.0E-1	5.6E-3	-	7.4E-4	1.1E-1
Cs-135	1.3E+0	5.5E-2	-	1.7E-2	1.3E+0
Ra-226	-	-	-	6.6E+1	6.6E+1
U-234	2.0E+1	-	-	-	2.0E+1
U-235	3.4E-1	-	-	-	3.4E-1
U-238	5.1E+0	-	-	-	5.1E+0
Np-237	3.4E+0	-	-	-	3.4E+0
Pu-239	4.7E+3	-	-	-	4.7E+3
Pu-241	2.0E+6	-	-	-	2.0E+6
Am-241	1.1E+4	-	-	4.9E+2	1.1E+4

*) latest data

Table III:
Issues determining the performance of the near-field of a repository

Main scenarios studied for post-closure analysis
- Groundwater transport through host rock - Groundwater through access tunnels - Severe glacial erosion - Human intrusion
Near-field processes considered
- Hydrology in and around excavations - Mechanical behaviour or host rock/engineered barriers - Evolution of near-field chemistry (pore water) - Degradation of waste, containers, backfill - Generation of gases (corrosion, biodegradation, radiolysis) - Microbial action - Solubility/speciation of radionuclides - Nuclide sorption on engineered barriers - Nuclide transport (diffusive, advective, gas-driven)

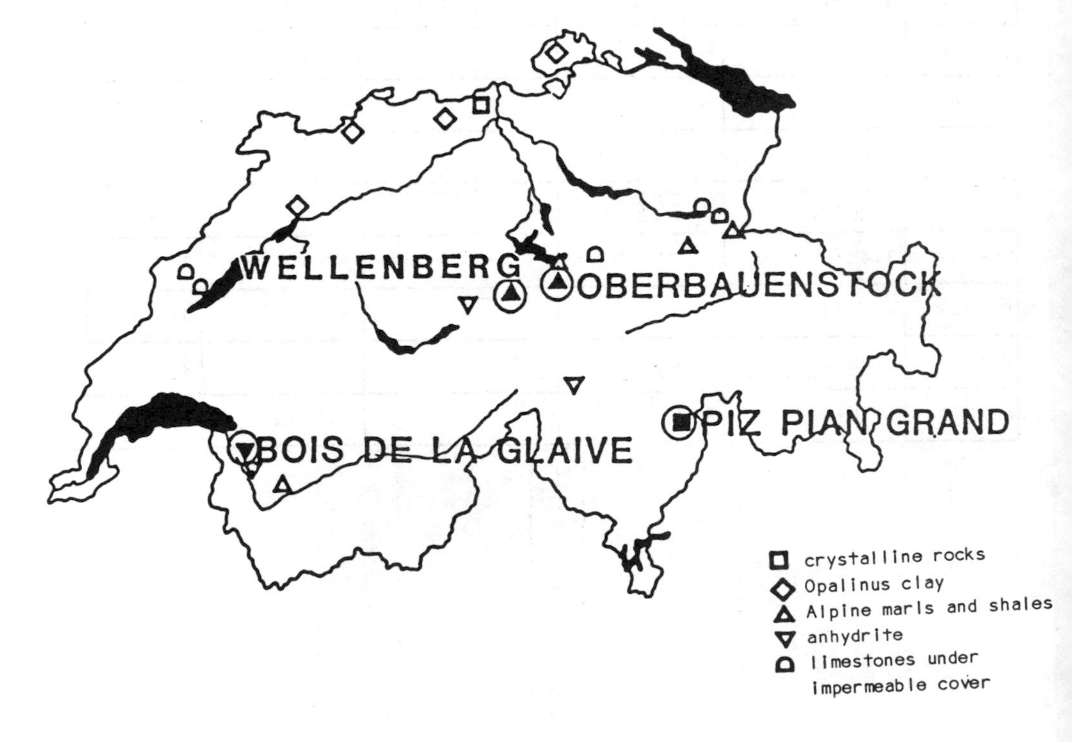

FIGURE 1: Potential L/ILW repository sites in Switzerland

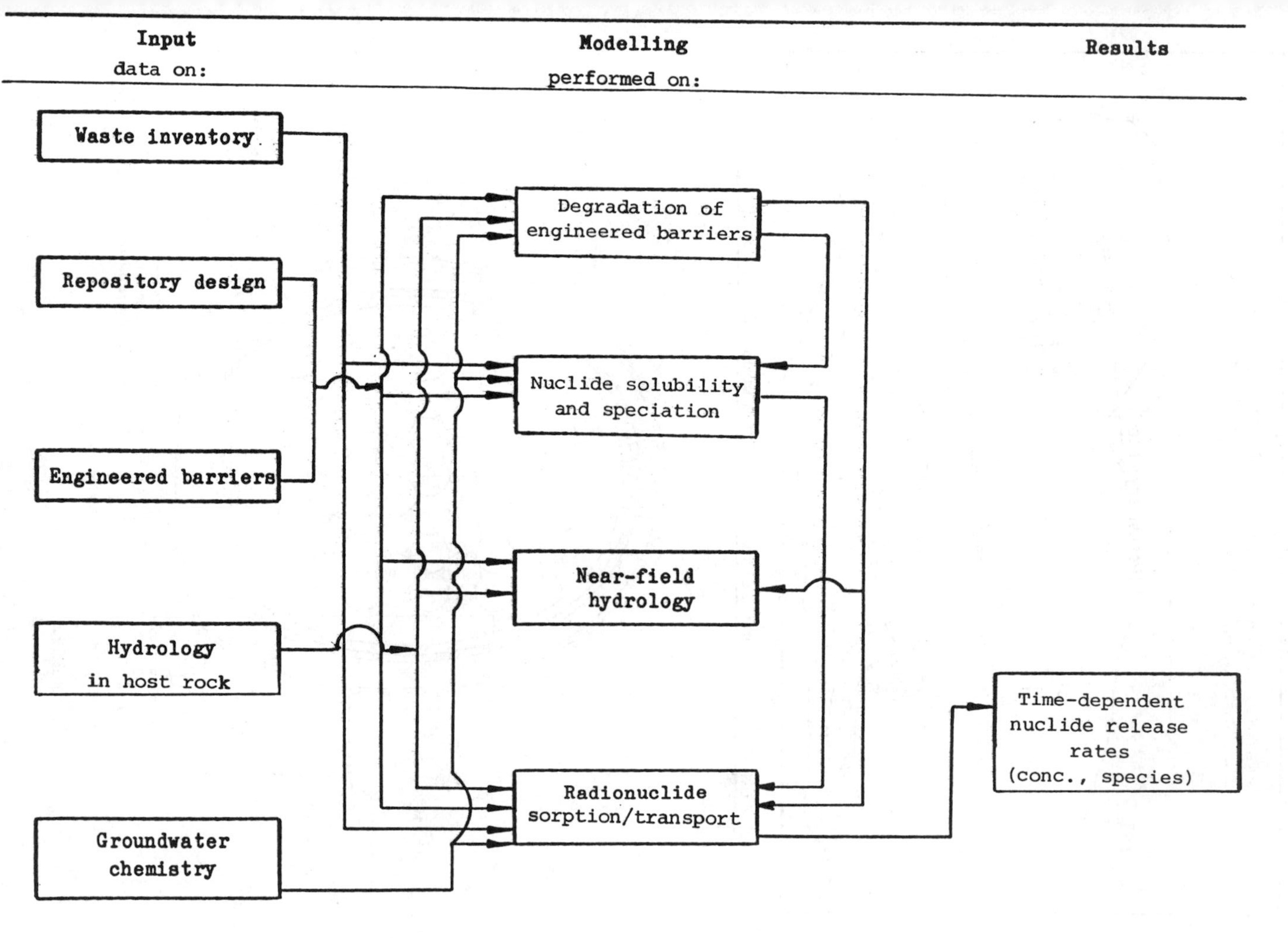

FIGURE 2: Near-field processes

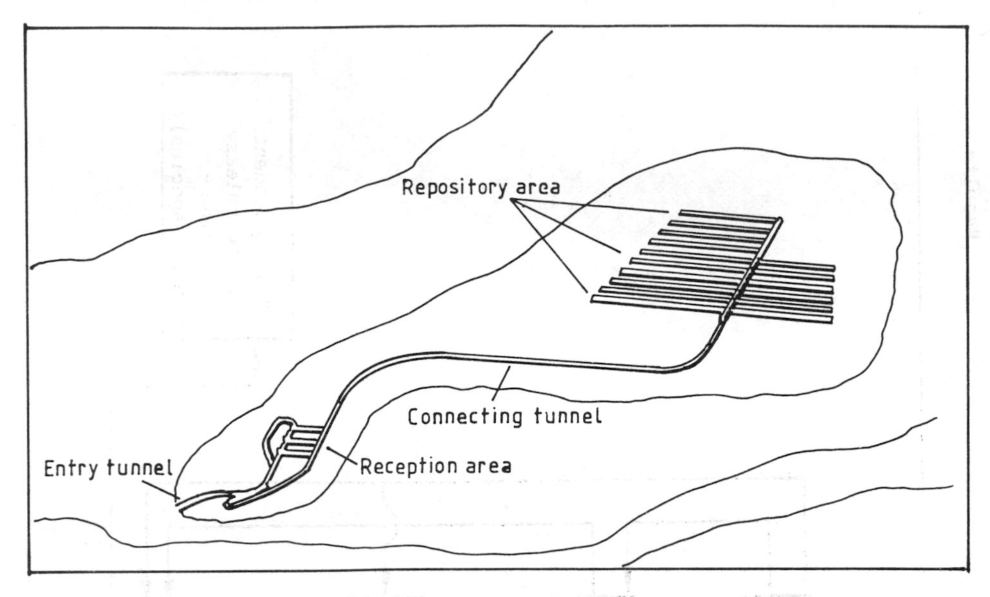

FIGURE 3: Schematic view of Oberbauenstock repository

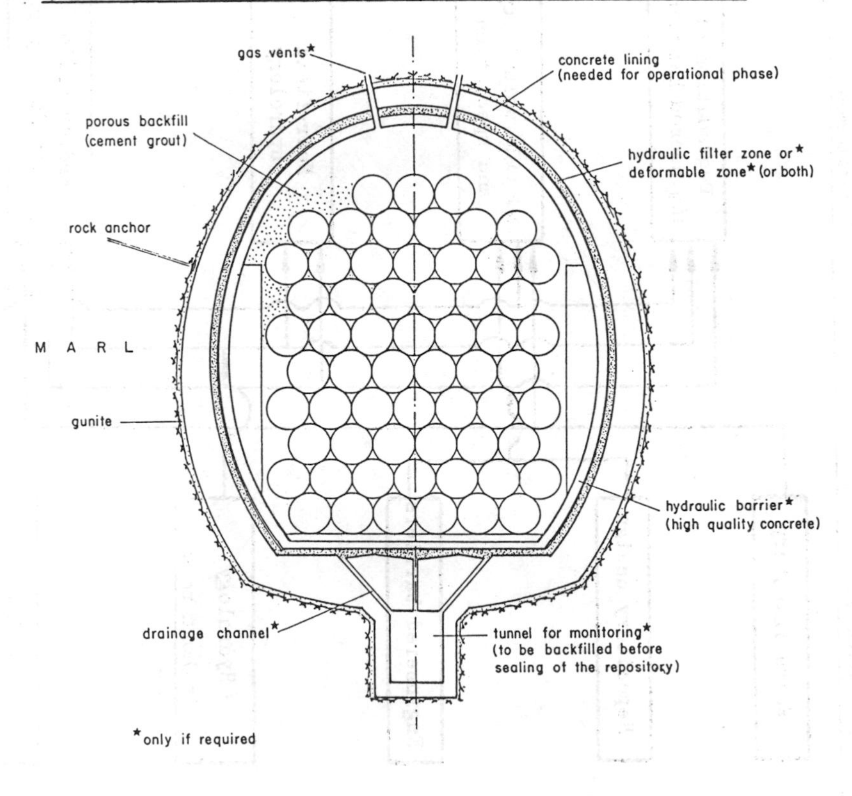

FIGURE 4: Possible near-field design for cavern

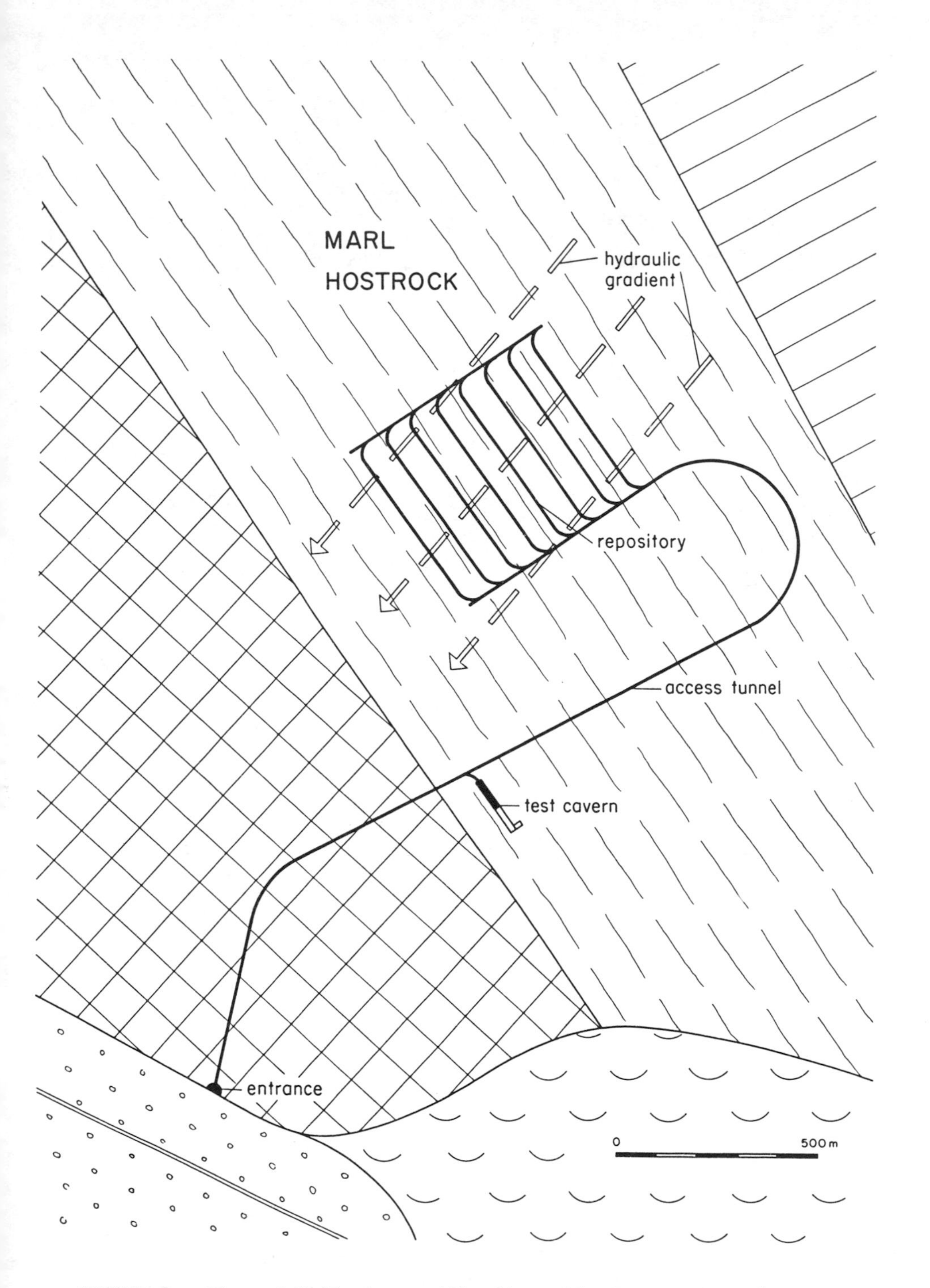

FIGURE 5: Plan of Wellenberg with alternative access tunnels

THE SWEDISH FINAL REPOSITORY FOR REACTOR WASTE (SFR) A SUMMARY OF THE SFR PROJECT WITH SPECIAL EMPHASIS ON THE NEAR-FIELD ASSESSMENTS

Jan Carlsson

Swedish Nuclear Fuel and Waste Management Co
Stockholm, SWEDEN

ABSTRACT

The first phase of the final repository for reactor waste (SFR) is scheduled for operation in April 1988. The construction work is finished and preoperational tests are in progress. Impact on the environment from SFR is analysed in a final safety report. This paper gives a summary of the design and performance of SFR. Assessments, made for the analyses of the long term safety, are given with special emphasis on the near-field. As a conclusion from the analyses, the dose commitment to the most affected individual during the post-closure period, has proved to constitute only an insignificant contribution to the natural radioactive environment of the area.

LE SITE SUEDOIS DE STOCKAGE DEFINITIF DES DECHETS DE REACTEURS (SFR) UN RESUME DU PROJET SFR AXE PRINCIPALEMENT SUR LA CARACTERISATION DU CHAMP PROCHE

RESUME

La première phase de stockage effectif des déchets de réacteurs (SFR) doit avoir lieu en avril 1988. La phase de construction est terminée et les tests préopérationnels sont en cours. L'impact des déchets SFR sur l'environnement est évalué dans un rapport final de sûreté. Cet article présente succinctement les caractéristiques et les performances de SFR. La sûreté à long terme est évaluée avec une attention toute particulière portée sur le rôle du champ proche. En conclusion, les analyses montrent que la dose maximale susceptible d'atteindre un individu pendant la période suivant la fermeture ne constitue qu'une contribution insignifiante à la radioactivité naturelle du site environnant.

1. SITING

1.1 Basic assumptions

Due to natural conditions in Sweden, it was early decided that the final disposal of radioactive waste should be underground in rock caverns. It was also a primary requirement that the repository should be located adjacent to existing nuclear facilities, i.e. the Nuclear power plants or the research centre Studsvik.

1.2 Siting of SFR

The sites mentioned above were evaluated on the basis of available data concerning the geological situation and other information of importance for site selection. When all aspects were considered, Forsmark came out as the best site for SFR. The rock mass in the area consists mainly of crystalline gneiss-granite.

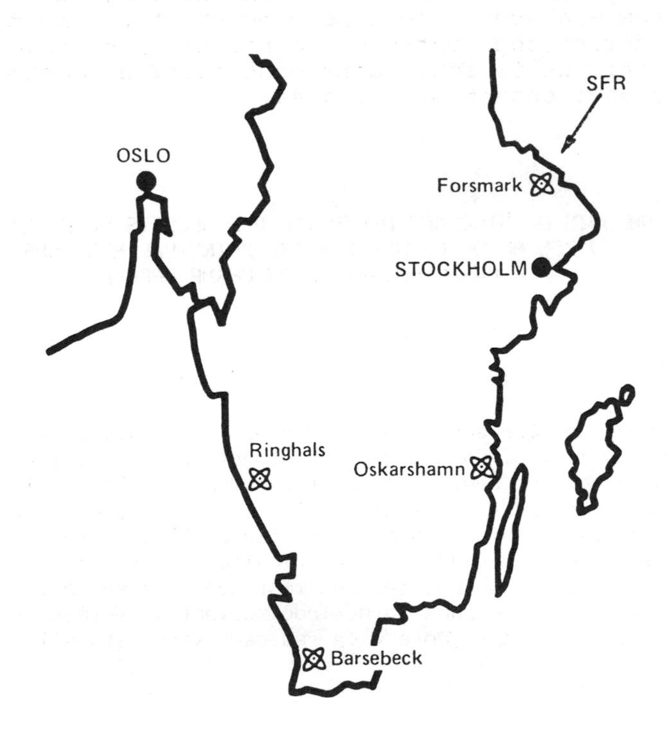

Figure 1 Location of SFR to the Forsmark region

2. GENERAL DESIGN

2.1 Design basis

The SFR has been designed to make possible a simple and controllable as well as a safe disposal of the low- and intermediate level radioactive material arising from the operation of the Nuclear power plants in Sweden.

The design and location should insure such isolation of the waste from the biosphere that the dose effects would not exceed the design limit, 0.1 mSv/a in the immediate vicinity of the facility.

Safety during the post-closure period should not be dependent on checking or corrective measures.

2.2 General design of the repository

The repository is located to the bedrock under the Baltic Sea, with a rock cover of about 60 m. This location ensures a very small hydraulic gradient and thereby the groundwater flow is low in the repository area. It also ensures that no one will drill for drinking water in the vicinity of the repository for at least 1000 years. After that time the land uplift, if continued with today's rate of 6 mm/a, would have rised the shallowest sea bed formations above the sea level.

As long as the Sea is recipient for nuclides released from the repository the dilution capacity is very high.

The SFR has various storage chambers, Figure 2, with different barriers, depending on the waste to be disposed of. The function of the engineered barriers is to limit the release of radioactivity to the groundwater.

The most active waste will be deposited in concrete silos surrounded by a clay barrier with low permeability. This ensures a nuclide transport governed by diffusion. A silo is 50 m high, 28 m in diameter and situated in a 70 m high cylindrical rock cavern.

160 m long rock caverns will be used for the less radioactive waste. The design of the caverns is dependent on the type and dose rate of the waste packages. The release of radionuclides from these caverns is mainly by the groundwater transport through the caverns.

The total amount of waste that is calculated for disposal in SFR is 90000 m^3. This is the volume of treated and packed reactor waste from the Swedish program up to the year 2010. The SFR will be built in two phases. The first phase consists of one silo and four rock caverns with a total storage volume of 60000 m^3. The second phase, including one silo and one or two rock caverns, is planned to be build at the end of the 1990s.

The layout of the tunnels and caverns allows future extension of the repository to accomodate waste from decommissioning of nuclear power plants.

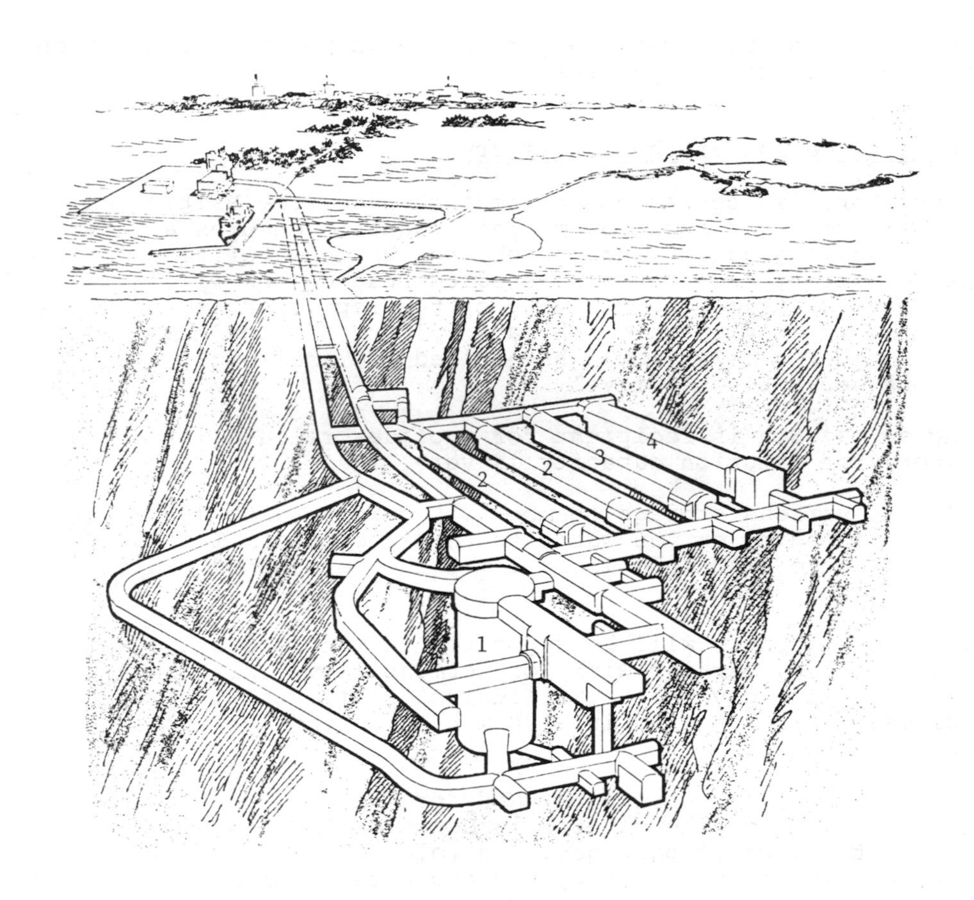

Figure 2 General layout of tunnels and caverns, phase 1

1= Silo, 2= BTF, 3= BLA, 4=BMA

3. LICENSING

3.1 Nuclear Waste Management

According to Swedish law, the primary responsibility for the realization of safe management and final disposal of the radioactive waste lies with the owners of nuclear utilities. This responsibility also includes financing of the total costs.

The four power utilities in Sweden, producing electricity in nuclear power plants are Forsmarks Kraftgrupp AB, Vattenfall (Swedish State Power Board), OKG AB and Sydsvenska Värmekraft AB.

These companies are joint owners of SKB, the Swedish Nuclear Fuel and Waste Management Company. SKB's functions are to plan, build, own and operate systems and facilities for the transport and disposal of spent nuclear fuel and radioactive waste.

SKB's work in this field is overseen by the Swedish Nuclear Power Inspectorate, SKI, and the National Institute of Radiation Protection, SSI.

The financing of future activities is ensured by a found which is financed by a fee per kWh of nuclear generated electricity production. The fee is based on a plan and a cost calculation for future activities presented every year by SKB. The supervision and reviewing authority is the National Board for Spent Nuclear Fuel, SKN.

The total cost for the management and disposal of spent fuel and radioactive waste as well as decommissioning of the power stations has been estimated to be about SEK 0.020 per kWh.

3.2 Licensing procedure

In March 1982, SKB applied for a license to build and operate the SFR. The application was based on a preliminary safety report, which was scrutinized by the Swedish authorities, SKI and SSI. In late June 1983 the Government granted a license according to the application.

The license was subject to certain stipulations. The most important are:

- SKB shall furnish information to the public on the progress of the project and safeguards against the release of radioactivity from the repository.

- A comprehensive quality control program backed up by a test and verification program shall be carried out during the construction phase.

- Further studies of gas production reactions shall be conducted, as well as a study of the gas transport capacity of the rock.

- SKI shall scrutinize the design and construction work and, if necessary, issue further requirements.

- Before commissioning, SKB has to apply for a license to operate the facility. The application shall be based on a final safety report.

- Final sealing will require a special license based on a reevaluation of the safety assessment.

3.3 State-of-the-art

The construction work on SFR phase one has come to completion. One silo and four rock caverns as well as auxiliary buildings and systems are under final preoperational tests.

A final safety report has been completed and SKB has in early October 1987, based on that report, applied for a license to operate SFR, phase one. The application is scrutinized by SKI and SSI and the license is foreseen to be given in April 1988.

4. SAFETY ASSESSMENTS

4.1 Sequence of Events (Scenarios)

After the operational waste from the Swedish reactor program has been disposed of in SFR, it is assumed that the repository will be sealed. According to current plans, this would take place in 2013, three years after the enclosure of the nuclear power plants. In the safety analysis it is assumed to take place immediately after enclosure, 2010. It has not been taken into consideration the possibility of keeping the repository opened and drained due to further utilization of SFR, e g for decommissioning waste.

When the sealing takes place, the pumping is interrupted and the repository is filled with water. After beeing filled with water, the hydraulic gradients will be restored in the repository area and it is now possible for the groundwater to move through the caverns. The clay barrier surrounding the silo prevent water from flowing through.

Based on hyrology calculations the groundwater is assumed to flow vertically through fissures in the rock and also through the caverns. Dissolved isotopes are assumed to follow the groundwater as it moves to the recipient, the Baltic Sea. To reach the groundwater from the silo, all transport is governed by diffusion.

This situation will be changed by time. One reason is that the engineered barriers will get weaker. Another is that the land rise in the area will elevate the sea bed so that it will be above sea-level and form a fresh water based ecological system.

The safety analyses are divided into two time periodes. The first covering the period during which the sea bed above SFR is still covered by the brackish waters of the Baltic Sea, the Salt Water Period. The second time period covers the time after the drying out of the sea bed and formation of the fresh water-based ecological system, the Inland Period.

In the analyses the Salt Water Period is assumed to last 2500 years.

The recipient for the radionuclide release from SFR is during the Salt Water Period the Baltic Sea and during the Inland Period a small lake and drinking water wells which, due to the bottom conditions, are anticipated to lie north of the repository.

4.2 Waste categories and activity contents

The waste that will be stored in the SFR consists mainly of ion-exchange resins and filter material from different water treatment systems. Other waste categories are contaminated components and material, trash and ash from the incineration of combustible waste. Before transport to the SFR, the waste is treated and packed at the reactor plants into the following main packages:

- Concrete moulds (1.2 m cube) with resins solidified in cement
- Steel boxes (1.2 m cube) with resins solidified in cement
- Steel drums (200 l) with bituminized resins
- Steel drums (200 l) with resins solidified in cement
- Steel drums (200 l) with trash and metal scrap
- Concrete tanks with dewatered resins (3.3*1.3*2.3 m)
- 20' or 10' containers

The division of waste into different chambers is based on activity content considerations as well as the stability and long term behaviour of the waste.

The waste with the highest activity content is allocated to the silo. This is mainly medium level resins and metal scrap. In the cavern for medium level waste, BMA, is low level spent ion-exchange resins and trash and scrap with a surface dose rate less than 30 mSv/h deposited. Concrete tanks with dewatered resins are allocated to a special cavern, BTF. Finally the very low level waste, transported in standard containers without any shielding, will be disposed of in the cavern for low level waste, BLA, together with the containers.

Before a waste package is disposed of in SFR, it has to be proven suitable for that particular chamber (the Silo or one of the caverns, BMA, BTF and BLA). The approval of a waste package for disposal is based on a so called Waste Type Description, WTD. There is one WTD for each type of waste (today approximately 30) to be deposited in SFR. The WTD's have the status of safety reports.

In summary this gives the following distribution of the total activity of 10^{16}Bq (at the year 2010) between the different repository parts:

Silo = 92% ; BMA = 6% ; BTF = 1.4% ; BLA $<$ 1%

Dominating nuclides are Cs-137, Co-60, Ni-63 and Sr-90. After decay during 1000 years the nuclides of highest contents are Ni-59, Pu-239, Pu-240, Tc-99, Am-241 and C-14.

Operational experiences of the nucler power plants gives, so far, a margin of a factor 3 for fission products, a factor 5 for activation products and at least one order of magnitude for the transuranics.

4.3 Engineered and Natural Barriers

The need of barriers against transport of radionuclides from the repository differs between the different parts of the repository. The rock chambers with the highest activity content need the most extensive barrier system while other rock chambers do not need any engineered barriers at all.

The most important barrier is the host rock with a hydraulic conductivity of 10^{-8} to 10^{-7} m/s. Taken the low hydraulic gradient into consideration (0.001 m/m) this gives a very low rate of water replacement in the repository.

Looking at the different chambers for waste disposal one will find the following engineered barriers:

SILO:
- The waste package (cement or bitumen as matrix and often a concrete box as packaging)
- Internal concrete walls in the silo
- Porous backfill concrete inside silo
- The cylindrical outher concrete wall of the silo
- Bentonite clay as backfill between the concrete wall and the rock
- Shotcreted host rock

BMA:
- The waste package (same as for the silo)
- Compartment walls, floor and lid
- Backfill material (sand)
- Shotcrete on the rock surface

BTF:
- Concrete tank walls
- Concrete backfill
- Shotcrete on rock walls
- Concrete floor

BLA:
- Waste package (only applicable to some waste types)
- Shotcreted host rock
- Concrete floor

The barrier functions can be summarized as follows:

- Transport restrictions inside the waste matrix
- Transport restrictions through the package walls
- Sorption on minerals (cement and ballast) inside the repository
- Low rate of water flow through the chamber
- Nuclide transport governed by diffusion through the concrete walls
- Uptake of radionuclides in the groundwater flowing through or around a chamber
- Transport restrictions in the rock
- Sorption in the rock

All these barrier functions do not apply to all the different chambers in the repository. As will be mentioned below all barrier functions has not been taken into consideration in the transport calculations, either because they have no importance or due to lack of data.

4.4 Assessments for the Near-Field Analyses

For the calculations of the transport of radionuclides from the repository, a referens case has been developed for each chamber and time period. The choise of referens case has been made so that dominating transport and release mechanisms are considered in a reasonable pessimistic way. In the final safety report for SFR numbers of variations has been analysed in order to quantify the importance of different assumptions and conservatisms introduced in the simplified referens calculations. In this paper only the referns case is considered.

Common assumptions for the calculations are:

- All activity, from 90000 m^3 waste, is concentrated to one chamber of each type (one silo, one BMA, one BTF and one BLA) and is evenly distributed through the waste. The distribution of activity between the different chambers are the same as mentioned in section 4.2.

- All the radioisotopes are considered capable of beeing instantaneously dissolved in the pore water of the waste matrix.

- Ground water flow has an upward direction during the Salt Water

Period. It is also assumed that the water flow takes place in a strongly canalized form. This imply that no sorption in the rock mass has been considered during this period.

- The Inland Period gives an increse in the ground water flow of a factor 10. The flow direction is now horizontal and at least 1000 m in length. Sorption has been taken into account in this calculations.

- The calculations for the Inland Period assume that no release of radionuclides have occured the first 2500 years, only decay has been accounted for.

- Sorption of radionuclides on concrete and bentonite has been considered but with different capacities for the two time periodes analysed.

- The transport of radionuclides from the repository has been calculated separately for every compartment in order to be able to take into account the differenses in design and function.

In order to give an idea of how the analyses has been carried out, the assumptions for the cavern for medium level waste (BMA) is given in more detail:

Waste packages are steel drums, concrete moulds and steel boxes disposed of in big concrete cells. Out of the barriers mentioned above only the concrete construction in the cells is given any transport restrictions.

BMA, Salt Water Period:

- Transport of nuclides from the compartment takes place through flow.

- All ground water passing through the cavern also passes through the compartment, through fractures in the concrete walls.

- The water flow rate is 32 m^3/a which gives a water replacement time of 400 years.

- Diffusion into and through the compartment walls are considered.

- The effective diffusivity in concrete walls are $3*10^{-11}$ m^2/s and distribution coefficients for aged concrete is used.

- The waste and the conditioning material is assumed to be so well mixed that initial sorption equilibrium is obtained.

- The material inside the compartment is modelled as a stirred tank and has no transport resistance.

- Radionuclides released from the compartment are assumed to be mixed with the water in the void space in the cavern outside the compartment and than following the ground water flow to the environment. The water replacement time for the void space is approximately 550 years

BMA, Inland Period

- Transport of radionuclides from the repository is goverend by the ground water, flowing through the cavern.

- The water flow is 320 m^3/a, which leads to a replacement time of 100 years

- All activity is still in the cavern, only decay during 2500 years is considered

- The whole cavern is modelled as a stirred tank.

- Sorption data for degraded concrete has been used.

4.5 Estimated dose commitment

The dose commitment from SFR, calculated for the Salt Water Period and the Inland Period, is given in Figures 3 and 4 respectively. From the figures it is seen that the individual dose commitment is well below the design goals.

The accumulated population dose commitment integrated over 10000 years is small compared with the population dose expected to arise from the normal operation of the nuclear power plants until the year 2010.

Bearing in mind the effects on the environment of the SFR plant even in the case of analyses performed in a pessimistic manner, have proved to fall far below current design goals for other plants in the nuclear power cycle, and only constitute an insignificant contribution to the natural radioactive environment of the area, the safety of the plant must be regarded as adequate.

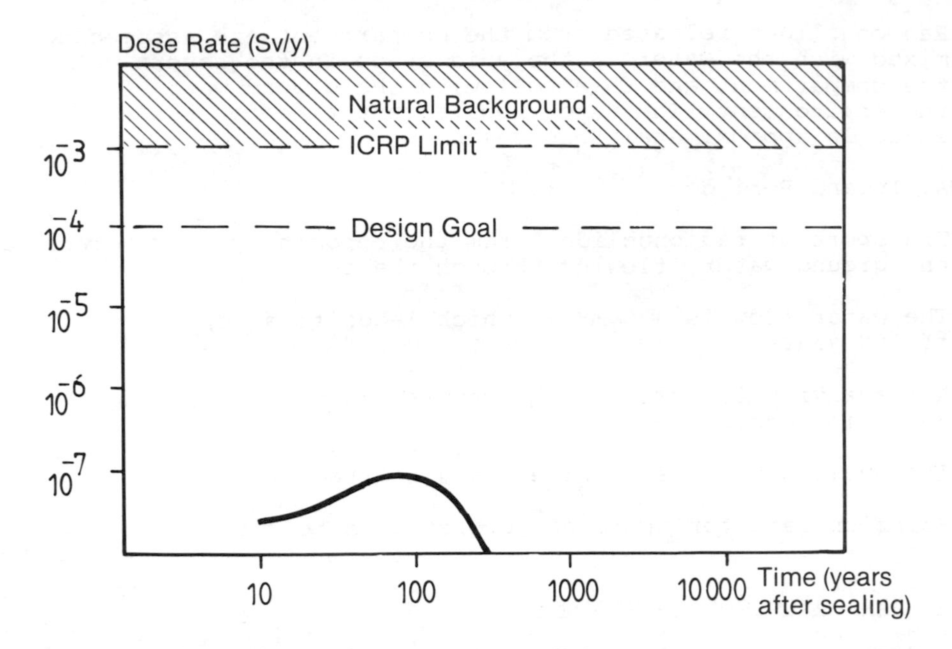

Figure 3 Calculated dose rate, to the most exposed individual during the Salt Water Period.

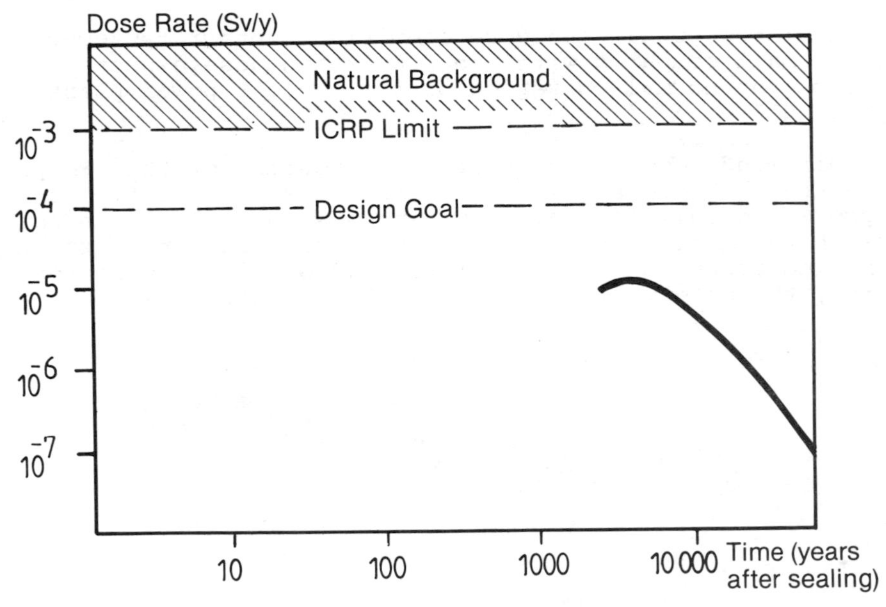

Figure 4 Calculated dose rate, to the most exposed individual, during the Inland Period.

SESSION II

GAS PRODUCTION AND GAS RELEASE

SEANCE II

PRODUCTION ET LIBERATION DES GAZ

Chairman - Président

P. ZUIDEMA

(Switzerland)

GAS PRODUCTION IN THE SFR REPOSITORY AND ITS POSSIBLE CONSEQUENCES FOR CONTAMINANT RELEASE

Luis Moreno and Ivars Neretnieks
Department of Chemical Engineering
Royal Institute of Technology
S 100 44 Stockholm, SWEDEN

ABSTRACT

Iron in the silo may corrode by reacting with water, forming hydrogen gas. The hydrogen pressure will increase with time and expel out potentially contaminated water. The internal pressure in the silo may grow to more than 0.5 to 1.0 MPa. At this pressure, the concrete walls of the silo will crack, forming fissures through which the gas and water may flow faster. To avoid this faster outflow of water, it is proposed that the top of the silo should be constructed with venting channels, which allow the gas to escape at a lower internal overpressure. This will considerably reduce the pressure in the silo and the outflow of contaminants from the silo will be less. By further surrounding the concrete boxes containing the waste with a low-capillarity, high-permeability concrete, less water will be exchanged with the interior of the boxes. A central case was defined where the parameter values were chosen such that they would represent a reasonably conservative case. Calculations were made for various combinations of values of the parameters.

LA PRODUCTION DE GAZ DANS LE DEPOT SFR ET SES CONSEQUENCES POSSIBLES POUR LE REJET DE CONTAMINANTS

RESUME

Le fer dans le silo peut se corroder par réaction avec de l'eau en formant de l'hydrogène gazeux. La pression de l'hydrogène augmentera avec le temps et expulsera de l'eau susceptible d'être contaminée. La pression interne dans le silo peut s'accroître au point de dépasser 0,5 à 1,0 MPa. A cette pression, les parois de béton du silo se lézarderont, formant des fissures à travers lesquelles les gaz et l'eau pourront s'écouler plus rapidement. Afin d'éviter cet écoulement plus rapide de l'eau, il est proposé que le haut du silo soit doté de canaux d'éventage qui permettront aux gaz de s'échapper à un niveau de surpression interne moins élevé. Cela réduira considérablement la pression à l'intérieur du silo et l'écoulement de contaminants à partir du silo sera moindre. En entourant en outre les compartiments de béton, qui contiennent les déchets, par un béton à faible capillarité et à perméabilité élevée, de moindres quantités d'eau seront échangées avec l'intérieur des compartiments. On a défini une hypothèse médiane pour laquelle les valeurs des paramètres ont été choisies de telle sorte qu'elles représenteraient une hypothèse raisonnablement empreinte de conservatisme. On a effectué des calculs pour diverses combinaisons de valeurs de ces paramètres.

INTRODUCTION

A repository for reactor waste (SFR) is built for the final disposal of low- and intermediate-level reactor waste. The repository is located under the sea in crystalline rock, the distance between the sea bed and the top of the repository is 50 m. The waste containing most of the activity will be deposited in the silo repository. This waste will consist primarily of solidified ion-exchange resins.

The silo repository is located in a cavern excavated in rock, and the space between the rock and the concrete silo is filled with bentonite. The interior of the silo is divided into vertical cells. The ion-exchange resins will be mixed with cement and deposited in concrete boxes. The boxes will be deposited in these cells and then surrounded by a porous concrete. When the repository is sealed, a concrete lid will be cast on top of the silo and evacuation pipes will be built through the lid to evacuate gas produced mainly by corrosion. A sand layer will be placed on top of the concrete lid as a distribution layer. Finally a layer of a sand-bentonite mixture will be deposited on top of the sand (Figure 1).

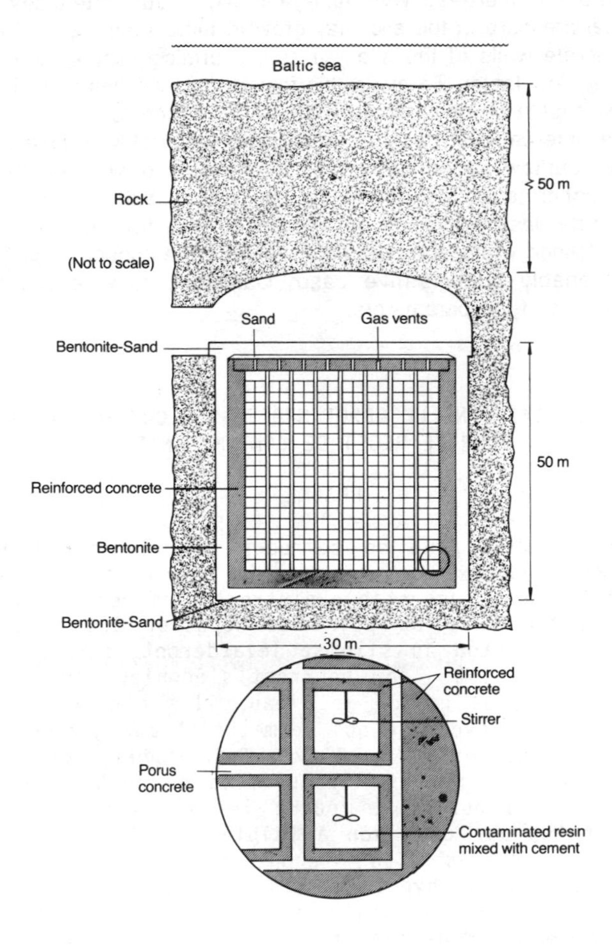

Figure 1 Outline of the repository (Not to scale)

GAS PRODUCTION

In the silo repository, iron is used as a reinforcement in the concrete in the walls of the silo and in the boxes containing the mixture of ion-exchange resins and cement. This iron will corrode. In the first stage this process is aerobic. After the repository is sealed and oxygen in the air trapped in the silo is consumed, the process then becomes anaerobic. The anaerobic corrosion of the iron may be written as:

$$3Fe(s) + 4H_2O \longrightarrow Fe_3O_4 + 4H_2$$

At the pH which exists in the water in the concrete (about 13.5) the corrosion rate is estimated to be about a few µm/yr [1]. Values of 3.0 µm/yr are used as the reference case. The expected gas production in the silo repository is then 1200 Nm^3/yr.

The production of inert gases resulting from radiolysis, may be important during the first 20-30 years. In the initial period this evolution may reach about 100 Nm^3/yr [2]. Biological degradation is not probable in the boxes containing the iron exchange resins due to the high pH. Biological degradation is possible if barrels with bitumen are deposited in the silo.

Calculations showed that for the smallest assumed corrosion rate and the highest assumed diffusion rate, the gas produced can not be transported from the concrete silo only by diffusion. The transport of hydrogen from the silo will thus have to occur by flow also.

The amount of iron in the silo is expected to be about $1.5 \cdot 10^6$ kg with a exposed surface of about 87,000 m^2. For a corrosion rate of 3 µm/yr the time for consuming all the iron in the silo would be about 2,500 years.

TRANSPORT MECHANISMS

If the water in the pores of the concrete in the silo is stagnant, the radionuclides may still escape by diffusion. The stagnant water may be set in motion by a differential pressure over the silo. Another cause is the internal overpressure resulting from the hydrogen gas produced by corrosion. The gas produced in the silo has to be eliminated to avoid a high pressure in its interior.

Water will flow into the silo during the initial period when the natural water pressure around the silo is being restored. The gas production will increase the pressure in the silo, and when it exceeds the external pressure, water will flow from the silo until steady state is reached.

Some of the mobile water coming from the silo may be contaminated by radionuclides. The contaminants may be transport by convection, i.e. dissolved in the water which is expelled from the silo. They also may be transported out by diffusion through the concrete walls, bentonite, and sand-bentonite.

For gas flow, capillarity must be accounted for. Capillary effects become important in materials with a very fine pore structure such as bentonite and concrete.

The finer pores will suck in water even against considerable gas pressure in the pores. Thus incompletely saturated bentonite and concrete will tend to suck up the water from the surrounding rock. The suction capillary pressures may be large (many tens of atmospheres). The process works in the other way also. Once the material is filled with water, a considerable pressure is needed to expel the water from the pores. For a given pressure (difference), steady state will be reached when just enough of the larger pores are empty and allow the gas to escape. Depending on the pore size distribution and the structure of the material the amount of water expelled may vary considerably.

The gas flow through water saturated porous media is possible if the critical pressure is overcome and some water is expelled. Experiments with bentonite [3] have shown that the critical breakthrough pressure depends on the degree of compactation of the bentonite. The amount of water expelled can be on the order of a few percent but may also be considerably lower.

Contaminants may be transported by diffusion in the water in the pores. For stationary transport through a porous structure the rate of transport, N, may be determined by Fick's law

$$N = -D_e A_T \frac{dC_p}{dz}$$

where D_e is the effective diffusivity, A_T the cross-sectional area, C_p the concentration in the pore water, and z is distance.

Some radionuclides may be sorbed or ion-exchanged on the surfaces in the pores in the concrete and bentonite. The migration of sorbing substances is then retarded. An apparent diffusion coefficient may be defined as $D_a = D_e/R$, where R is the retardation factor. This is the ratio between the time needed by a sorbing substance to the time needed by a nonsorbing substance to move a given distance.

RESULTS FROM PRELIMINARY CALCULATIONS

The water in the pores in the resins-cement mixture is contaminated. The water in others places in the silo (concrete walls) is not contaminated originally. To keep the contamination of the groundwater by radionuclides low, it is important to separate the contaminated water from the noncontaminated water. A way of doing this is to surround the boxes containing the radioactive waste by a concrete (or other material) with a high permeability and low capillarity.

The first property allows the water expelled from the cell walls (not contaminated) to flow along the zone with high permeability and does not force water flow through the concrete boxes containing the radioactive waste. The second property allows the gas to flow through the silo. The low capillarity zones will also avoid suction of water from the boxes. The use of porous concrete in the silo allows for the separation between contaminated and noncontaminated water in some situations.

In these calculations, first the transport by diffusion of the contaminants from the interior of the concrete boxes to the water in the pores of the surrounding porous concrete was calculated. Thereafter the effects of the evolving gas on the pressure in the silo were determined. The gas will expel water from the porous concrete in the silo once the pressure has increased above the surrounding pressure in the rock. The gas can not escape as yet from the silo even in those parts of the silo where the gas is in direct contact with the walls of the silo. This is because the capillary pressure must be overcome. The internal overpressure must be more than about 1.5 MPa for this to happen. The silo is not designed to contain such internal overpressure, the walls in the silo will have to form small cracks to let the gas out. Based on these calculations it was realized that it might be beneficial to design gas escape passages in the top of the silo to ensure that the build-up of a high pressure and the resulting large expulsion of water never takes place. The calculations showed that if the pressure in the interior of the silo is great a large volume of water will escape from the silo. This water may be contaminated to some extent. On the other hand, if escape tubes are located in the lid, the contamination from the silo may be strongly decreased because no large overpressure builds up to drive water out of the silo.

CALCULATIONS OF THE ESCAPE OF CONTAMINANTS

Processes involved

When the silo is sealed, the pressure is 0.1 MPa absolute (1 atmosphere) in the interior of the silo and in the surrounding rock nearest the silo. Groundwater flows into and restores the natural pressure around the repository. A part of the pores of the concrete in the silo is filled with air. The oxygen is consumed by means of aerobic corrosion. The volume of pores filled with gas is not important. Part of the remaining gas may be dissolved in the water when the pressure is increased. When the water table in the surrounding rock reaches the bottom of the silo, the groundwater begins to enter the silo. This water transport takes a considerable time because the permeabilities of the bentonite surrounding the silo and the sand-bentonite on the top as well as the lateral concrete wall of the silo are very low.

During this period the oxygen is consumed, and then hydrogen starts forming. Two different cases may be discerned. In the first the production of hydrogen may start early and a pressure may build up so that the silo will never become entirely water saturated before pressure equilibrium has been attained. In this case there will only be inflow of water into the silo. No contaminants escape by flow. In the second case it was assumed that the silo is first filled entirely with water before the gas production becomes important. When the gas production starts and bubbles are formed, an equivalent amount of water will be expelled from the pores in the materials of the silo. The water and gas will escape through the vents in the lid of the silo. The gas formed in the boxes and in the internal walls will collect in the porous concrete, whereas the gas formed in the outer walls of the silo may also move towards the rock.

If the water expulsion takes place before the mobile pore water in the porous concrete has been contaminated, no contaminant will escape by flow. Otherwise the escaping water will carry contaminants to the top of the silo. As the gas production rate probably will change with time some intermediate case is more probable.

The contaminated water which has reached the sand and sand-bentonite layers on the top of the silo will flow through these barriers. These will provide additional delay and also a dilution by mixing with the pore waters of these barriers. If the amount of expelled water is less than the pore water volume in the sand and sand-bentonite barriers, the contaminant front will only flow part way through before water movement stops. Further contaminant movement will take place by diffusion

Escape of contaminant from the concrete boxes to the mobile water

The transport of contaminants from the box containing the ion-exchange resins to the water in the pores of the porous material around these boxes takes place by diffusion and possibly by flow due to gas generation in the boxes. It is assumed that all contaminants (radionuclides) are very soluble in water and that they do not sorb on or ion exchange with the solid materials in the silo. This is a very conservative assumption for many species.

The gas production around the stirrer, used to mix the resins and the cement, will increase the pressure in the interior of the boxes. The gas produced around the iron reinforcement in the walls of the boxes will flow out to the porous concrete or into the boxes. This transport will occur through fissures or microfissures in the concrete wall.

The results show that when equilibrium is reached between the concentration in the concrete boxes and the porous concrete, about 50 % of the radionuclides will reside in the mobile water. A concentration of about half the original maximum nuclide concentration will be reached in 20 - 100 years. The first figure applies if the walls crack at t = 0. The impact of the gas produced around the stirrer in the interior of the boxes is negligible.

Transport of contaminants out of the silo

The water in the porous concrete surrounding the boxes is much more mobile than the water in the other parts of silo. When gas production starts and bubbles form, part of this water will be displaced by gas. This water will flow upward through the porous concrete, out through the escape tubes in the concrete lid, and into the sand and sand-bentonite layers covering the lid. When the gas reaches the sand and sand-bentonite layers some pore water will be displaced by the gas as the latter forms channels through the layers. The contamination front will be slightly (in the order of %) advanced due to this process.

The expected amount of escaping water is on the order of 150 m^3 in all. The pore water volume in the sand and sand-bentonite layers is larger, about 300-400 m^3. In an ideal plug flow situation, the contaminated water would travel only part way through the layers and the contamination front would not reach the rock.

Inhomogeneities of the layers will disperse the front, however, and some contaminants may escape. Later, molecular diffusion will lead to an ever-increasing amount of the contaminants reaching the rock.

The contaminants in the water in the silo may also escape by molecular diffusion through the vertical walls and bottom of the silo and continue out through the bentonite layer. This process is practically independent of the escape through the top of the silo. The two escape rates are additive.

When so much water has been expelled that gas channels have been formed, they will permit the gas to escape as it produces and then no more water will be expelled.

The pressure drop through the tubes in the concrete lid may be changed by varying the number of tubes, the diameter of the tubes, or the type of filling material. For this reason the pressure drop through them may be maintained sufficiently low.

The water flow

The water expelled from the sand-bentonite bed will form a network of gas channels. The amount of water that must be expelled before gas can start flowing is determined by the volume of gas bubbles formed in the silo. In the following it is assumed that 1.0-3.0 % of the pore volume will be gas-filled. This is based on laboratory measurements [4]. The value of 2 % is used as a central value for the calculations.

The pressure difference in the silo above the hydrostatic pressure varied from 7.4 kPa for low gas production and high hydraulic conductivity to 1840 kPa for a high gas production rate and low hydraulic conductivity.

These results indicate that a high packing density of the sand-bentonite bed may produce a pressure that is too high in the interior of the silo. This high pressure may cause damage to the external walls of the silo. This situation will occur if the high gas production starts everywhere at the same time. But this situation is not very probable. For this reason another more gradual increase has also been considered.

It was assumed that the gas production increases linearly from zero to its maximum value during a certain period of time, e.g. 20 years. It is reasonable to assume that anaerobic corrosion starts first in those sections of the silo which have been water-filled from the beginning, i.e. near the bottom of the silo, and then proceeds upwards as the silo is filled with water.

It is seen that the maximum pressure, in general, is less in this case. But in one case it is above the failure pressure of the silo. In all cases considered the times to expel the water are relatively short compared to the decay times of the radionuclides.

Contaminant escape with water expelled from the silo and the sand and sand-bentonite layers.

The water flowing from the silo may carry out radionuclides. The concentration of contaminants in the water expelled from the silo depends on when the gas production in the silo starts. In the calculations it is assumed that nuclides have escaped from the boxes and have contaminated all the water in the silo.

The contaminants transported by the water flow will be dispersed when water flows through the sand and sand-bentonite beds. This will permit some contaminants to arrive at the outlet before all the pore water in the sand and sand-bentonite beds has been expelled. Although the pore volume is about 370 m^3 and the water volume, which is being displaced, is considerably less, mixing of the two water volumes by dispersion will occur and let some contaminants escape. So the higher the dispersion, the more contaminants will escape.

It is assumed that the gas flow starts after a certain water volume (1.0, 2.0, or 3.0 % of the water volume) is expelled from the silo through the evacuation tubes located in the lid. The gas flow starts when some of the larger channels have been emptied. The gas penetration continues with the expulsion of water and increasing transport capacity of the channels.

For a porous medium there is a critical value for the pressure needed to open the channels for gas flow. Above this value the gas may start to flow through the sand-bentonite bed. Water is expelled from some channels. If the pressure is decreased again, the channels are closed, and to start the gas flow again the critical pressure must be exceeded. The opened channels are kept open if the pressure is reduced to below the critical pressure. A second critical value may be defined below which the channels are closed. Gas is enclosed in the sand-bentonite bed in this case.

For a sand-bentonite bed with 10 % bentonite the critical opening pressure is in the interval 10 to 50 kPa [3]. The above values are rough estimates only. The final results are, however, not very sensitive to them.

Gas flow expels a certain water volume from the sand-bentonite bed. Simulations were made assuming that 1 to 10 % of the water in the pores is expelled (3.4 - 34 m^3).

The contaminants which remain in the sand and sand-bentonite beds after the start of gas flow may be transported out by molecular diffusion. The outflow rate of contaminants by this mechanism is low if compared with the outflow rate in the initial period (during water flow and the start of gas flow). After about 200 years 60 % of the contaminants in the sand and sand-bentonite beds have diffused out. In the calculations it was assumed that the concentration in the rock is zero. This implies that all contaminants which reach the rock are quickly transported away. This is a conservative assumption.

Pressure within the silo

The silo is surrounded by bentonite clay outside the vertical walls and by a sand-bentonite mixture on top and bottom of the silo. When the bentonite is fully water saturated, all the voids are filled with water. Gas has to displace the water in the voids before it can start flowing. Because the pores are very small, considerable capillary forces will have to be overcome before any gas can penetrate the pore structure. This critical pressure increases with increasing compaction of the bentonite.

At gas pressures higher than the critical pressure the gas starts to flow. The larger pores are then blown free of water. Only the larger pores form a connected gas network. Almost all of the water is still strongly bound in the smaller pores. When the gas pressure decreases, the system closes to gas flow again. The pressure at which the system closes is of course lower than the opening pressure. It can be envisaged that there will be pressure oscillations because of this process. First, the pores open at P_{op}. The gas escapes until the pressure has dropped to P_{cl}. The pressure builds up again to

P_{op}, etc. In the silo where there are inhomogeneities in the clay and the packing density varies somewhat from place to place, there will probably be some places that are constantly open. They will let just enough gas pass to compensate for the production. The pressure will then stabilize at a constant level.

Because there is an overpressure within the silo, water will also flow out through the walls and bottom. The gas which is produced will rise through the large pores of the porous concrete. These voids will make up a few percent, at the most, of the volume of the porous concrete. The rest of the pores are filled with water which is held by capillary forces. In such a case an overpressure can be maintained within the silo without any flow of either gas or liquid through the walls. The expected overpressure is reasonably small, about 0.5 - 2.0 m water head and the gas cushion formed will be contained in the sand, the gas escape tubes in the top of the silo, and in the upper part of the silo itself. No additional expulsion of water owing to overpressure in the silo is expected.

Some calculations have been performed on the water velocity in the walls if there were to be a pulsating pressure. The water flowrate through the lateral wall depends on the values of the hydraulic conductivity used in the calculations. For P_{cl}= 10 kPa and P_{op} =50 kPa, the contaminant front in the concrete wall will oscillate within a maximum amplitude of the inward and outward flow of about 1 mm.

SUMMARY OF RESULTS FOR THE CENTRAL CASE

The contaminants from the boxes diffuse out to the mobile water in the silo at such a rate that 50 % of the final concentration is reached in about 100 years.

Two percent of the water volume is expelled during a short time once the gas production starts. This water, which is assumed to be contaminated, displaces pore water from the sand and sand-bentonite layers on the top of the silo. Due to dispersion effects, about 0.30 % of the contaminants are carried through the sand-bentonite layer to the water in the rock.

Subsequent gas production will expel some water from the sand-bentonite layer. This will add another 0.04 % to the contamination.

Diffusion from the sand-bentonite layer will transport another 1.4 % of the contaminants to the rock in the period up to 600 years. Note that no decay has been accounted for so far.

In parallel to the escape mechanisms described above, contaminants escape by diffusion through the walls of the silo. This adds another 0.2 % to the contamination during the first 600 years.

Accounting for radioactive decay changes the picture considerably. Results show that the activity outside the silo by all transport mechanisms is never more than 0.03 % of the total activity deposited in the silo. Radioactive decay is discussed some more in the next section.

DECAY AND RETARDATION OF THE RADIONUCLIDES

During the first 10 years there is little decay of the radioactivity. In the period from 10 - 100 years the activity decays by nearly one order of magnitude. During the period from 100 - 1000 years it decays further by one and a half order of magnitude. Five nuclides dominate the activity in the silo: ^{137}Cs, ^{60}Co, ^{63}Ni, ^{3}H, and ^{90}Sr. After this period the activity is dominated by ^{59}Ni, ^{14}C, ^{240}Pu, and ^{239}Pu. These will have decayed to insignificance after 10^5 - 10^6 years.

During the first 600 years the activity decays by about 2.5 orders of magnitude, and this will considerably influence the amount of activity which reaches the mobile water in the rock.

Previously it has been assumed that the radionuclides behave as if they were very soluble and as if they are not retarded by any sorption in the concrete or in the clay.

Cesium, ^{137}Cs, is not expected to be retarded significantly in the concrete [5]. It will, however, undergo considerable retardation in the clay [6]. In compacted bentonite a retardation factor of 2800 has been observed. In the less compacted bentonite outside the silo and in the sand-bentonite mixture a considerably smaller retardation can be expected. Still, it may be expected that the transit time through the barriers will be on the order of several hundreds of years. Cesium will then decay considerably (several orders of magnitude) due to the retardation effects.

^{60}Co will decay by about 2 orders of magnitude during the first 30 years. During this time no escape is envisaged in practice.

Tritium, ^{3}H, will decay by nearly one order of magnitude in 30 years but will not be retarded in any of the barriers.

^{90}Sr can also be expected to have a retardation factor of several hundred in the bentonite barrier since it has a retardation factor of many thousand (5800) in compacted bentonite [6]. It will thus decay significantly in the barrier. Furthermore, ^{90}Sr can be expected to be retarded considerably in the concrete due to solubility limitations in the calcium carbonate-rich concrete. This can not be quantified at present. Of the more long-lived nuclides, ^{59}Ni may well be hindered from escaping due to solubility limitations, and probably also ^{14}C if it exist in the carbonate form.

It is thus very conservative to neglect the impact of retardation and solubility in the safety analysis.

SENSITIVITY TO ASSUMPTIONS AND PARAMETER VALUES USED

In this section the influence of the different assumptions made is studied. In the calculations of contaminant transport through the silo lid, it has been assumed that the water expelled was fully contaminated. The contaminant concentration depends on when the gas production starts. An early start means a low concentration in the water expelled from the silo because the concentration in the water in the pores of the porous concrete is low. A late start of gas production means that more contaminants may have escaped from the boxes before the water is expelled, but on the other hand more of the radioactivity has decayed during the longer time span.

Values for the different parameters which influence the transport processes were estimated. In some cases the uncertainty in the values is very large and the intervals used have been wide. Two orders of magnitude between the minimum and maximum values are not uncommon.

A reference case has been defined. In this case reasonably probable, with a bias to conservative, values of the parameters were used. In some cases these values were chosen as the mean between two extreme values.

DISCUSSION AND CONCLUSIONS

It is difficult to make an accurate analysis of the consequences of the hydrogen production caused by the corrosion of the iron because of the large variability of data.

The data on the hydraulic and capillary properties of the concrete have been estimated but the confidence in these data is low. The corrosion rate is estimated from experimental data under different conditions.

The preliminary calculations showed that for most of the combinations of parameter values studied, the walls of a nonvented silo will crack due to a build-up of internal pressure. For this reason it was proposed that the silo should be constructed in such a way as to allow the gas to escape at lower pressures.

The use of evacuation channels in the silo lid avoids the development of large overpressures in the silo. If the overpressure in the silo is low, the water transport through the lateral concrete wall and the bentonite will be very low. By surrounding the concrete boxes with a high-permeability, low-capillarity concrete, water will not be drawn from the boxes by capillarity and water will not be forced into or out from them by a pressure gradient over the silo. The water will flow around the boxes instead.

In the reference case the critical mechanism for the escape of contaminants to the rock is the diffusion from the sand and sand-bentonite layers on top of the silo. These layers have been contaminated by the water expelled from the porous concrete in the silo. This water in turn has been contaminated by the diffusion from the boxes. There are some data available on Cesium and Iodide diffusion in relatively young (1-2 years) concrete samples [7]. Cesium was found to have an effective diffusivity on the order of $4{\cdot}10^{-14}$ m^2/s. Accounting for the slight sorption of Cesium the apparent diffusivity, D_a, was estimated to be about $1{\cdot}10^{-14}$ m^2/s. Iodide diffusion was estimated to be of the same magnitude. It can be expected that all small ions and molecules will have an effective diffusivity on the order of $1{\cdot}10^{-14}$ m^2/s in this kind of concrete.

The process of how the water or gas may be transported from the locations outside the silo through the rock has not been studied. There are two critical aspects to be considered: the water flow from the silo at the start of the gas flow and the gas flow itself. In the first case, 144 m^3 of water may be expelled from the silo during a time of about 1-10 years. In the second case, about 1200 Nm^3 of gas per year must be evacuated from the cavern at the top of the silo to the bottom of the sea above. A gas cushion may be formed at the ceiling of the cavern. The thickness of this cushion depends on the permeability of the rock surrounding the silo. Braester and Thunvik [8] calculated the amount of the gas transported from a low-level waste repository. In the calculations a fracture frequency of 5-10 fractures per meter was assumed. If the same assumptions are made it is found that the rock surrounding the silo may remove the gas with a relatively low overpressure.

REFERENCES

1 Hansson, C.M.:"Hydrogen evolution in anaerobic concrete resulting from corrosion of steel reinforcement", The Danish Corrosion Centre, 14 p, Denmark, 1984.

2 Wiborgh, M. and L.O. Höglund:"Gasbildning i SFR", SKBF/KBS SFR 83-03, Slutförvar för reaktoravfall, arbetsrapport, Stockholm, 1983. (In Swedish)

3 Pusch, R.:"SFR-Buffertar av bentonitebaserade material i siloförvaret. Funktion och utförande", SKBF/KBS SFR 85-08, Slutförvar för reaktoravfall, arbetsrapport, Stockholm, 1985. (In Swedish)

4 Moreno, L. and M. Cederström:"Permeability of porous concrete", Teknisk PM , SFR/SKB, 1985.

5 Allard, B., L. Eliasson, S. Höglund, and K. Andersson:"Sorption of Cs, I, and actinides in concrete systems", SKBF/KBS TR 84-15, Stockholm, Sweden, 1984

6 Neretnieks, I.:"Diffusivities of some constituents in compacted wet bentonite clay and the impact on radionuclide migration in the buffer", Nuclear Technology, 71, p 458, 1985.

7 Andersson, K., B. Torstenfeldt and B. Allard:"Sorption and diffusion studies of Cs and I in concrete", SKBF/KBS TR 83-13, Stockholm, 1983.

8 Braester, C. and R. Thunvik:"An analysis of the conditions of gas migration from a low-level radioactive waste repository", SKBF/KBS TR 83-21, Stockholm, 1983.

IMPACT OF PRODUCTION AND RELEASE OF GAS IN A L/ILW REPOSITORY
A SUMMARY OF THE WORK PERFORMED WITHIN THE NAGRA PROGRAMME

P. Zuidema*/ L.O. Höglund**
*Nationale Genossenschaft für die Lagerung radioaktiver Abfälle
Nagra, 5401 Baden/Switzerland
**Kemakta Konsult AB
11228 Stockholm/Sweden

ABSTRACT

In a repository for low- and intermediate-level radioactive wastes, gases will be formed due to corrosion of metals, microbial degradation of organic materials and radiolytic decomposition of water and organic materials. The predominant source of gas is calculated to be anaerobic corrosion of metals, particularly iron.

Gas pressure will build up in the near-field until it is released through the system of engineered barriers into the geosphere at a rate equivalent to the production rate. Excessive gas pressures may damage the engineered barriers if no precautions are taken. Radionuclide transport both through the host rock and near-field may be influenced by such gas releases. Water will be displaced and local hydrology will be altered. The significance of these alterations are site-specific; theoretical studies as well as field investigations are underway to clarify the rôle of the different processes involved.

INCIDENCE DE LA PRODUCTION ET DU REJET DE GAZ DANS UN DEPOT DE DECHETS DE FAIBLE ET DE MOYENNE ACTIVITE -- RESUME DES TRAVAUX EXECUTES DANS LE CADRE DU PROGRAMME DE LA CEDRA

RESUME

Dans un dépôt destiné à des déchets de faible et de moyenne activité, des gaz se forment par suite de la corrosion des métaux, de la dégradation microbienne des matières organiques et de la décomposition radiolytique de l'eau et des matières organiques. La principale source de gaz est constituée, d'après les calculs, par la corrosion anaérobie des métaux, en particulier du fer.

La pression des gaz augmentera dans le champ proche jusqu'à ce qu'ils soient libérés dans la géosphère à travers le système de barrières ouvragées à une vitesse équivalente au taux de production. Des pressions excessives des gaz peuvent endommager les barrières ouvragées si aucune précaution n'est prise. Le transport de radionucléides au travers tant de la roche receptrice que du champ proche, peut être influencé par de tels rejets de gaz. Il y aura déplacement d'eau et modification de l'hydrologie locale. L'importance de ces modifications est propre à chaque site ; des études théoriques de même que des recherches sur le terrain sont en cours afin d'élucider le rôle des différents processus en jeu.

1 INTRODUCTION

In order to estimate the gas formation in a L/ILW repository and to model the subsequent transport of the gas through the host rock, the physical and chemical conditions in the repository and the surrounding rock must be known. Therefore, information is required on:

- Waste inventory (nuclides and chemical composition of all types of waste) with information also on matrices and containers
- Description of and data on the relevant gas formation processes
- The repository design which specifies system, dimensions, materials and their properties
- Site-specific data on the host rock properties, water chemistry, etc.

The work described here consists mainly of the studies performed for Project Gewähr 1985, in which the potential site at Oberbauenstock /1/,/2/,/3/,/4/,/5/ was chosen as an example. The estimation of gas formation and the models to describe the release of gas are presented in more detail in /6/.

The wastes foreseen for the L/ILW repository can be divided into four main groups with respect to their origin:

- operational wastes
- reprocessing wastes
- decommissioning wastes
- MIR wastes (**m**edicine, **i**ndustry and **r**esearch).

A condensed description of the different materials in the wastes from a 240 GWe years energy scenario is given in Table 1 (/4/,/7/).

2 GAS FORMATION

Gas in a L/ILW repository is mainly formed by three different processes:

- Corrosion of metals present in the wastes, in the waste containers and in the reinforcement of the storage containers and the repository structures
- Microbial degradation of organic constituents of the wastes and the waste matrices
- Radiolytic decomposition of water and organic materials.

Of these three processes, the contribution from radiolytic decomposition has been calculated to be small and will not be discussed further.

Table I: Summary of total amounts of materials originating from waste and conditioning material (tonnes)

Conditioning material	Operational waste	Reprocessing waste	Decommissioning waste	MIF waste	Whole repository
Concrete	54,730	9,120	118,280	8,410	190,600
Bitumen	300	3,100	-	-	3,400
Plastics	100	-	-	-	100
Waste					
Steel	2,800	8,300	39,700	-	50,800
Al/Zn/	-	600	6	-	600
Salt/concentrates	300	1,800	8,500	-	10,600
Ashes	200	-	100	-	300
Glass	400	-	-	-	400
Ion exchange resins	12,200	100	-	-	12,300
Concrete	20	-	4,300	-	4,300
Cellulose	-	5,300	-	-	5,300
Plastic	5	600	3,000	-	3,600
Other organics	0	1	-	-	1
Other solids	3,000	-	2,400	3,300	8,700
Waste packages					
Steel drums	9,500	4,300	-	1,200	15,000
Zn coating	41	-	-	140	181
Reinforcement					
In concrete containers	-	-	-	-	22,100
In the lining	-	-	-	-	40,100

Table II: Theoretical amounts of gas formed by complete corrosion of different groups of metal items in the repository

Origin of metal	**Total theoretical volume m^3 (STP) of gas**
Steel drums in concrete	$7.2 \cdot 10^6$
Steel drums in bitumen	$0.4 \cdot 10^6$
Steel drums in plastic	$0.05 \cdot 10^6$
Reinforcement in drums	$0.39 \cdot 10^6$
Steel waste	$27 \cdot 10^6$
Aluminium and zinc	$0.74 \cdot 10^6$*
Reinforcement-concrete containers	$12 \cdot 10^6$
Reinforcement-repository lining	$21 \cdot 10^6$
Whole repository	$69 \cdot 10^6$

* 50% of this volume is assumed to be produced before storage

2.1 Corrosion of metals

Corrosion of metals is the most important mechanism for gas formation in a L/ILW repository. Large amounts of metals will be present:

- The wastes contain steel, aluminium and zinc
- Steel drums are used as containers for many of the wastes
- The concrete containers used as overpack for some of the wastes (e.g. decommissioning waste) are reinforced with steel
- The lining and other structures of the repository contain steel reinforcement.

During the initital period after closure of the repository, aerobic conditions exist because large volumes of air will be present in pores and voids in the repository. Aerobic corrosion of iron will not yield any gas as a reaction product. The oxygen will be consumed both by corrosion and by microbial activities. As the groundwater in the surrounding rock is free of oxygen, no oxygen supply will exist and the environmental conditions in the repository will change and become anaerobic and other electron acceptors (e.g. SO_4, NO_3, etc.) will be used. Only when these are consumed (or unavailable) the redox potential will drop sufficiently for reduction of water. Then corrosion with formation of hydrogen is assumed to begin. In the calculations it was assumed that anaerobic corrosion of iron will form $Fe(OH)_2$ which will eventually be transformed into Fe_3O_4. The corrosion rate of iron in concrete at temperatures less than 50°C and atmospheric pressure is expected to be in the range of 10^{-4} to 10^{-3} mm Fe per year. This corresponds to a hydrogen evolution of 0.02 to 0.2 mol per m^2 and year /6/. The thermodynamics of the system imply hydrogen equilibrium pressures far in excess of repository hydraulic pressures /6, 18/; therefore, despite the build-up of hydrogen pressure, corrosion will not stop and it is still uncertain how far an increasing hydrogen pressure, the formation of hydrogen bubbles and the growth of corrosion product layers might reduce the corrosion rate.

In connection with the Nagra HLW programme, experimental work has been performed on corrosion of iron /8/,/9/,/10/. These and other experiments, together with engineering judgement, allow the general conclusion for the L/ILW programme that the above mentioned range of corrosion rates is reasonable.

Corrosion of other metals can also be important. Aluminium is, under normal conditions, protected against corrosion by a layer of oxides. At high pH, aluminium oxide is soluble and the protection is considerably reduced, which explains the high corrosion rates reported in literature /11/.

To calculate the amount of hydrogen generated in a repository under anaerobic conditions the following information is required:

- The corrosion rate expressed in mass per unit surface area and year
- The amount of hydrogen generated per unit mass of corroded metal
- The surface area of corroding metals.

For the inventory defined in Table I steel waste and steel reinforcement are the most important potential sources of hydrogen. With respect to the rate of gas generation, the significance of the different sources is summarised in Figure 1. During the first 20 years, aluminium may be of predominant importance; however there is considerable uncertainty on this point. After this period, the predominant source of gas will be the corrosion of steel drums and reinforcement bars. The rate of gas production of steel waste, which continues over a long time period, is low. This is due to the small surface area of this steel.

2.2 Microbial degradation

Many of the organic and inorganic materials in the repository are potential sources of energy to microbes through oxidation. To allow for oxidation of these energy sources, however, oxidants, nutrients and water are needed. Their availability can restrict microbial degradation /12/.

Specialised microbial groups are known to live at extreme conditions of temperature, pH, pressure, salinity and irradiation /13/. The most important constraint on microbial growth in the Swiss L/ILW near-field is likely to be the hyperalkaline conditions of the cement pore water but recent work has shown that relevant microbial groups (e.g. sulphate reducing bacteria) can be found in natural water of similar chemistry /14/. The L/ILW near-field is sufficiently heterogeneous that a wide range of microenvironments may exist, each of which could host a different range of microbes utilising different substrates and producing different by-products.

For the calculation of the amount of gases generated by microbial degradation, the following simple global parameters were defined:

- The total amount of compounds which will undergo degradation
- The gas formation rates (moles of total gas produced annually per kg substance)
- The distribution between different gaseous products.

The most important products formed by microbial degradation are organic macromolecules, methane (CH_4), hydrogen (H_2) and carbon dioxide (CO_2).

The carbon dioxide formed will react with $Ca(OH)_2$ dissolved in the porewater and precipitate as $CaCO_3$. The total quantity of CO_2 expected is, however, in excess of the dissolved $Ca(OH)_2$ in the porewater. It is, therefore, to be expected that part of the CO_2 will react with the $Ca(OH)_2$ solid phase in the concrete and cause changes in the matrix properties. These changes may result in decreased porosity and permeability and will also eventually lower the pH.

The gas formation rates used were derived from the literature /15, 16/. With respect to gas formation rates, cellulose material is the predominant source. The contributions from bitumen, plastics and ion exchange resins are minimal by comparison.

2.3 Overall gas formation

The total gas formation rates (the sum of gases from corrosion, microbial degradation and radiolytic decomposition) are depicted in Figure 2. There is significant uncertainty about these rates. Therefore a minimal and a maximum case have also been calculated (Figure 3). Of the three gas production reactions (corrosion, degradation, radiolytic decomposition), the contribution from radiolytic decomposition is minimal. The contribution of gas from corrosion of aluminium is calculated to be very high during a very short initial period. Then, during the first 1'000 years microbial degradation predominates as a source of gases; here, degradation of cellulose is calculated to be very high during a short time period after closure of the repository (Figure 4). In the long term, steel corrosion is the dominant source.

For the gas migration calculations, the information from Figure 2 is simplified and the following assumptions are made (/6/, /17/):

- during the first 20 years approx. $5.7 \cdot 10^4$ m^3 (STP) of gas are produced per year. This is equivalent to approx. $2.5 \cdot 10^6$ moles H_2/a.
- during the following 730 years approx. $2.5 \cdot 10^4$ m^3 (STP) of gas are produced per year. This is equivalent to approx. $1.1 \cdot 10^6$ moles H_2/a.

3 GAS TRANSPORT

The gas which evolves will escape through the near field and subsequently through the far field. Different flow mechanisms are possible depending upon gas production rates and material properties. In general, it can be said that a system will be established with gas pressures such that the gas production rates are in equilibrium with the release rates. The following release mechanisms are considered to be possible /17/:

- Diffusion of gas dissolved in the groundwater
- Advective transport of gas dissolved in the groundwater
- Flow of gas bubbles
- Displacement of groundwater with subsequent gas flow.

3.1 Release from the near field

Rough estimates of advective transport of hydrogen by the groundwater flowing through the repository have been made. They show that, with a solubility of $2.4 \cdot 10^{-2}$ g/l (at repository pressures) and a groundwater flow of approx. 10 l/m^2 per year (with $k=3 \cdot 10^{-10}$ m/s and a hydraulic gradient of approx. 1), only $2.4 \cdot 10^{-1}$ g/m^2 per year (approx. $1.1 \cdot 10^4$ moles H_2/a for the whole repository) can be released from the near field. That means that advective transport cannot balance the production rates /17/.

The importance of diffusion compared with advection is given by the Peclet number ($Pe=UZ/D_e$). Even when assuming a diffusivity of $D_e=10^{-2}$ m^2/a (assuming a porosity of 3 % and a geometry factor of 1), a water flow rate of U=0.3 m per year and a migration length of Z=10 m, the Peclet number is larger than 10 and therefore release by diffusion is considered not to be important /17/.

The small transport rates by advection and diffusion imply that transport of gas dissolved in water cannot balance the production rates and that another process must be responsible for the release of gas from the near field. Displacement of porewater with subsequent gas flow is considered to be the most probable release mechanism of gas from the near field /6/. It is envisaged that a gas pressure will be built up until the water in some of the pores will be blown out and then gas flow can start /6, 18/. The gas pressure has to overcome not only the groundwater pressure but also the capillary pressures to displace the water (see Figure 5). Because capillary pressure is directly related to the pore size, the importance of the pore size distribution is obvious; in Figure 6 the pore size distributions and corresponding capillary forces are depicted for concrete with different W/C ratios /19/. This figure shows that high differential pressures (approx. 2 MPa) are required to overcome the capillary forces. These pressures result in stresses in the engineered barriers in the order of the ultimate tensile strength of concrete (approx. 2-3 MPa) and hence fracturing cannot be excluded. However, if one assumes that a significant amount of (small) fractures exists (e.g. due to shrinkage of the concrete during setting) then displacement of water with subsequent gas flow should be possible through these fractures at considerably lower pressures.

When the porewater has been displaced, a (gas) pressure gradient is needed to allow the necessary gas flow to occur through the open pores and fractures, respectively. For a gas production rate of $5.7 \cdot 10^4$ m^3(STP)/a it should be possible to release the gas through a fractured concrete with about 5-10 fractures per m, the fractures having an aperture of approx. 10^{-5} m /6/.

The properties of concrete relevant for the release of gas are not well defined and there exists some concern that in the required high quality concrete the pressure differences required to displace the water might fracture the concrete. Uncontrolled fracturing will affect significantly the initially good performance of the engineered barriers. Therefore, the provision of gas vents through the lining in combination with a porous backfill are currently being considered.

In designing a gas release system the main problem is the trade-off between assuring acceptable low pressure differences by choosing material having low capillary pressures (unavoidably connected with increased permeabilities) and the potentially increased radionuclide release rates due to the increased permeability. In this connection the problem of clogging of these vents due to precipitates should also be mentioned. The final selection of the backfill material is still open and possible problems with its long-term performance are currently under investigation.

Although concrete is currently the main material proposed for the engineered barriers in a L/ILW repository, bentonite is still considered as an option. In connection with Nagra's HLW programme, some experiments have been performed for bentonite to determine at what overpressures gas flow starts to take place. It was observed, that gas bubbles can be released through conductive passages of the clay, if a critical pressure is exceeded. Laboratory investigations were performed to determine the gas pressure required to overcome "capillary retention". It was observed that the critical pressure is normally close to the swelling pressure /20/.

3.2 Transport of gas through the host rock

The gases produced in the repository can theoretically be transported through three different zones in the rock - through the rock matrix, through the fractures and through the decompressed zones around shafts or tunnels. The transport capacity of the different zones is determined by their conductivity and the cross section available for transport /6/.

In the vicinity of the repository it is not possible to transport all the gas in a dissolved form. After the gas has been released from the engineered barriers it will accumulate in the decompressed zone and thus a gas cushion will form. The gas pressure will increase (resulting in an increasing height of the gas cushion) until the pressure reaches a level that allows the gas to displace the water in the larger fractures of the host rock. As a first approximation, it was assumed that transport would be restricted to a set of parallel fractures extending from the repository to the surface. The overpressure needed to displace the water from the fractures is equal to the capillary pressure - therefore, water will be displaced from the fractures with the greatest aperture first (Figure 7). The overpressures (expressed in terms of the height of the gas cushion) at steady state gas flow are depicted in Figure 8 as a function of the conductivity of the rock matrix and of the fracture zones.

The calculations showed that under certain circumstances the gas cushion will displace contaminated water from the repository downwards into the host rock.

This conceptual model used for the calculations with a set of parallel fractures implies that gas channels will evolve that extend from the repository to the surface. This model will at least under certain circumstances be over-conservative in that divergent flow is neglected. When allowing for divergent flow it can be envisaged that at a certain distance from the repository the gas concentration could be low enough to allow the gas to be dissolved completely in the water.

Currently, work is underway to model gas transport through the host rock in a two-dimensional vertical cross section that explicitly allows divergent flow /21/. In this model the following processes are considered:

- Two-phase flow; i.e. displacement of water by gas
- Dissolution of gas in the water.

Both in the simplified one-dimensional model as well as in the more sophisticated two-dimensional model the gas overpressures needed to overcome "capillary retention" in the host rock play a very important rôle. Although there exists from gas reservoir engineering a broad body of experience on the behaviour of gas entrapped by water, and empirical relationships between conductivity and capillary pressure have been established (see e.g. /22/), the evaluation of site-specific data is considered to be necessary. Therefore, during the first phase of site investigations at the potential site at Oberbauenstock a first simple gas test was performed. The zone tested contains a fracture zone of moderate permeability. Gas was injected at a constant rate into the borehole and it was possible to measure the threshold pressure at which the displacement of water by the gas started. Three injection tests with different flow rates were performed; for all three tests the "capillary retention" turned out to be only moderate (in the order of a few bars) /23/. Such pressures should not lead to severe problems, and further site investigations will show whether the tested interval is representative of the whole formation.

4 CONCLUSIONS

Significant amounts of gas will be generated in a L/ILW repository. The gases as such will have no detrimental effect on the safety of the repository. Indirectly, however, the gas generation can influence the release of radionuclides from near field into the geosphere as well as transport of nuclides through the geosphere. The following aspects are of major importance:

- Due to corrosion and degradation gas may be produced at such rates that it will not be possible to dissolve the gas in the (flowing) groundwater. Therefore, a gas phase will develop and a gas pressure will build up. This pressure will act as a driving force to the gas and will, under certain circumstances, influence near-field release of nuclides as well as local hydrology and geosphere transport.

- Under certain circumstances the engineered barriers could be damaged due to excessive gas pressure differences and thereby their function to act as a transport barrier will be diminshed. To rule out this possibility the option of engineered gas vents is currently considered. With these gas vents significant build-up of pressure differences can be avoided.

- Release of gas through the geosphere depends mainly upon the properties of the fracture zones. First results from field investigations at the potential site of Oberbauenstock indicate that it should be possible to release the gas through the geosphere without excessive pressure build-up. However, it remains to be shown how far these data are representative.

ACKNOWLEDGEMENTS

The authors take pleasure in acknowledging the contributions of all the persons and organizations involved in the Swiss gas evaluation programme.

REFERENCES

/1/ Project Gewähr 1985: "Nuclear Waste Management in Switzerland: Feasibility studies and safety analyses", Nagra NGB 85-09 (1985).

/2/ Projekt Gewähr 1985: "Endlager für schwach- und mittelaktive Abfälle: Bautechnik und Betriebsphase", Nagra NGB 85-06 (1985).

/3/ Projekt Gewähr 1985: "Endlager für schwach- und mittelaktive Abfälle: Das System der Sicherheitsbarrieren", Nagra NGB 85-07 (1985).

/4/ Projekt Gewähr 1985: "Endlager für schwach- und mittelaktive Abfälle: Sicherheitsbericht", Nagra NGB 85-08 (1985).

/5/ T. Schneider, S. Kappeler: "Geowissenschaftliche Grundlagen des Sondierstandortes Oberbauenstock", Nagra NTB 84-20 (1984).

/6/ M. Wiborgh, L.O. Höglund, K. Pers: "Gas formation in a L/ILW repository and gas transport in the host-rock", Nagra NTB 85-17 (1985).

/7/ "Inventar und Charakterisierung der radioaktiven Abfälle in der Schweiz", Nagra NTB 84-47 (1984).

/8/ B. Knecht, C. McCombie: "High-level waste overpack for final storage in the Swiss granite bedrock: Material selection, design and characteristics", Proc. of the Int. Topical Meeting on High-level Nuclear Waste Disposal, ANS, Pasco/WA (1985).

REFERENCES (cont.)

/9/ J.P. Simpson, R. Schenk, B. Knecht: "Corrosion rate of unalloyed steels and cast irons in reducing granitic groundwaters and chloride solutions", Proc. of 9th Int. Symp. on the Scientific Basis for Nuclear Waste Management, MRS-85, Stockholm (1985).

/10/ R. Schenk: "Experimente zur korrosionsbedingten Wasserstoffbildung in Endlagern für mittelaktive Abfälle", Nagra NTB 83-16 (1983).

/11/ H.H. Uhlig: "The Corrosion Handbook", John Wiley & Sons Inc., New York (1948).

/12/ H.A. Grogan, J. West: Personal communication (1987).

/13/ I.G. McKinley, J.M. West, H.A. Grogan: "An analytical overview of the consequences of microbial activity in a Swiss HLW repository", Nagra NTB 85-43 (1985).

/14/ A.H. Bath et al: "Trace element and microbiological studies of alkaline groundwaters in Oman, Arabian Gulf - a natural analogue for cement pore-waters", Nagra NTB 87-16 (in preparation).

/15/ M. Molecke: "Gas generation from transuranic waste degradation, data summary and interpretation", SAND 79-1245 (1979).

/16/ M. Molecke: "Degradation of transuranic-contaminated wastes under geologic isolation conditions", IAEA-SM-246/37 (1980).

/17/ A. Rasmuson, M. Elert: "The influence of gas flow on solute transport", Nagra draft report (1987).

/18/ I Neretnieks: "Some Aspects of the Use of Iron Canisters in Deep Lying Repositories for Nuclear Waste", Nagra NTB 85-35 (1985).

/19/ B.K. Nyame, J.M. Illston: "Capillary Pore Structure and Permeability of Hardened Cement Paste", 7th Int. Congress on the Chem. of Cement, VI-181, Paris (1980).

/20/ R. Pusch, L. Ranhagen, K. Nilsson: "Gas Migration through MX-80 Bentonite", Nagra NTB 85-36 (1985).

/21/ K. Pruess: Personal communication (1987).

/22/ M.A. Ibrahim, M.R. Tek, D.L. Katz: "Threshold Pressure in Gas Storage", AGA Monograph (1970).

/23/ T. Küpfer: Personal communication (1987).

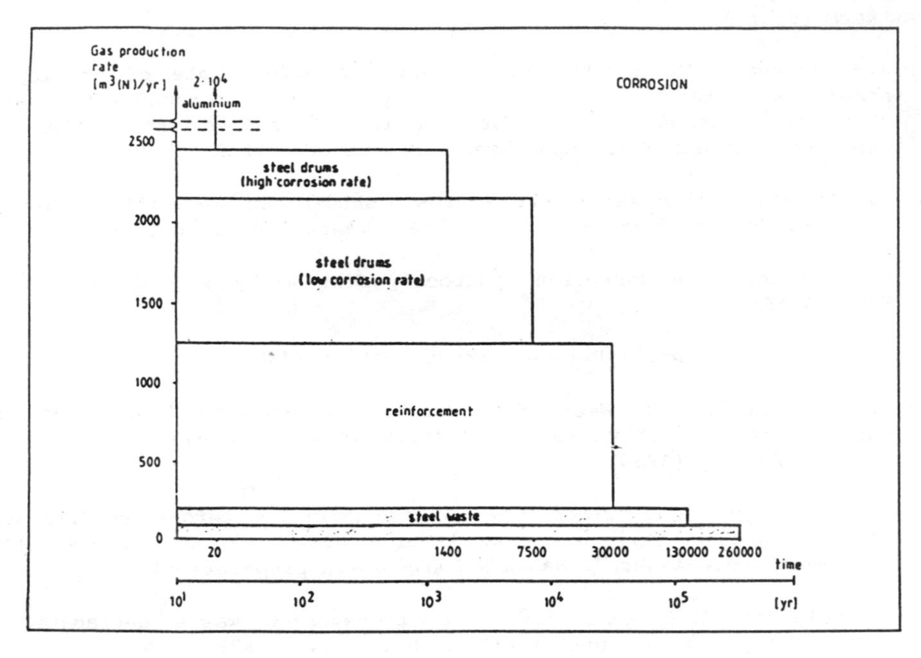

Figure 1: Annual rate of gas formation (STP) for different metals in the waste, base case

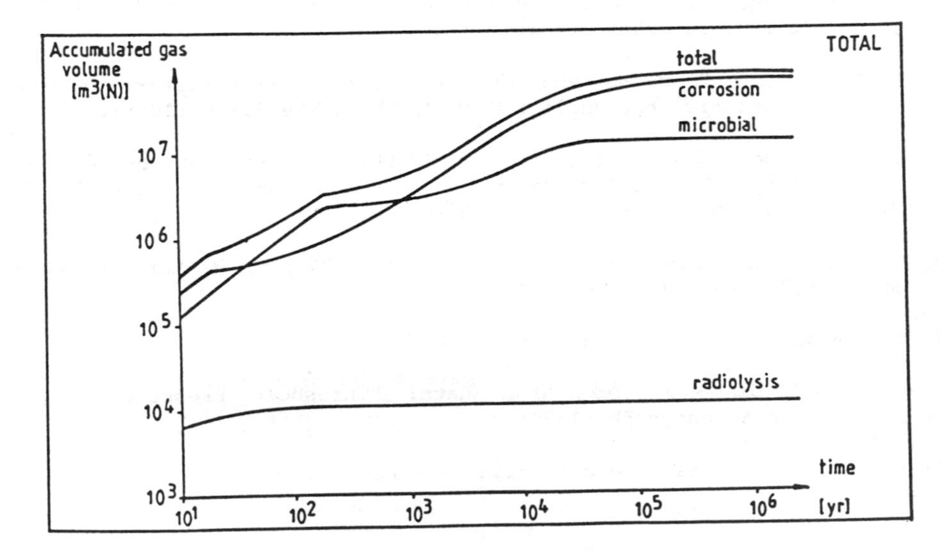

Figure 2: Accumulated gas volumes (STP) for the different gas-forming processes, base case

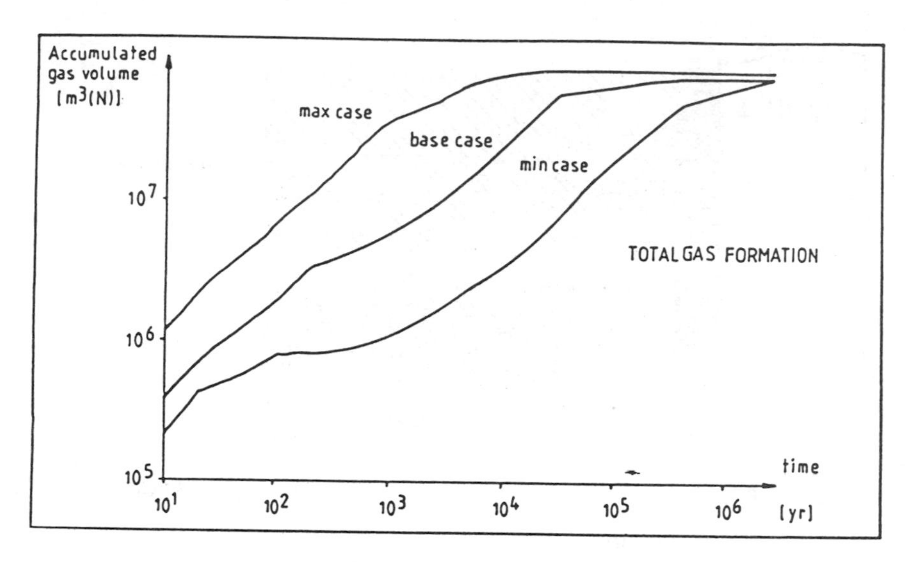

Figure 3: Accumulated gas production (STP) in the repository for different cases

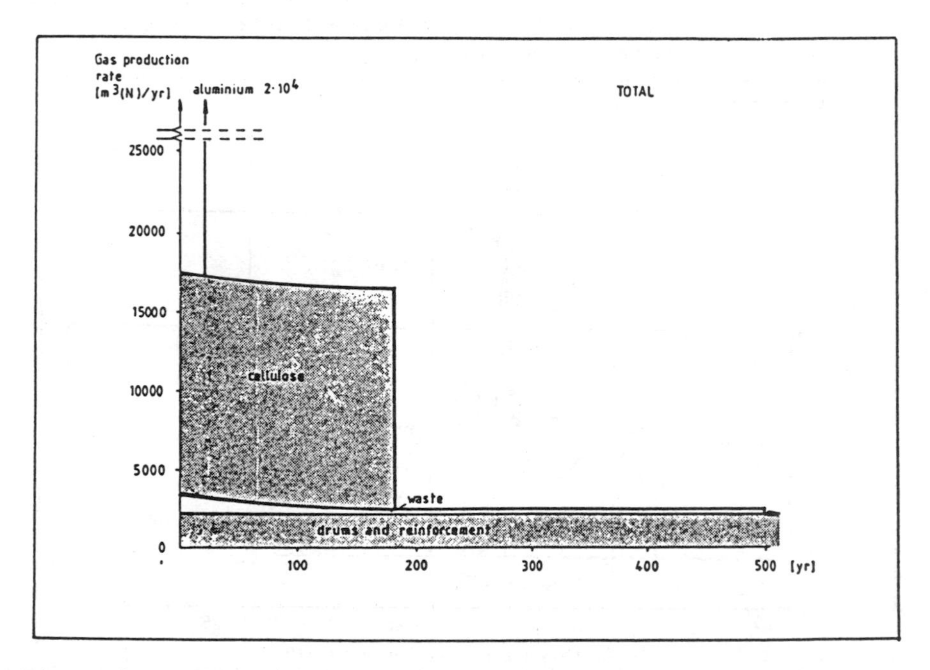

Figure 4: The contribution from different materials to the total rate of formation (STP) of insoluble gas during the first 500 years

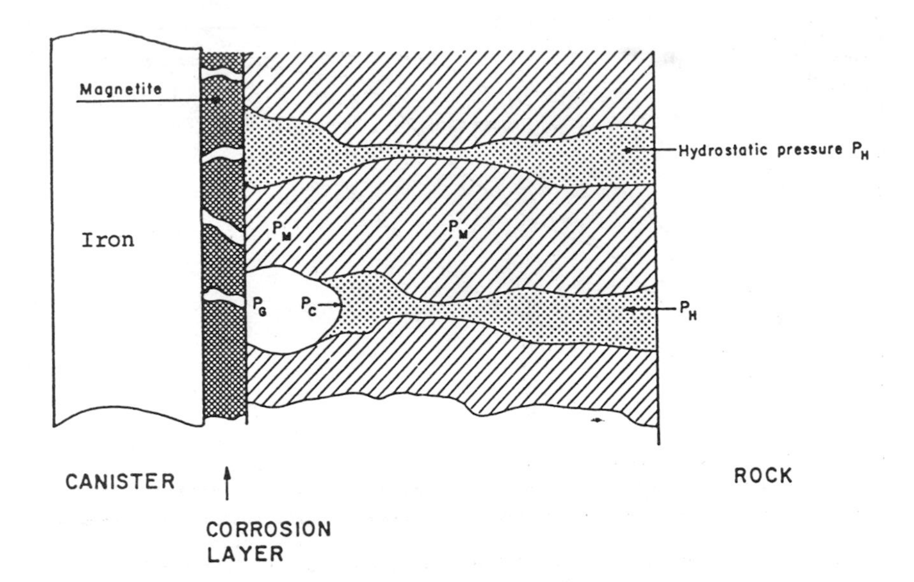

Figure 5: Schematic illustration of the parameters relevant for the displacement of water by gas. P_C: capillary pressure ($P_C = P_G - P_H$); P_G: gas pressure; P_M: rock stress; P_H: hydrostatic pressure

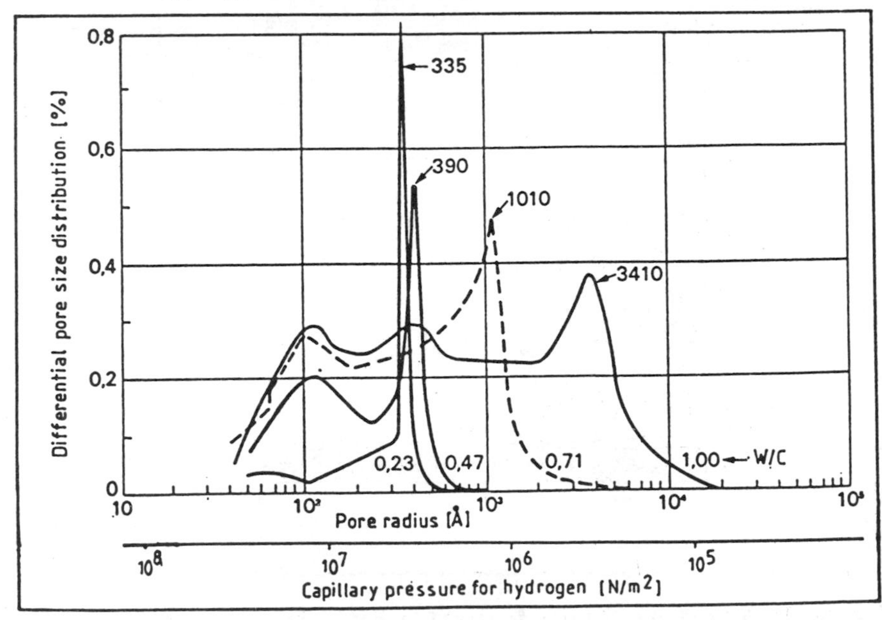

Figure 6: Differential pore size distribution for concretes with different w/c ratios. Numbers on the peaks indicate the maximum continuous pore radius (Å) /19/

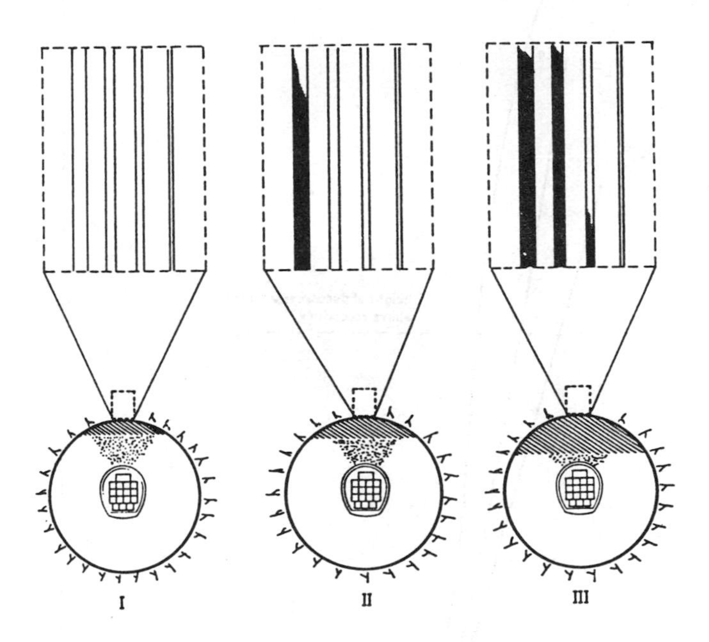

Figure 7: Schematic illustration of the accumulation of gas in the decompressed zone (I), the displacement of water from fracture zones as the gas pressure increases (II), and the steady-state situation (III)

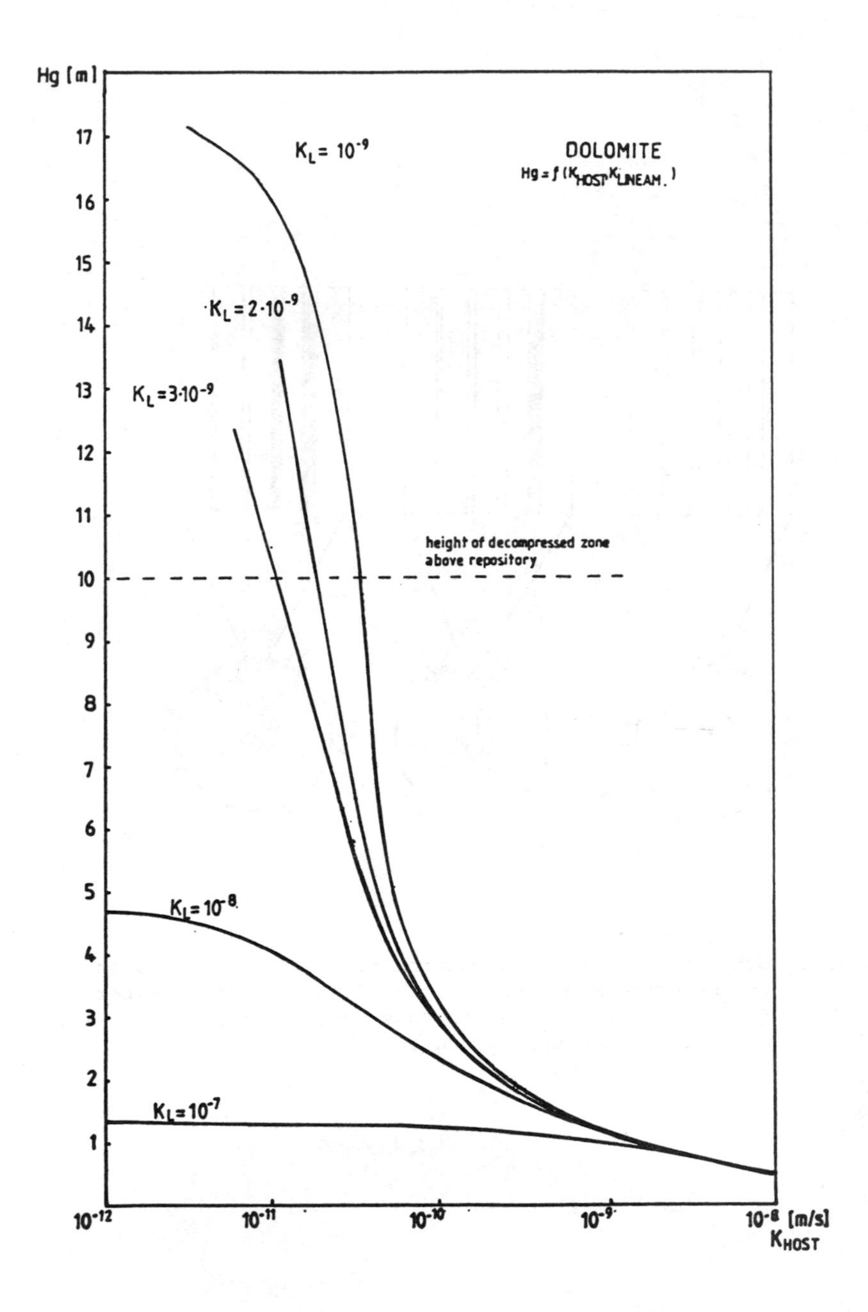

Figure 8: Calculated gas cushion height, Hg, as a function of the hydraulic conductivities in the rock matrix and in the fracture zones at steady-state conditions

GAS GENERATION AND MIGRATION IN WASTE REPOSITORIES

D. A. Lever and J. H. Rees

Theoretical Physics and Chemistry Divisions
Harwell Laboratory, Oxon, OX11 0RA, U.K.

ABSTRACT

Recent estimates of gas generation in repositories suggest that substantial volumes of gas will be produced on long time-scales. Metals such as steel corrode very slowly and produce hydrogen. Microbiological degradation of organic materials in the waste may produce substantial volumes of carbon dioxide and methane in the anaerobic conditions expected in the repository. Radiolysis of water in cement and organic materials is estimated to produce substantially less gas than corrosion or microbiological degradation. Finally, some long-lived nuclides (^{238}U, ^{232}Th, ^{235}U) have radioactive gaseous daughters (^{222}Rn, ^{220}Rn, ^{219}Rn), which could escape with other gases, together with minor amounts of other active and toxic gases.

In this paper, we review (i) the gases that could possibly be produced in a repository, (ii) the volumes and rates of generation of the main inactive gases, and (iii) the volumes and rates of generation of their main active derivatives. Then we discuss the ways these gases move through the repository materials and into the surrounding geology.

PRODUCTION ET MIGRATION DES GAZ DANS DES DEPOTS DE DECHETS

RESUME

Des estimations récentes de la production de gaz dans des dépôts laissent penser que des volumes notables de gaz seront produits à des échelles de temps prolongées. Des métaux, tels que l'acier, se corrodent très lentement et produisent de l'hydrogène. La dégradation microbiologique des matières organiques contenues dans les déchets peut produire des volumes substantiels de dioxyde de carbone et de méthane dans les conditions anaérobies escomptées dans le dépôt. La radiolyse de l'eau dans le ciment et les matières organiques produit, estime-t-on, notablement moins de gaz que la corrosion ou la dégradation microbiologique. Enfin, certains nucléides à vie longue (^{238}U, ^{232}Th, ^{235}U) ont des produits de filiation gazeux radioactifs (^{222}Rn, ^{220}Rn, ^{219}Rn), qui pourraient s'échapper avec d'autres gaz, ainsi qu'avec de faibles quantités d'autres gaz radioactifs et toxiques.

Dans la présente communication, les auteurs examinent (i) les gaz qui pourraient éventuellement être produits dans un dépôt, (ii) les volumes et les vitesses de production des principaux gaz inactifs, et (iii) les volumes et les vitesses de production de leurs principaux dérivés actifs. Ils examinent ensuite les manières dont ces gaz se déplacent à travers les matériaux du dépôt et dans le milieu géologique avoisinant.

1. INTRODUCTION

There are a number of strands in the assessment of the impact of gas generated in a repository. First, the total volume of gas and the rate at which it is produced have to be evaluated. If the volumes are large, then the way the gas escapes from the repository and through the surrounding geology has to be investigated, and the impact on other migration pathways has to be assessed. To evaluate the direct radiological impact, the proportion of the major gases that are active has to be calculated. Hazards from toxic and flammable gases have also to be considered.

These issues have been receiving attention in a number of countries. Biddle et al.[1] have recently presented a comprehensive assessment of the volumes of gas that will be produced in low- and intermediate-level waste repositories. Wiborgh et al.[2] have undertaken a thorough review of the whole problem for the Swiss programme. They have examined generation, transport and the possibility that a gas cushion will be formed around the repository. Moreno and Neretnieks[3] have looked at the ability of gas to escape through repository materials. They highlighted the possibility that concrete structures may crack at the sort of pressures that may be experienced. Water and gas may then flow more quickly along these cracks. To avoid this, they propose the construction of channels of high permeability materials at the top of the repository, to allow the gas to escape at lower pressures. A number of authors have examined the movement of gas through the surrounding geology. Pusch and co-workers[4,5] have undertaken experiments on the migration of gas through bentonite clay. They conclude that it is possible that gas could percolate through a number of narrow passages that form when the gas pressure is sufficiently high. Thunvik and Braester[6,7] and Pahwa[8] have modelled gas migration in hard fractured rocks. Both studies conclude that if gas moves through geologies of this type, then the transport times will probably be order tens of days rather than years.

In the remainder of this paper, we review the gases that could possibly be produced in a repository, the volumes and rates of generation of the main inactive gases and the volumes and rates of generation of their active derivatives. Then the ways these gases move through the repository materials and into the surrounding geology are discussed. As an example, a shallow repository in clay for cemented low-level waste will be discussed. Such repositories were being investigated by UK Nirex Ltd. However since a change of national policy was announced earlier this year, shallow repositories are no longer being considered. Many of the basic ideas carry over to deeper repositories for cemented low- or intermediate-level wastes, although the ambient pressure of deep repositories is substantially higher.

2. GAS GENERATION

2.1 Introduction

Gases will be evolved in repositories in three main ways[1]:

i) Corrosion of metals generating hydrogen under the anaerobic conditions prevailing in a repository.
ii) Microbiological degradation of organic wastes, particularly paper and wood. This is expected to occur in regions that locally are not at a high pH when the repository is closed, and may therefore take

place in the centre of drums of compacted waste. Microbial activity is expected to be much lower in wastes at the high pH generated either by wastes coming into contact with cement water when they are grouted into steel boxes or by saturation with high pH water after closure of the repository and breaching of the steel containment by groundwaters. It could take hundreds of years to reach this second situation, by which time the susceptible wastes may have been consumed. Carbon dioxide and methane are produced under the anaerobic conditions assumed.

iii) Radiolysis of water in cement, producing hydrogen, and of organic materials yielding hydrogen as the main product, with lesser amounts of carbon dioxide, carbon monoxide and various simple hydrocarbons.

In the initial study of gas production[1], it was found that radiolysis was much less important than corrosion and microbial decomposition for cemented intermediate- or low-level waste, and this route is not discussed further here.

Other gases, either toxic or active, may be formed in addition. Possible gases are listed in Table I, where they are shown in approximate order of decreasing importance. This work is limited to consideration of hydrogen, carbon dioxide and methane because they are formed in bulk. Radon, which is evolved through the decay of uranium and thorium, is also potentially radiologically important. However, the radon isotopes are both soluble and short-lived. So we expect the most significant path for radon to be transport of the long-lived radionuclide to the near-surface region by groundwater, then decay to radon and transport as a gas into, for example, houses. Other active gases are expected to be formed in very low quantities, while several of the toxic gases, for example H_2S, H_2Se, $Ni(CO)_4$ and SnH_4 are expected to react in the repository. Carbon dioxide is also expected to react with cement, giving calcium carbonate. It is nevertheless considered further in this section because it is expected to be one of the major gases formed.

2.2 Volumes of gas evolved

To calculate the volumes and rates of production of gases, we use shallow disposal of cemented low-level waste in clay as an example. The volumes here are based on recent calculations[1] but the precise volumes are rather larger as the waste volume and compositon have been updated. The important repository parameters, such as repository volume and mass of metals, are given in Table II. Allowance is made for corrosion of the steel packing received from waste producers, but not for any additional packaging or for steel reinforcement in the repository.

(a) Corrosion

The bulk of the metals considered here are various steels; the volume of hydrogen that will be generated at S.T.P. will be about 10^9 m^3.

(b) Microbial degradation

Again basing the calculations on [1], a maximum volume at S.T.P. of about 3 10^6 m^3 each of carbon dioxide and methane will be formed from the paper and wood expected to be in the repository. If any of this material is brought to a high pH before or soon after closure of the repository, the volumes would be less.

2.3 Rates of evolution

The rates of evolution are based on those in[1], although as mentioned earlier, the precise rates of evolution are rather larger as the waste volume and composition have been updated. The results are summarised in Figure 1, where the accumulated volumes of the main gases are plotted.

(a) Corrosion

The data given in [1] indicate that hydrogen will be evolved at a steady rate of about 8 10^4 m^3 yr^{-1}. The rate of corrosion depends on the surface area of the steels. The rate quoted is an average of values obtained assuming all the waste steel to be in sheet form and then in massive form.

(b) Microbial degradation

The rate of production of both methane and carbon dioxide is estimated from[1] to start at a rate of 10^4 m^3 yr^{-1}. It falls off approximately exponentially over a time-scale of a few hundred years.

2.4 Rates of evolution of active gases

The proportion of the active component of the gas has been derived from the appropriate ratio of the quantity of the active isotope to the inactive one of the element in question. The results are summarized in Table III.

(a) Tritiated hydrogen

Tritium will be in dynamic equilibrium with the water in the repository that causes corrosion, both porewater and water bound as part of the cement hydration products. From the repository parameters given in Table II, we estimate the volume of porewater (ϕ_R V_R) to be 5.2 10^5 m^3. There will be a smaller amount of water bound into cement hydrate phases, though this is not taken into account here. The estimate of tritium initially in the inventory is 23 TBq, and this decays relatively quickly because of the short half-life. So the initial proportion of the hydrogen in the repository that is ^3H is about 4 10^{-13}.

(b) Tritiated methane

As a worst case, it is assumed here that all the ^3H is present in paper and wood. The estimated mass of this in the waste is 10^4 te at repository closure. Taking 10% as the weight of hydrogen in these wastes gives 10^7 moles of H in them. So the initial ratio of tritium to hydrogen in the paper and wood is 2 10^{-9}.

(c) $^{14}CH_4$ and $^{14}CO_2$

If 50% by weight of the wood and paper is carbon, the 10^4 te of wood and paper in the repository contain 4 10^8 moles of C. The initial inventory of

^{14}C is 34 TBq, corresponding to 14 moles of ^{14}C. If we assume all of this is associated with paper and wood, the initial proportion of $^{14}CH_4$ in methane and $^{14}CO_2$ in carbon dioxide is 3.4 10^{-8}. As some of the ^{14}C will be associated with materials that are not susceptible to attack, these rates are over-estimates.

3. GAS MIGRATION

3.1 Introduction

In the previous section the volume of gas and the rate of production have been estimated. Next we have to consider what happens to that gas, whether it remains in the vicinity of the repository or whether it escapes and migrates to the surface. The conceivable scenarios are as follows:

i) the gas dissolves in the porewater, and then diffuses or advects away from the repository.

ii) the gas pressure builds up until it is sufficient to overcome the capillary pressure of the surrounding material. It then enters that porous material as a gas. It may move preferentially into more permeable strata, where the capillary pressure is lower.

iii) before the pressure reaches the capillary pressure, fractures and cracks could open up in the surrounding material and the gas could migrate through them. After the gas has escaped, the fractures may heal, or they may remain and offer a preferential water flow path.

iv) the gas could remain in the immediate vicinity of the repository, unable to escape, with the gas pressure slowly increasing.

v) the gas could expand into the geology around the repository and remain at the ambient pressure, but not escape.

vi) if the gas does remain in the vicinity, it could affect the near-field environment and significantly affect other processes, e.g. the rate of gas generation could be suppressed.

vii) the gas could escape through the surrounding geology to the surface.

It is important to determine which takes place and to evaluate if there are consequences for other aspects of the radiological assessment. In this section we examine the first scenario - the ability of gas to escape from the repository dissolved in porewater.

3.2 Modelling local diffusion away from the canister

First, we examine the ease with which gas moves through the concrete in the repository, by considering local diffusion away from a canister. At the canister wall corrosion is producing hydrogen. We compare the rate of hydrogen generation with the maximum diffusive flux of H_2 dissolved in the concrete porewater, assuming that the concentration at the canister wall is the solubility at the local pressure (c_s) and that it drops to zero over about half the canister spacing (ℓ). This is a very optimistic estimate, as it assumes the hydrogen can be carried away, and so in practice the flux will be less. If the maximum flux is less than the generation rate per unit area (R), then bubbles will form in the gap between the canister and surrounding backfill. So bubbles form if

$$R > D_i^{con}\ c_s/\ell \quad , \tag{1}$$

where D_i^{con} is the intrinsic diffusion coefficient of dissolved H_2 moving through concrete. This is a simplified form of the model being developed by Sharland et al.[9].

If bubbles do form, their subsequent behaviour is a complicated problem. Pressure will build up, but the gas will only move into the backfill if it can overcome the capillary pressure, which is the pressure required to overcome the surface tension forces so the gas can move up the capillaries. The simplest model for a circular tube capillary gives

$$p_c = 2\Gamma\cos\theta/r \quad ,$$

where Γ is the surface tension, θ is the contact angle and r the pore radius. So the smaller the pore, the larger the capillary pressure becomes. For concrete, the capillary pressure is estimated to be quite high, approximately 1.5 MPa[3]. However, it is possible that the concrete will not be able to withstand these pressures, and so fractures may form to allow gas to escape. Alternatively, as the pressure rises the gas solubility increases, and so an increased diffusive flux results. The pressure at which the balance between generation and diffusion takes place, should be compared with the capillary pressure and the pressure at which fracturing takes place. There is one further possibility to consider. The blanket of H_2 may slow down the corrosion rate, and consequently the rate at which H_2 is generated.

3.3 Modelling migration of dissolved gas through the geosphere

Next we examine whether the dissolved gas can migrate through the geosphere, by considering the quantity that can diffuse or convect away in the local groundwater.

The maximum quantity that can diffuse to the surface from a repository with horizontal cross-sectional area A at a depth d over a time t is

$$Q^d = A\ D_i^{clay}\ t\ c_s/d \quad , \tag{2}$$

whereas the maximum that can be convected away by the groundwater flow with Darcy flux q is

$$Q^c = A\ q\ t\ c_s \quad . \tag{3}$$

The best current estimates suggest that migration will be advectively dominated ($qd \gg D_i^{clay}$), and so Q^c will be much larger than Q^d. We can express Q^c in terms of the repository volume V_R (V_R = Ah, where h is the height of the repository) and the water travel time t_w($q = \phi_g d/t_w$, where ϕ_g is the porosity of the surrounding geology), giving

$$Q^c = dt\ \phi_g\ c_s\ V_R/ht_w \quad . \tag{4}$$

We expect Q^c to be an overestimate of the actual flux. As long as the repository retains its integrity and very low permeability, the flow will be around the repository rather than through it, and so the quantity of dissolved gas entering the groundwater will be limited by diffusion. If Q^c is larger than the quantity of gas produced, then the gas cannot migrate away from the repository as a dissolved species, and one of the other possibilities outlined in section 3.1 will come about.

3.4 Typical results

We now give some typical results for a number of cases. As an example, we consider shallow disposal of low-level waste in clay. Many of the ideas carry over to deeper disposal. The parameters used in this example are given in Table II.

(a) Diffusion of hydrogen away from a corroding surface

The rate of gas generation at a corroding surface is approximately[9]

$$R = 3\ 10^{-10}\ \text{mol.m}^{-2}\ \text{s}^{-1} \equiv 7\ 10^{-12}\ \text{m}^3\ (\text{gas:STP})\ \text{m}^{-2}\ \text{s}^{-1} \quad .$$

This is consistent with the rates quoted in section 2.3. We assume that the intrinsic diffusion coefficient for H_2 in concrete[9] is similar to that measured for dissolved ions or molecular oxygen and so we take a value of $D_i^{con} = 4\ 10^{-12}\ m^2\ s^{-1}$. We note that the intrinsic diffusion coefficient for H^+ ions is an order of magnitude larger than this, but this is due to the easy exchange mechanism with water molecules. A substantially higher diffusion coefficient for dissolved H_2 would increase the potential for H_2 to diffuse away. Finally we require the solubility of hydrogen. At atmospheric pressure, this is approximately[10]

$$c_s = 10^{-3}\ M \equiv 2\ 10^{-2}\ \text{m}^3\ (\text{gas:STP})\ \text{m}^{-3}(\text{water}) \quad .$$

This gives a maximum diffusive flux (1) of $8\ 10^{-14}\ m^3(\text{gas:STP})m^{-2}s^{-1}$, which is substantially less than R. So unless the repository is specially engineered, gas bubbles will form around corroding surfaces, and the pressure will start to build up.

(b) Convection of hydrogen through the surrounding geology

In section 2, the total volume and rate of production of hydrogen were estimated. These correspond to

$$V_{H_2} = 10^9\ m^3\ ,\ t_{H_2} = 10^4\ yr \quad .$$

So the estimate (4) of the quantity of dissolved gas that can be convected away gives

$$Q^c_{H_2} = 9\ 10^4\ m^3 \quad .$$

This is substantially less than V_{H_2}. As Q^c is regarded as an overestimate, it seems unlikely that the groundwater will be able to carry the gas away in solution.

(c) Convection of methane through the surrounding geology

Taking the average rate of production quoted in the previous section, the volume of CH_4 and the production time is

$$V_{CH_4} = 3\ 10^6\ m^3 \quad , \quad t_{CH_4} = 6\ 10^2\ yr.$$

The solubility of methane is[10]

$$c_s = 2\ 10^{-3}\ M \equiv 4.5\ 10^{-2}\ m^3(gas:STP)\ m^{-3}(water) \quad .$$

So the estimate (4) of the volume of gas that could be convected away is

$$Q^c_{CH_4} = 1.2\ 10^4\ m^3 \quad .$$

As this is less than V_{CH_4} above, it is unlikely that CH_4 can be carried away in solution, even if it can escape through the concrete. Furthermore, as mentioned above, the estimate Q^c is probably an overestimate of the quantity that can actually be carried away, particularly as on the short time-scales considered here, the repository will largely retain its integrity.

(d) Carbon dioxide

The solubility of carbon dioxide is[10]

$$c_s = 6\ 10^{-2}\ M \equiv 1.4\ m^3(gas:STP)m^{-3}(water) \quad ,$$

which is thirty times the solubility of methane. The volumes of carbon dioxide and methane are the same, and so the estimate of the volume of carbon dioxide that can be convected away is

$$Q^c_{CO_2} = 3.6\ 10^5\ m^3 \quad .$$

This is still substantially less than the volume that is generated.

However, carbon dioxide can react with the calcium hydroxide of the concrete. So we examine the quantity of concrete required to mop up the carbon dioxide. The volume of CO_2 that is generated ($3\ 10^6\ m^3$) corresponds to $1.3\ 10^8$ mol. To estimate the number of moles of Ca in the repository, we assume

i) half the repository is concrete,

ii) concrete has 400 kg m^{-3} cement,

iii) cement has 10 mol kg^{-1} of Ca.

Thus the total number of moles in the repository is $5.2\ 10^9$ mol. So 2.5% of the Ca must be available to react with the CO_2 generated as a gas.

Now this is certainly feasible. However, it is possible that if the epository is fractured, then $CaCO_3$ could seal the walls of the pores and fractures. If this occurred, the Ca in the bulk of the concrete would be protected from CO_2. It does not make any difference to the gas assessment, however, as the presence of CO_2 would lead to an increase of only a factor of 2 in the volume of gas generated by microbial degradation.

4. DISCUSSION

In this report, we have examined the volumes of gas that are expected to be produced in a repository. Although the total volumes are substantial (about 400 times the repository volume at S.T.P.), the gas is produced over very long time-scales (roughly 10^4 years). This corresponds to each waste package, on average, producing its own volume of gas every 10 years. The simple migration models, examined in the last section, suggest that the gas will not be able to escape dissolved in the porewater. So it will either build up in the environs of the repository or escape as gas. Which takes place will depend on the properties of the materials of the repository and the surrounding geology.

For LLW, the proportion of the gas that is active is very small, and so the volumes of active tritium, methane and carbon dioxide are all less than 10^{-3} m^3 yr^{-1}. However, the radiological impact of this gas has to be assessed if it escapes from the repository and returns to the surface. The results will be very different for different environments, e.g. marine, undevelopable terrain, and land that could be developed for housing. An assessment, which is not reported here[11], has been made for two scenarios. The first is for houses built at the edge of the site during the site use restriction period, and the second is for houses built over the site after the site use restriction has lapsed. The assessment concludes that, for the current inventory, the radiological risk is well below the U.K. regulatory target.

ACKNOWLEDGEMENTS

This work is funded by UK Nirex Ltd. We would like to thank Dr. A. Atkinson, Mr. P. Biddle, Dr. P. E. Rushbrook, Dr. P. W. Tasker and Mr. S. J. Wisbey for their advice.

REFERENCES

1. Biddle, P., McGahan, D., Rees, J.H. and Rushbrook, P.E.: "Gas Generation in Repositories", Harwell Laboratory Report, AERE-R.12291, 1987.
2. Wiborgh, M., Höglund, L.O. and Pers, K.: "Gas Formation in a L/ILW Repository and Gas Transport in the Host Rock", Nagra Report NTB 85-17, 1986.
3. Moreno, L. and Neretnieks, I.: "Gas, Water and Contaminent Transport from a Final Repository for Reactor Waste", SKB Progress Report SFR 85-09, 1985.
4. Pusch, R. and Forsberg, T.: "Gas Migration through Bentonite Clay", KBS Technical Report 83-71, 1983.
5. Pusch, R., Ranhagen, L. and Nilsson, K.: "Gas Migration through MX-80 Bentonite", Nagra Report NTB 85-36, 1985.
6. Braester, C. and Thunvik, R.: "An Analysis of the Conditions of Gas Migration from a Low-Level Radioactive Waste Repository", KBS Technical Report 83-21, 1982.

7. Thunvik, R. and Braester, C.: "Calculations of Gas Migration in Fractured Rock", SKB Progress Report SFR 86-04, 1986.
8. Pahwa, S.B.: "Gas Migration in Fractured Rock from a Low-Level Waste Repository using a Two-Phase Flow Code", INTERA Technologies Inc., Austin, Texas, 1986.
9. Sharland, S.M., Tasker, P.W. and Tweed, C.J.: "Evolution of the Eh in the Porewater of a Model Nuclear Waste Repository", Harwell Laboratory Report AERE R.12442, 1987.
10. Lange, N.A.: "Handbook of Chemistry", 14th Ed., McGraw Hill, 1985.
11. Hodgkinson, D.P. and Robinson, P.C.: "Nirex Near-Surface Repository Project: Preliminary Radiological Assessment: Summary", Harwell Laboratory Report NSS/A100, 1987.

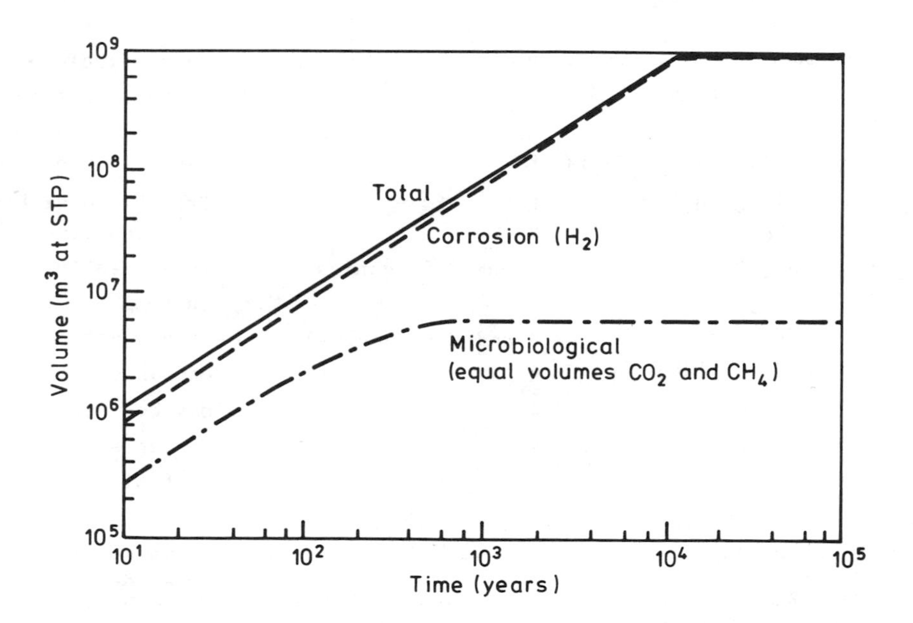

FIG. 1. BEST ESTIMATE OF THE VOLUME OF GAS PRODUCED AS A FUNCTION OF TIME

Table I Minor active and toxic gases that might form in a repository

Gas	Active isotope(s)	Method of formation*	Comment
Rn	$^{222}Rn, ^{220}Rn$	Decay of U and Th	Unreactive but soluble
Kr	^{85}Kr	Present in some waste	Unreactive
He	-	α-particles, decay of ^{3}H	Unreactive
Low mol. wt alkanes	$^{3}H, ^{14}C$	R and M (organics)	Stable
Low mol. wt alkenes	$^{3}H, ^{14}C$	R and M (organics)	Fairly reactive
Other organics	$^{3}H, ^{14}C$	R and M (organics)	-
CO	^{14}C	R and M (organics)	Fairly reactive
CH_3Cl	$^{14}C, ^{3}H, ^{36}Cl$	Mainly R	May be hydrolysed
CH_3I	$^{14}C, ^{3}H, ^{129}I$	Mainly R	May be hydrolysed
$(CH_3)_2Se$	$^{14}C, ^{3}H, ^{79}Se$	R	May be hydrolysed
H_2O	^{3}H	-	-
NH_3	^{3}H	R	Very soluble
$HSeCH_3$	$^{3}H, ^{79}Se, ^{14}C$	R	Probably unstable
HS CH_3	$^{3}H, ^{14}C$	R	Probably unstable
CH_3SnH_3	$^{121m}Sn, ^{126}Sn, ^{3}H, ^{14}C$	R	Probably unstable
CH_3PbH_3	$^{210}Pb, ^{3}H, ^{14}C$	R	Probably unstable
Cl_2	^{36}Cl	R	Very reactive
HCl	$^{3}H, ^{36}Cl$	R and C (organics)	Very reactive
I_2	^{129}I	R	Unstable
HI	$^{3}H, ^{129}I$	R and C (organics)	Very reactive
SnH_4	$^{121m}Sn, ^{126}Sn, ^{3}H$	R	Unstable
PbH_4	$^{210}Pb, ^{3}H$	R	May not exist
SbH_3	$^{125}Sb, ^{3}H$	R	Unstable
BiH_3	^{3}H	R	Unstable
$Ni(CO)_4$	$^{59}Ni, ^{63}Ni, ^{14}C$	C	Unstable
H_2Se	$^{3}H, ^{79}Se$	R	Forms metal selenides
H_2S	^{3}H	R	Forms metal sulphides

*R = Radiolytic, C = Chemical, M = Microbial

Table II Repository, inventory and groundwater parameters

Quantity	Symbol	Value
Repository volume	V_R	$2.6\ 10^6\ m^3$
Weight of metals	-	$2.5\ 10^6$ te
Weight of wood and paper	-	10^4 te
Inventory of ^3H	-	23 TBq
Inventory of ^{14}C	-	34 TBq
Repository height	h	5.7m
Repository depth	d	10m
Half canister separation	ℓ	1m
Repository porosity	ϕ_R	0.2
Groundwater travel time	t_w	$2.5\ 10^3$ yr.
Rock porosity	ϕ_g	0.25

Table III Summary of gas evolution rates

Year after closure	Rate of formation ($m^3 yr^{-1}$ at STP)						
	H_2	CH_3	CO_2	H^3H	$^{14}CH_4$	$CH_3{}^3H$	$^{14}CO_2$
0	$8\ 10^4$	$1\ 10^4$	$1\ 10^4$	$3\ 10^{-8}$	$3\ 10^{-4}$	$2\ 10^{-5}$	$3\ 10^{-4}$
50	$8\ 10^4$	$9\ 10^3$	$9\ 10^3$	$2\ 10^{-9}$	$3\ 10^{-4}$	$9\ 10^{-7}$	$3\ 10^{-4}$
100	$8\ 10^4$	$7\ 10^3$	$7\ 10^3$	$8\ 10^{-11}$	$2\ 10^{-4}$	$6\ 10^{-8}$	$2\ 10^{-4}$
300	$8\ 10^4$	$3\ 10^3$	$3\ 10^3$	$2\ 10^{-15}$	$1\ 10^{-4}$	$3\ 10^{-13}$	$1\ 10^{-4}$

SESSION III

CHEMICAL EFFECTS IN THE NEAR-FIELD

SEANCE III

EFFETS CHIMIQUES DANS LE CHAMP PROCHE

Chairman - Président

I. NIERETNIEKS

(Sweden)

PHYSICAL AND CHEMICAL ENVIRONMENT AND RADIONUCLIDE MIGRATION IN A LOW LEVEL RADIOACTIVE WASTE REPOSITORY.

by

John Torok and Leo P. Buckley

Waste Management Technology Division
Atomic Energy of Canada Limited
Chalk River, Ontario K0J 1J0

ABSTRACT

The expected physical and chemical environment within the low-level radioactive waste repository to be sited at Chalk River is being studied to establish the rate of radionuclide migration. Chemical conditions in the repository are being assessed for their effect on buffer performance and the degradation of the concrete structure. Experimental programs include the effect of changes in solution chemistry on radionuclide distribution between buffer/backfill materials and the aqueous phase; the chemical stability of the buffer materials and the determination of the controlling mechanism for radionuclide transport during infiltration.

MILIEU PHYSIQUE ET CHIMIQUE ET MIGRATION DES RADIONUCLEIDES DANS UN DEPOT DE DECHETS NUCLEAIRES DE FAIBLE ACTIVITE

RESUMÉ

Les auteurs étudient le milieu physique et chimique escompté dans le dépôt de déchets de faible activité qui doit être implanté à Chalk River, afin d'établir la vitesse de migration des radionucléides. Ils évaluent les conditions chimiques existant dans le dépôt du point de vue de leurs effets sur le comportement du tampon et la dégradation de l'ouvrage en béton. Les programmes expérimentaux couvrent l'effet des modifications de la chimie de la solution sur la distribution de radionucléides entre, d'une part, les matériaux constituant le tampon et le remblai et, d'autre part, la phase aqueuse, la stabilité chimique des matériaux utilisés pour le tampon, et la détermination du mécanisme régissant le transport des radionucléides au cours de l'infiltration.

INTRODUCTION

AECL has an on-going comprehensive program in low level waste management. It involves the characterization, processing and storage of low and intermediate level waste streams generated on-site at the Chalk River Nuclear Laboratories (CRNL) and shipped from waste generators including hospitals and universities across Canada. A program for the transition from storage to disposal of low level waste is well advanced [1]. It involves site selection, facility design, the assessment of safety and environmental protection, licensing and quality control. The program outlined in this presentation is focused on the chemical characterization of the repository and the transport of radionuclides through the engineered barriers. It is part of the safety and environmental assessment and also impacts on the design and optimization of the facility. Our plans call for the construction of IRUS (Intrusion Resistant Underground Structure), a concrete facility with an open bottom placed underground but above the water table in a stable sand dune.

A persistent problem in repository characterization is the large degree of uncertainty associated with the composition, physical and chemical properties of the various waste forms. Waste that is amenable to volume reduction and/or stabilization is treated in the Waste Treatment Centre at CRNL. A small fraction of other waste (<10%) will be placed in the repository without treatment. While the long-term performance of stabilized wastes, the ones incorporated into bitumen, plastic or concrete, has been extensively explored, the contents of miscellaneous and compacted waste are subject to large variations. Yet, the latter group of wastes are more likely to undergo significant chemical and physical changes soon after disposal and thus influence repository performance.

While there will be a variety of engineered barriers put in place to restrict water movement, to retard radionuclides and to reduce intrusion [2], the focus of this paper will be on the performance of granular materials to be placed at the bottom of the repository and around the various waste forms. The engineered barriers in the repository are designed to minimize the transport of critical radionuclides both in the expected normal and possible, but unlikely failure modes. The critical radionuclides range from the relatively short half-life radionuclide, tritium, whose transport is the same as the transport of water; to ^{137}Cs, ^{90}Sr and ^{60}Co radionuclides whose movement can be effectively retarded by ion exchange media. The performance and stability of ion exchange media, however, is dependent on both the concentration of dissolved radionuclides and the water chemistry. Radionuclide migration through the layers of granular material must be evaluated using a wide range of potential water chemistries in the repository. Thus, the physical and chemical environment expected in the low-level radioactive waste repository at CRNL is being studied to establish the rate of radionuclide migration and to predict the extent of their movement. The conditions in the repository should remain quite dry and with little movement of the water. However, two improbable extremes, flooding or infiltration, are also being evaluated to compare radionuclide migration with that expected under normal operating conditions.

PHYSICAL AND CHEMICAL ENVIRONMENT IN THE REPOSITORY.

Normal and Failure Modes

The physical and chemical environment and radionuclide migration of the facility after closure is being evaluated in both the normal mode, where the facility meets its design criteria and in different failure modes. The two most plausible failure modes which have been considered are: flooding due to elevation of the water table; and, infiltration of rain water or melting snow due to the cracking and/or collapse of the repository roof. Schematic diagrams of the repository in these three conditions are illustrated in Figure 1. It is expected the repository as designed with its various protective water-shedding layers will keep water out for a long period of time. The determination of this time span is being addressed through concrete durability studies.

Chemical Interactions in the Repository

A simplified illustration of the interaction of waste forms, the chemical and physical processes taking place in the repository and their ultimate effect on water chemistry and radionuclide transport is illustrated in Figure 2. Only waste forms that have a significant effect on chemical conditions are included. These include compacted and miscellaneous waste, metals and wastes stabilized with bitumen, plastics or concrete. All of these events are not likely to be encountered until there are sufficient quantities of water entering the repository either by infiltration or by flooding.

The major chemical processes involve carbon dioxide generation by biological activity, carbonic acid formation, when the carbon dioxide is dissolved in water, and the reaction of carbonic acid with concrete waste forms, concrete rubble and the concrete structure of the repository. The products of this reaction are calcium and magnesium carbonates and bicarbonates. The generation of the latter will increase the salt content of the aqueous phase. Fatty acids and complexing agents are also products of the biological degradation process. Biological degradation together with the corrosion of metals such as carbon steel create a reducing environment in the repository. Physical processes include the leaching of soluble chemicals from stabilized and unstabilized waste and the leaching of calcium and magnesium hydroxide from concrete. The water chemistry can also be altered by interaction with the buffer and backfill materials and any high surface area waste in the repository.

The physical and chemical processes the concrete is subjected to not only affect the water chemistry in the repository, but what could be more important, they can influence the durability of the concrete structure.

SELECTION OF BUFFER AND BACKFILL MATERIALS

Desirable Hydraulic Properties

Current plans for the low level radioactive waste repository at CRNL call for the incorporation of two different types of engineered barriers, each designed to maximize the retardation of specific radionuclides. These barriers include: buffer material which will be put in place at the bottom of the repository as the repository is constructed; and, backfill material to be placed around the waste forms as the repository is filled.

Sand was chosen as the backfill material. In the dry normal repository environment, it is an ideal barrier to moisture movement due to its large pore structure and minimal capillary pore water content. Relative to finer structured soils, sand has a high hydraulic conductivity at high water content prevalent during infiltration, when short residence time for water is desirable, and low conductivity in the dry state, where minimal transport of the radionuclides is desirable.

A buffer zone, approximately 0.5 to 1.0 m thick will be placed at the bottom of the repository, immediately below the waste forms. Candidates for the buffer are clinoptilolite and/or clay mixed with sand. Clinoptilolite, a natural molecular sieve, was chosen as one of the materials because of its high selectivity for the critical cationic radionuclides. The other major candidate, a marine sediment clay (Dochart clay), is available locally and consists of in decending quantity: rock flour, chlorite, smectite and illite clay. Physical properties of clay/sand mixtures have been fully evaluated [3].

Desirable Buffer Properties.

The chemical environment in the repository can have a profound effect on the long term reliability of the buffer layer to retard the migration of radionuclides. The chemical environment in the repository is difficult to predict because of the broad range of wastes placed in the repository, and their poorly defined composition. Chemicals leached from the waste can have a dual effect on the buffer. They can influence the distribution of radionuclides between the solid and aqueous phases; and/or the chemicals can cause the gradual selective destruction of ion exchange sites of the buffer by slowly dissolving or by altering the skeletal structure of the adsorbent. Two approaches are taken to mimic the chemical environment in the repository. In the first approach, extremes in pH values are used since the expected uneven distribution of acid and base sources could result in localized high and low pH values in the buffer layer. This chemistry change can have an important effect on the adsorption properties and long term stability of the buffer. The potential range of pH conditions are expected to be broad in the repository due to acids and bases that form part of the low level waste load in the repository, from short chain fatty acids generated by biological degradation processes and from the leaching of sodium -, potassium -, and calcium hydroxides from concrete. The other approach is to subject the buffer and backfill materials to waste leachates. In the IRUS repository, compacted wastes with low levels of radioactivity will be placed in packages of low durability and thus will be the most susceptible to leaching.

EXPERIMENTAL PROGRAMS

The objectives of the experimental programs are to identify the controlling mechanisms of radionuclide transport within the repository in normal and failure repository conditions; and to quantify the rates of water and radionuclide transport both in the laboratory and scaled-up field experiments.

Effect of Environment on Buffer Performance

The potential for pH value fluctuations in the repository is significantly higher than in nature. Acids and bases in the waste will affect the pH of the leachate. Other potentially more dominant processes for acid and base generation are the anaerobic degradation of organic materials producing short chain fatty acids, and hence creating an acidic environment in the pH range 4.3 to 5. Water in contact with concrete will generate a leachate having a high pH, in the range of 11 to 12.

The candidate buffer materials, Dochart clay and clinoptilolite, were subjected to leachants in the pH range 2 to 11 at room temperature and at 60°C. Temperature and pH conditions were more extreme than expected in the repository (8 °C) to accelerate the chemical reactions. The conditioning solutions were acetic acid, the product of anaerobic degradation; potassium and calcium hydroxide, the main constituents of short- and long-term concrete leachate respectively. Several approaches were used to define the nature and extent of clay degradation. The most quantitative approach was to analyze the composition of the spent leaching solution for the chemical components of the buffer materials. The initial leachant solution, containing mostly ion exchanged cations, was discarded and only the subsequent leach solutions that more closely reflect the dissolution of the mineral skeleton were analyzed.

Another approach was to compare the distribution coefficients of the key radionuclides adsorbed on buffer materials that were leached at different pH values. Major changes in distribution coefficients may be attributed to the preferential dissolution of specific adsorption sites or the destruction of the ion exchange structure. To determine the structural changes, X-ray diffraction (XRD) was employed and Scanning Electron Microscopy (SEM) was used to identify changes in morphology. The total moles of cations in the leachate solution as a function of pH is presented in Figure 3 for both candidate buffer solutions. As the pH is reduced, a significant increase in dissolved cation concentration is apparent between pH 4 and pH 2 for both candidate buffers, especially for Dochart clay.

The distribution coefficient of the three key radionuclides was determined for all of the pH conditioned samples at the same pH as the conditioning environment. The distribution coefficient values were plotted as a function of conditioning pH and are presented in Figures 4 and 5 for clinoptilolite and Dochart clay respectively. The relatively small fluctuations in distribution coefficient values with changes in pH for clinoptilolite is remarkable considering the large pH range covered. It indicates that major changes in the adsorption sites did not take place even though some of the skeletal structure was dissolved at pH 2 (See Figure 3). Much larger changes in distribution coefficients with changes to pH values are apparent for Dochart clay. An order of magnitude reduction of the cesium distribution coefficient at pH 2 may be attributed to the removal of adsorption sites, the dissolution of disordered clay particles and/or the dissolution of the iron coatings on the mineral particles. Further work is required with Dochart clay between pH 2 and 4 to determine the long term stability of this candidate and the reason for the drop in the cesium distribution coefficient at the low pH value of 2.

Work has commenced on generating dissolution rate data between pH 2 and 4 to facilitate the estimation of maximum buffer degradation rates in the repository.

Activity Transport Mechanisms in the Buffer Layer.

To simulate on a small scale the combined leaching of waste forms surrounded by buffer and backfill materials and the subsequent transport of radionuclides through the buffer layer, a series of small scale lysimeters was constructed. Details of the equipment, procedure and most of the results were elaborated before [4] and thus only a short description is presented here. Waste forms, approximately 5 cm diameter, 5 cm high were either sodium phosphate immobilized in bitumen or shredded, compacted paper and plastic trash. All waste forms were tagged with ^{137}Cs, ^{85}Sr, and ^{60}Co radioisotopes. The waste forms were placed in the centre of a 30 cm diameter, 30 cm high container and were surrounded by the candidate buffer materials: 95/5 weight ratio of sand and clay or sand and clinoptilolite. Water was allowed to infiltrate through the lysimeter. The radionuclide content of the effluent water, and at the end of the experiment the soil below the waste form , were determined. Calculations based on the results of earlier experiments indicated that the retardation coefficients for ^{137}Cs and ^{85}Sr were significantly lower than obtained in batch experiments [4]. Particulate transport was suspected to be responsible for some of the apparent discrepancy. The following experiments confirmed that particle transport can be of major importance in activity transport. In two experiments the waste form in the lysimeter was replaced by a thin layer of neutron-activated sand/clay mix. Since neutron activation resulted in the generation of radionuclides in the bulk material, any activity release would be primarily due to particle transport. In Table I the fraction of activity released into the effluent water over that leached from the waste form (or introduced as activated clay) is compared. The average values for ^{85}Sr and ^{137}Cs in the waste form lysimeters are very similar to the average values for four radionuclides in the activated clay lysimeters. Identical values would not be expected since source geometry and the activity distribution between various clay particles are governed by different mechanisms in the two types of experiments. The much higher ratios for ^{60}Co in the waste form lysimeters indicate low distribution coefficients for this isotope, hence predominantly solution, rather than particle, transport.

Two experiments performed, where the buffer material was 95% sand/5% clinoptilolite, gave additional evidence of particle transport. The clinoptilolite particle size was in the same range as the sand particles, producing a buffer with hydraulic properties very similar to that of sand. Water fed to one of the lysimeters contained sodium hypo-chlorite to destroy bacterial cultures that may contribute to the particle transport of the activity. Clinoptilolite, a natural molecular sieve, contained some bentonite clay. In a similar way to the sand/clay lysimeters, the radionuclide released with the effluent water was much higher than expected from calculations based on distribution coefficients. Activity release from the sodium hypo-chloride containing lysimeter was actually higher than from the other control lysimeter suggesting that transport by bacteria was not a dominant activity transport mechanism.

The radionuclide concentration in horizontal slices of the buffer layer below the waste form was obtained and normalized to the radionuclide inventory in the waste form. In Figures 6 and 7 the results are plotted as a function of the distance from the waste form. Note the very similar values obtained for ^{85}Sr and ^{137}Cs. This could only be explained if we assume identical distribution coefficients for these two radionuclides, a very unlikely coincidence. Particulate transport by the clay component released by the clinoptilolite is a more reasonable explanation. Because of the high surface area of colloid particles dispersed in the soil pore water, they may preferentially adsorb the radionuclides released from the waste form, and the colloid particles are then transported by the infiltrating water. Work is continuing on gathering further evidence on the contribution of colloid particles to radioactivity transport.

SUMMARY AND CONCLUSIONS

The construction of a low level waste repository is planned at the Chalk River Nuclear Laboratories. A buffer layer, approximately 0.5 m thick will be placed at the bottom of the repository to retard the migration of cationic radionuclides, such as ^{90}Sr, ^{137}Cs and ^{60}Co. The chemical environment in the repository is being assessed and is being simulated in the laboratory to evaluate the long term performance of candidate buffer materials. Both candidate materials, Dochart clay and clinoptilolite, are stable in the pH 4 to 11 region. The dissolution of the structural matrix accelerates when the pH is lowered from 4 to 2. Experiments with laboratory scale lysimeters suggest that the movement of particles control the transport of ^{85}Sr and ^{137}Cs in the candidate buffer materials when water is infiltrating at moisture conditions very close to saturation.

REFERENCES

[1] Dixon, D.F., ed. "A Program for Evolution from Storage to Disposal of Radioactive Waste at CRNL", Atomic Energy of Canada Limited Report AECL-7083, October 1985.

[2] Buckley, L.P., "The Influence of Engineered Barriers on the Disposal of Low-Level Radioactive Waste", Atomic Energy of Canada Limited Report AECL-9611, September 1987.

[3] Buckley, L.P., Arbique, G.M., Tosello, N.B., Woods, B.L., "Evaluation of Backfill Materials for a Shallow-Depth Repository", Atomic Energy of Canada Limited Report AECL-9337, November 1986.

[4] Buckley, L.P., Tosello, N.B., Woods B.L., "Leaching Low Level Radioactive Waste in Simulated Disposal Conditions", presented at the 4th International Hazardous Waste Symposium, Atlanta, Georgia, May 2 - 6, 1987.

TABLE I

COMPARISON OF RADIONUCLIDE TRANSPORT THROUGH SAND / CLAY BUFFER

Wasteform in Lysimeter (bitumen encapsulated or compacted)			**Activated Clay in Lysimeter**
Activity in effluent water * / Activity leached from wasteform			Activity in effluent water ** / Activity in neutron activated clay
^{85}Sr	^{137}Cs	^{60}Co	
0.015	0.016	0.19	0.027

Buffer: 95% sand, 5% clay;
Thickness of buffer below the wasteform (or activated clay): 5 cm.
Average water infiltration velocity: 0.34 cm/h

*(Average of 7 lysimeters) ** (Average of 4 radionuclides in 2 lysimeters)

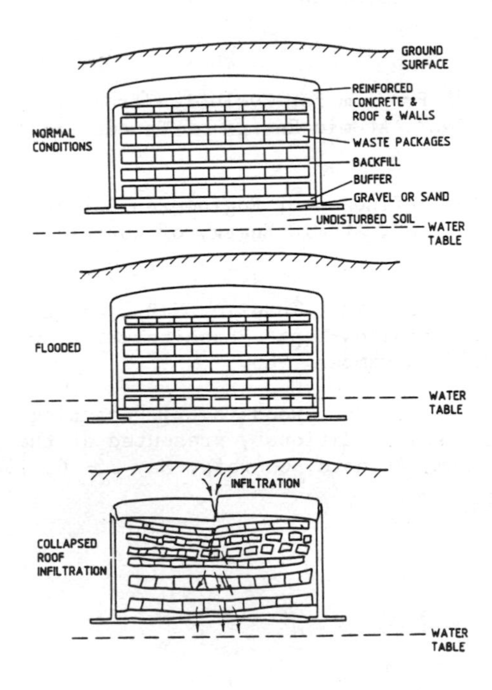

Figure 1
Repository Conditions

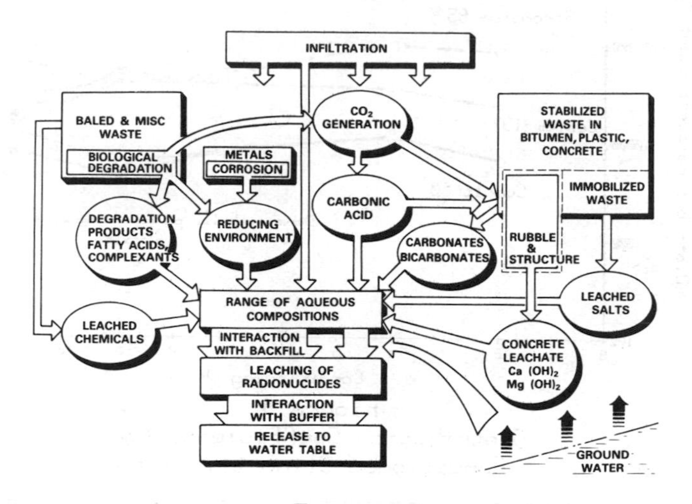

Figure 2
Aqueous Chemical Environment

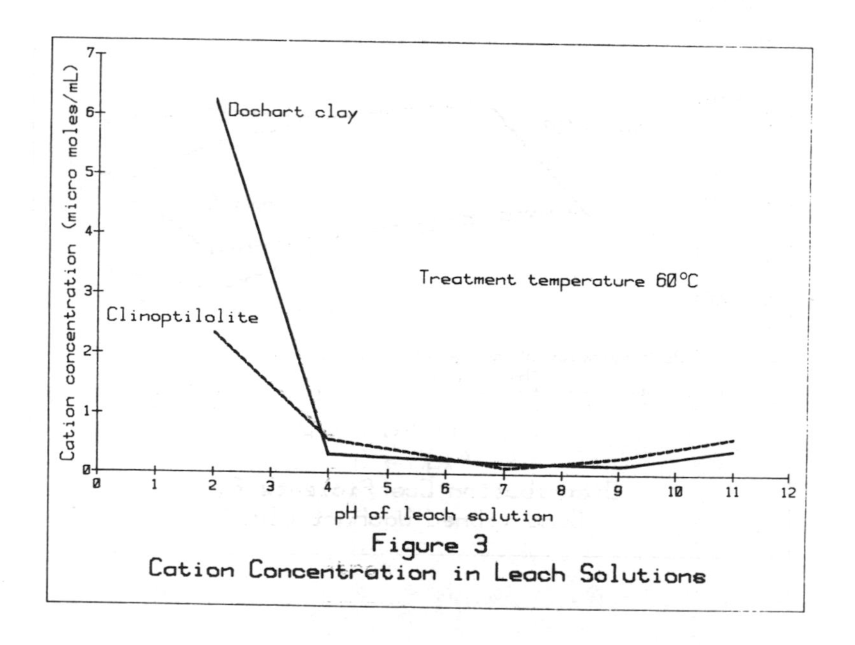

Figure 3
Cation Concentration in Leach Solutions

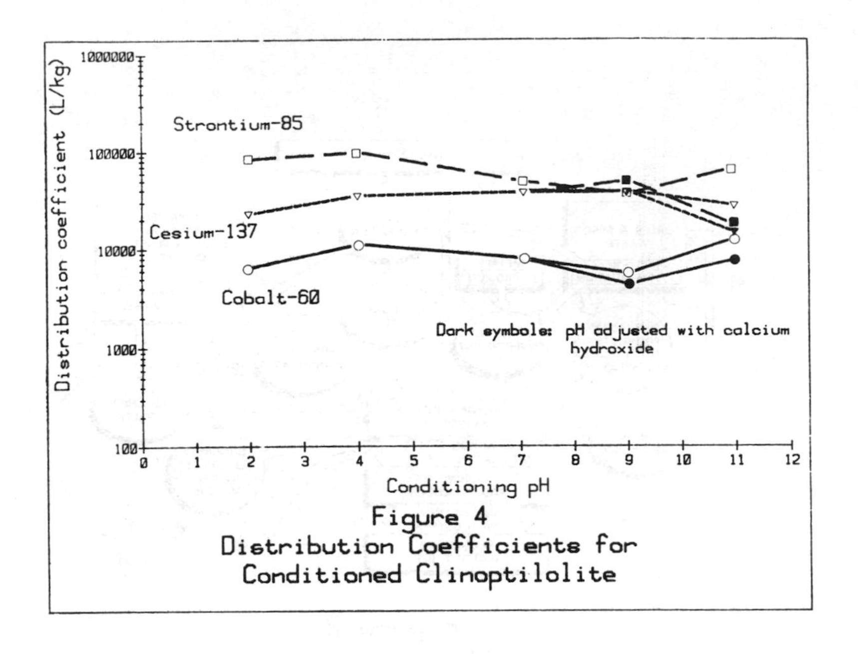

Figure 4
Distribution Coefficients for Conditioned Clinoptilolite

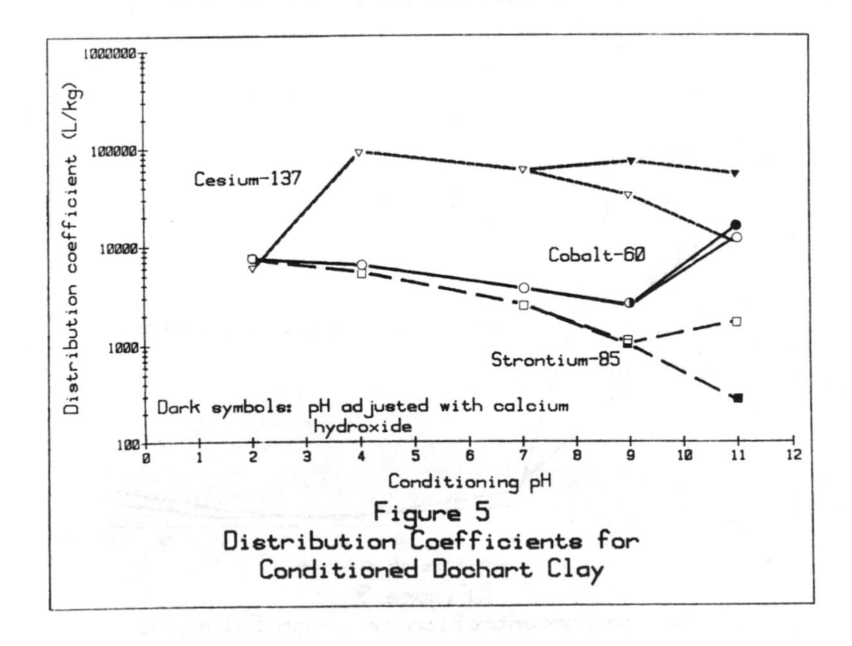

Figure 5
Distribution Coefficients for Conditioned Dochart Clay

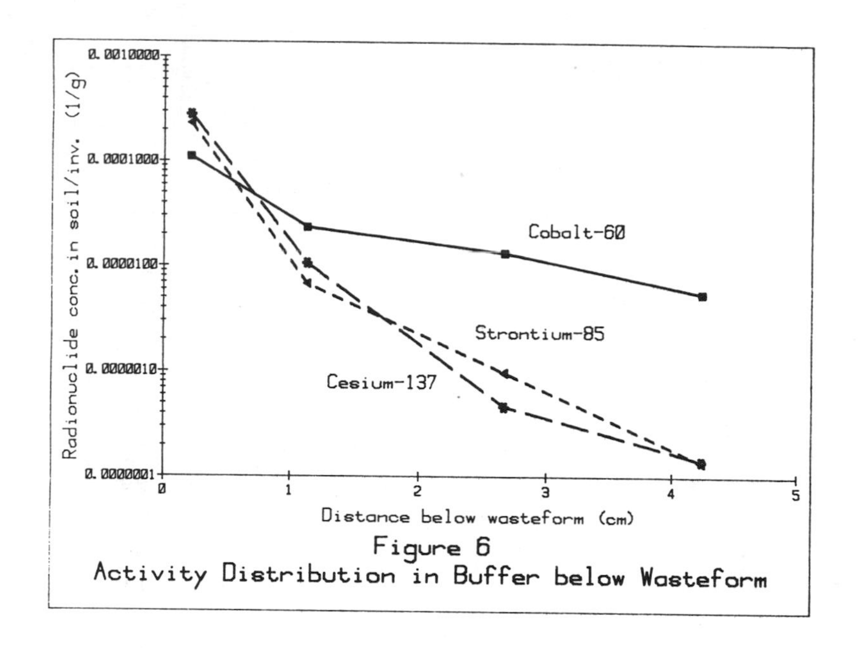

Figure 6
Activity Distribution in Buffer below Wasteform

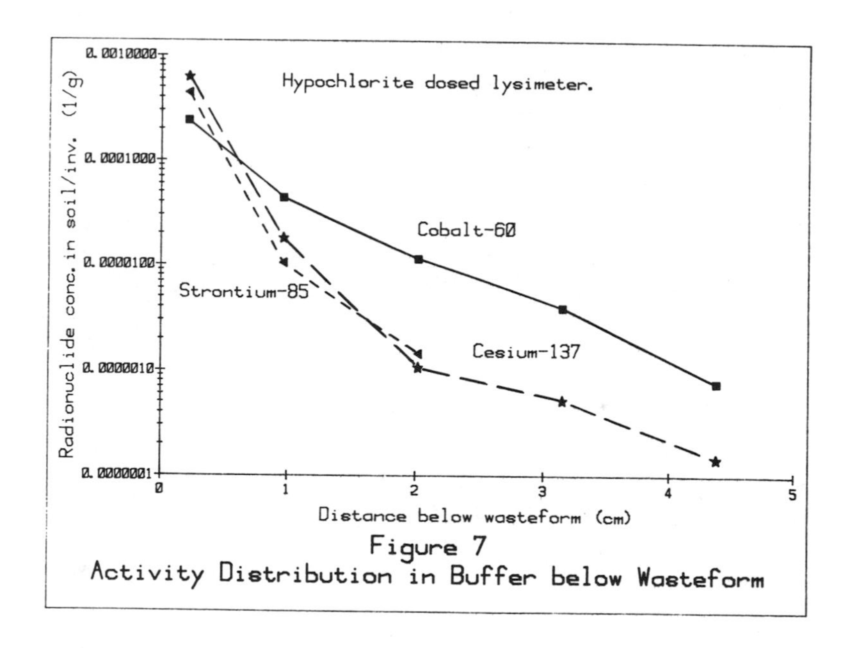

Figure 7
Activity Distribution in Buffer below Wasteform

EXPERIMENTAL AND MODELLING STUDIES OF THE NEAR-FIELD CHEMISTRY FOR NIREX REPOSITORY CONCEPTS

A.Atkinson, F.T.Ewart, S.Y.R.Pugh, J.H.Rees, S.M.Sharland, P.W.Tasker and J.D.Wilkins

UKAEA, Harwell Laboratory,
Oxford, OX11 0RA, U.K.

ABSTRACT

A research programme is described which is designed to investigate the chemical conditions in the near field of a concrete based repository and the behaviour of the radiologically important nuclides under these conditions. The chemical conditions are determined by the corrosion of the iron components of the repository and by the soluble components of the concrete. Both of these have been investigated experimentally and models developed which have been validated by further experiment. The effect of these reactions on the repository pH and Eh, and how these develop in time and space have been modelled using a coupled chemical equilibrium and transport code. The solubility of the important nuclides are being studied experimentally under these conditions, and under sensible variations. These data have been used to refine the thermodynamic data base used for the geochemical code PHREEQE. The sorption behaviour of plutonium and americium, under the same conditions, have been studied; the sorption coefficients were found to be large and independent of the concrete formulation, particle size and solid liquid ratio. Recent experimental results from sorption/exchange experiments with lead and 14–carbon are also reported. The programme has also investigated experimentally the possible perturbation of the repository chemistry by microbial action and by natural and added organic material. A final set of experiments combine all the repository components and the waste in a long term equilibration experiment.

ETUDES EXPERIMENTALES ET DE MODELISATION DE LA CHIMIE DU CHAMP PROCHE SE RAPPORTANT AUX CONCEPTS DE DEPOTS DE LA NIREX

RESUME

Les auteurs décrivent un programme de recherche conçu en vue d'étudier les conditions chimiques prévalant dans le champ proche d'un dépôt fondé sur l'emploi de béton, et le comportement, dans ces conditions, des radionucléides qui revêtent de l'importance du point de vue radiologique. Les conditions chimiques sont déterminées par la corrosion des composants en fer du dépôt et par les composants solubles du béton. Les uns et les autres ont été étudiés expérimentalement et on a mis au point des modèles qui ont été validés par de nouvelles expériences. L'effet de ces réactions sur le pH et l'Eh régnant dans le dépôt, et la manière dont ils évoluent dans le temps et dans l'espace ont été modélisés à l'aide d'un programme de calcul associant équilibre chimique et transport. La solubilité des principaux nucléides fait actuellement l'objet d'études expérimentales dans ces conditions et en présence de variations notables. Ces données ont été utilisées pour affiner la base de données thermodynamiques utilisée pour le programme de calcul géochimique PHREEQE. On a étudié le comportement du plutonium et de l'américium du point de vue de la sorption dans des conditions identiques ; on a constaté que les coefficients de sorption sont importants et indépendants de la composition du béton, de la taille des particules et du rapport liquide/solide. Il est également rendu compte des résultats récents obtenus à partir d'expériences de sorption/échange portant sur le plomb et le carbone 14. Dans le cadre de ce programme, on a également étudié, au plan expérimental, la perturbation éventuelle de la chimie du dépôt par l'action microbienne et par des matières organiques naturelles et ajoutées. Une dernière série d'expériences associe tous les composants du dépôt et les déchets dans le cadre d'une expérimentation de l'équilibrage à long terme.

Introduction

Solid radioactive waste disposal studies are divided into two broad categories: those which address the host geology – the far field, and those which address the vault and the waste form – the near field. It is the study of the near field which is discussed in this paper. Presently, the main thrust of the radioactive waste disposal research in the U.K. is directed towards the disposal of low and intermediate level wastes. There are a number of repository design concepts under consideration, but for the purpose of the research programme described here there are only minor differences resulting from the design variations. Basically, these designs consist of a waste material which is immobilised in a cementitious matrix and packed in a steel drum. These drums are then stacked in the repository and backfilled with a cementitious grout. The repository structure itself consists of a large concrete vault which resides in a cavity excavated from the host geology.

The objectives of the near field research programme are to obtain experimental data which will show how radionuclides will behave in the repository environment, to obtain an understanding of the processes which will control the aqueous concentration of the important radionuclides and to construct a model based on this understanding, which will enable the behaviour of the repository to be predicted over extended timescales.

Near field conditions

Work on the near-field conditions themselves is needed both to provide general understanding of the complex environment and to suggest the appropriate chemical conditions for the experimental study of the solubility and sorption of the radionuclides. The system will evolve both spatially and in time and modelling plays a vital role in extrapolating over the long periods involved. We can envisage several processes contributing to the evolution of the chemical conditions. First, there will be a period of resaturation of the repository during which groundwater seeps in from outside. An important requirement of the host geology will be to restrict the water flow to low levels, and linear flow rates of $10^{-10} m.s^{-1}$ are found to be feasible.

This water will equilibrate with the concrete which will determine the pH within the near-field. Corrosion of the steel canisters and of reinforcing steel will occur. This generates corrosion products and determines the time at which water will contact the waste. It is likely that the corrosion products (dissolved iron and hydrogen) will determine the redox conditions, Eh, in the repository. In addition, the production of both solid oxides and gas may affect the structural integrity of the near-field. Over long timescales, the concrete pore water will diffuse and flow into the surrounding geology. This will lead to changes in the concrete composition and hence in the pH within the repository. These effects will be compounded if aggressive ions are present in the pore water of the surrounding geology and transport into the repository to attack the concrete. The near-field leachate will also influence the surrounding geology, e.g. clay, possibly causing both physical and chemical changes and creating a disturbed zone around the repository. The slow movement of water and solute species that results from the low flow rate means that chemical equilibrium will be established in the near field and also at each point in the migration path from the near field into the host geology. The research programme therefore addresses equilibrium chemistry. All these topics are the subject of current research in the U.K.

Concrete Chemistry and pH Evolution

The leaching of cement and the composition of the pore water has been studied in some detail by Atkinson[1]. A thermodynamic model of the CaO-SiO_2–H_2O system has been developed to aid the understanding of the experiments. Using this equilibrium model, Atkinson was able to estimate the time evolution of the pH in the repository using a simple mass transfer calculation. Experimental leaching studies were carried out in which granules of hydrated sulphate resisting portland cement (SRPC) paste were contacted with successively increasing volumes of demineralised water for one week. The leachates were analysed for pH and aqueous concentrations of calcium, silicon and aluminium, and the results compared with the predictions of the model, for an assumed groundwater flux density of $10^{-10} ms^{-1}$, as shown in Figure 1 . The discrepancies between the model and the experimental data have been shown to be due to the development of concentration gradients in the samples and consequent incomplete equilibrium being attained in the leaching experiment. The results confirm the view that the pH is controlled by the $CaO-SiO_2-H_2O$ system and that the aluminium and iron containing phases play only a secondary role. It is thus concluded that the initial high pH of about 13 is due to the presence of alkali metal oxides and hydroxides which are leached first from the repository. After a few thousand years the pH settles at 12.5, determined by solid $Ca(OH)_2$. This is maintained in this model for the following 10^5 years until the calcium silicate hydrate (CSH) gel phases start to dissolve incongruently. Over the next million years, the pH falls to 10.5 while the calcium to silicon ratio drops to 0.85. Thereafter, the pH is held at 10.5 while the CSH gel dissolves congruently. This is greatly in excess of the likely structural lifetime of the backfill and indicates that the chemical conditions may act as a retarding barrier even after the physical integrity of the concrete cannot be relied upon.

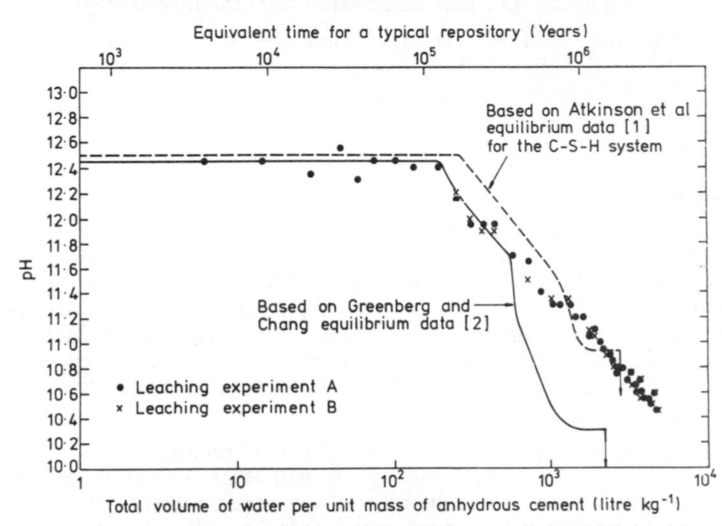

Figure 1. Evolution of the pH of concrete pore water

The Evolution of the Groundwater around a Repository.

A complete understanding of how the chemical environment develops requires us to understand the coupling between the chemical and transport processes. To this aim, we have developed a computer program that in its original version combines one-dimensional diffusion and electromigration of aqueous species with local chemical equilibration. The code called CHEQMATE (CHemical EQuilibrium with Migration And Transport Equations) incorporates the geochemical program PHREEQE[3]. PHREEQE uses thermodynamic data and solves the coupled equations describing chemical equilibria, mass balance and electroneutrality. This enables the calculation of the pH, Eh, concentration of dissolved elements and their aqueous speciation, the saturation state of the solution with respect to mineral phases and the amounts of precipitation and dissolution.

The CHEQMATE program has been used to model the perturbation of the groundwater chemistry within the host rock by the presence of the engineered barriers of the repository. Such changes may have important effects on properties such as nuclide solubilities and sorption characteristics. After construction, water will enter the repository from the surrounding geology. Over a period of time a chemical equilibrium will be achieved in this and any existing pore water from the construction process, under the influence of both the minerals within the backfill material and the degradation processes of other components of the repository (such as corrosion of the metallic containers and reinforcements). It is likely that the chemical composition of the backfill pore water will become significantly different to that of the natural groundwater, leading to a marked chemical discontinuity at the edge of the backfill. Ionic migration processes such as diffusion will tend to level out such variations over a period of time, leading to the progression of a plume of calcium hydroxide from the edge of the repository into the pores of the surrounding rock, preceeding the release of radionuclides. The progress of this plume, and later of the nuclides, will also be influenced by any flow of groundwater through the backfill and rock. An ingress of alkalinity may result in important changes to the chemical and physical properties of the host geology; the cation-exchange capacity within the mineral structure may be affected which will influence the sorption characteristics of the nuclides and also the permeability and porosity of the rock may be changed which could affect the migration of nuclides out of the repository. Such changes could clearly influence a repository assessment.

A model of the groundwater around the repository has necessitated various extensions to the original CHEQMATE code. These include the addition of advection as an ionic transport mechanism so the effect of a groundwater flow can be modelled and also the addition of a change in physical properties of the porous medium within a CHEQMATE section, such as at the backfill/rock interface. The code has separately been extended to model a spherical geometry to investigate the dilution of various ionic concentration fronts spreading out from the repository. The preliminary model has been applied to a concrete backfill embedded in a clay geology[4]. (Although a site for a low– and intermediate– level waste has not yet been decided in the U.K., many rock formations contain substantial amounts of clay minerals).

The model consists of a grid of cells extending across the concrete/clay boundary with one cell of concrete and the rest clay. The thermodynamic model of clay used in the base case was developed using the PHREEQE code to reproduce experimental results from titrations of clay against alkali. The model comprises an idealised clay mineral equilibrated with calcite and water and models the buffering capacity in terms of sequential substitution of cations at the clay surface by calcium ions[5]. As a first stage, it is assumed that calcium hydroxide equilibrated with water provides a reasonable approximation for concrete pore water as far as alkalinity and ionic strengths are concerned. During the calculation it is further assumed that the concrete cell does not become depleted of calcium hydroxide and effectively acts as an infinite source of high pH solution.

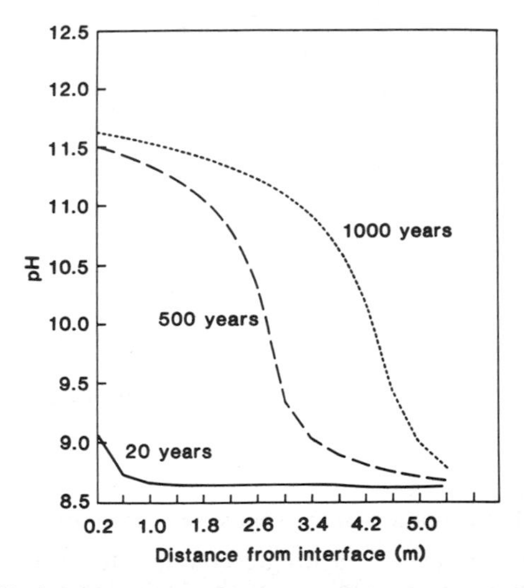

Figure 2. pH Profile in clay for set of base case parameters at 20, 500 and 1000 years

The natural chemistry of the clay exhibits a resistance to changes in pH as the highly alkaline solution enters from the backfill pores. Although this mechanism will not be effective indefinitely (and eventually the clay porewater adjacent to the repository achieves a high pH), the model suggests that for a typical set of physical and chemical parameters, the buffering action has a strong influence on the chemistry over the timescales associated with the possible first release of radionuclides (about 1000 years). Figure 2 shows the pH profile in the clay at 20, 500 and 1000 years. The clay starts at a pH of approximately 8.5 with only the first two cells showing a rise in pH at 20 years. At 500 years the high pH plume has progressed further into the grid as ion exchange sites are progressively saturated and by 1000 years, a pH of over 11 extends over 4 m into the clay. Sensitivity analysis performed with the various models suggest that perturbations to the natural chemistry within the host rock groundwater will be most delayed when there is a slow flow of groundwater through the system or near the corners of the repository where dilution effects are greatest.

The repository Eh

The oxidation potential of the near-field water is another critical chemical parameter and is likely to be determined by the corroding steel containers. We can apply the CHEQMATE program to construct a model of the repository consisting of a regular array of carbon steel canisters embedded within a concrete backfill with a mean spacing of 1.2m. We make the pessimistic assumption that the porewater is initially saturated with oxygen and is hence highly oxidising. Since we wish to neglect the influence of the backfill in reducing the Eh of the system, we model it initially as $Ca(OH)_2$ which produces a realistic pH. Corrosion of the containers occurs consuming oxygen and generating hydrogen and iron oxides. The corrosion products diffuse from the canister surface and gradually the whole chemical environment changes. Figure 3 shows the development of the oxidation potential between a pair of canisters for our base case calculations.

The calculations predict that canister corrosion is sufficient to establish reducing conditions within the repository in approximately 100 years after saturation. Sensitivity analysis to these parameters yields a lower limit of about 50 years and a maximum of 150 years. In all cases, reducing conditions are established very quickly close to the container surface which is important with regard to nuclide solubilities.

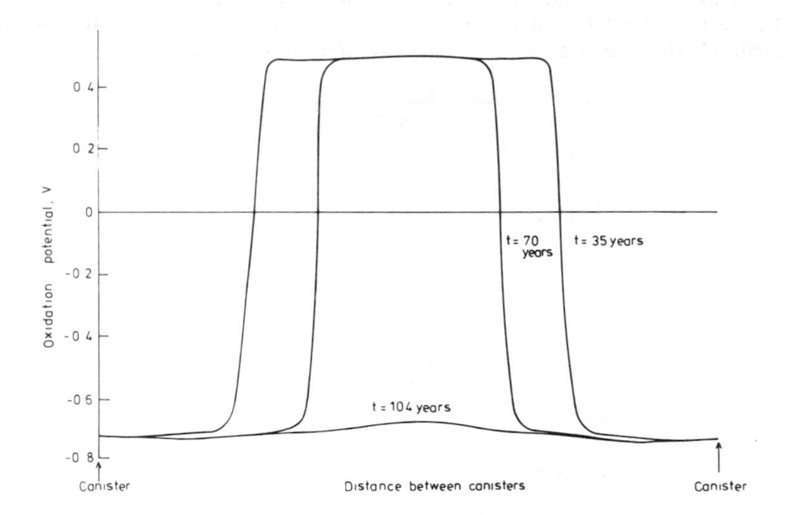

Figure 3. The evolution of the repository oxidation potential

The calculations indicate that the aqueous hydrogen concentration controls the Eh, but of course, the effective Eh may be limited by the kinetics of the hydrogen redox reactions. Under these conditions the dissolved iron controls the Eh and reducing conditions are still established close to the cans in a short time. In our base case calculations hydrogen bubbles form when the partial pressure of hydrogen exceeds the ambient pressure. We have not considered the subsequent motion of gas out of the repository. This may have important consequences for repository design but does not directly influence our conclusions on Eh development. Fairly large quantities of hydrogen gas are generated by the corrosion process. Our model predicts a rate of generation of 0.27 litres per year per m^2 of canister surface. The principal solid corrosion product is magnetite and the solid volume increase due to corrosion is fairly small (about 10^{-7} m^3 per year per m^2 of container surface). Nevertheless it will be important to assess the influence of all the corrosion products on the integrity of the concrete backfill and repository design. The CHEQMATE model allows a substantial sensitivity analysis to be made with respect to the assumptions and choices in the model. These have been discussed in detail elsewhere[6].

The main picture that is emerging from the studies of the near-field chemical environment is that a high pH, low Eh chemistry will be established in the pore water fairly quickly after resaturation of the repository. These conditions are likely to persist for time that is long compared with the structural lifetime of the repository components.

Nuclide behaviour

The elements which are being studied in this programme are those which have been shown to be significant by the assessment modelling of the repository. Since these models necessarily consider only a simplified chemistry, the list of significant nuclides depends strongly on the values of the simplified parameters which are used. The behaviour of elements in the near field will be dominated by their limiting solubility and by sorption from the solution on to the sorbents available in the near field. For material of relatively high inventory and low solubility, the solubility limit may well be reached while the higher solubility nuclides may have concentrations determined by their sorption. The experimental programme has therefore studied the solubility under high pH conditions and sorption on to cements and concretes, with the objective of combining these effects in the near field model. Further studies have been mounted to investigate the possible modification of the expected behaviour caused by the presence of complexing molecules, both naturally occurring and arising from the emplaced waste and to study the possibility of synergistic effects, not expected from the above studies, by mounting long-term equilibration experiments which contain all the major constituents of the repository.

Solubility

The solubility of a number of important radioelements has been determined over a range of pH values between 9 and 13 in waters which had been pre-equilibrated with hardened cement. The details of the experimental methods used in these determinations have been reported in references 7 and 8. The trends in the solubility with pH were modelled using the PHREEQE geochemical code, thus predicting both the speciation and the solubility. Comparisons of the predictions from modelling and the experimental results have enabled the thermodynamic database to be both refined and validated. As an example of this work Figure 4 shows the comparison of the predictions, using the final database, and the experimental results for the case of americium. It can be seen that the agreement is good thus giving confidence in the use of the code for further predictions.

The americium solution species are dominated by the hydroxy carbonate complexes although the solid americium hydroxycarbonate is replaced by solid americium hydroxide at pH values greater than pH 9. In this region the predicted solubility is quite sensitive to the solid phase selected for the modelling. Since there is no experimental evidence for a levelling out of the solubility curve at high pH, it is concluded that the hydroxide solution species gives precedence to the hydroxycarbonate in the range of pH values studied. Attempts were made to reproduce the experimental data by modelling the americium carbonate and hydroxide species reported by Vitorge [9] but it was not possible to obtain any similarity between the modelled and experimental data. The present thermodynamic data [10], refined during these experiments from those of Kim [11] are preferred.

Sorption

The objective of the experimental sorption programme is to obtain quantitative information on the sorption of radionuclides onto the solid phases likely to be encountered in the near field of of a repository. The experimental programme involves both batch sorption and combined diffusion-sorption studies; the former technique produces sorption coefficients relatively quickly and, under some conditions, sorption isotherms can be constructed, the latter produces more realistic conditions but may require extended timescales to produce results.

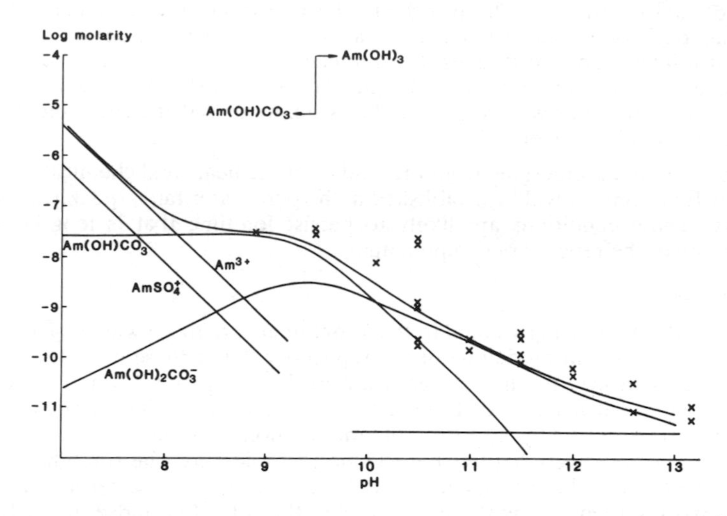

Figure 4. Americium solubility and speciation Eh 200mV, carbonate 3.10^{-5}M

Throughout this discussion we will refer to the sorption coefficient, or Rd, which is the concentration of the nuclide in the solid divided by the concentration in the solution. When these concentrations are expressed in terms of mass and volume respectively then the dimensions of Rd are $length^3 \cdot mass^{-1}$. In the examples quoted here the units will be $ml.g^{-1}$.

In batch studies of the sorption of plutonium and americium the experimental approach has endeavoured to maintain water chemistry conditions that will mimic those within the repository and to preserve the pore structure of the concrete by using cast coupons of concrete. These coupons were 2cm in diameter and 3mm thick and were cast from ordinary Portland cement (OPC), OPC with blast furnace slag additive (BFS), OPC:BFS 1:3, and OPC with pulverised fly ash (PFA) additive, OPC:PFA 1:3. Based on the premise that the frequently discussed solid– liquid ratio effect is primarily due to concentration changes during the sorption experiment, we have endeavored to minimise these by maintaining the aqueous concentration of the actinide within reasonably close limits. These concentrations were set to be approximately five to ten times lower than the solubility limits determined above.

Table I
Americium Sorption on Cement – Effect of Particle Size

Rd/(ml/g)	Particle Size
7×10^3 7×10^3	coupon coupon
4×10^4 6×10^4	1–2 mm 1–2 mm
3×10^4 2×10^4	125–250 μm 125–250 μm

Some experiments were mounted to study separately the effects of solid liquid ratio and of concrete particle size. The solid liquid ratios were varied between 1:40 and 1:400 and the concrete coupons were used intact and crushed into two size groups, 1.0 to 2.0 mm and 125 to 250 μm. Examination of the concrete samples by water vapour sorption showed that the surface area was unaffected by the crushing within the sample to sample variation of 50–80 m^2/g. It was found that the sorption occurred more rapidly on the crushed samples than the coupons, which we attributed to the slow diffusion of the sorbate into the pores of the concrete coupon.

The coupons were sectioned after some four months exposure to the plutonium solution and analysed for plutonium penetration, by alpha autoradiography. It was clear that the plutonium had penetrated no more than a few hundred microns into the concrete. Tables I and II show a selection of the results obtained; they have been grouped to allow the effects of particle size and solid liquid ratio to be shown. The results obtained on the coupons have been calculated for the measured penetration depth only. It is clear that, under these conditions the effect of particle size is negligible and the solid liquid ratio has no influence on the observed Rd.

Table II

Plutonium Sorption on Cement – Effect of Solid–Liquid Ratio

Rd/(ml/g)	Solid–Liquid Ratio
1×10^4	1:40
1×10^4	1:40
2×10^4	1:40
1×10^4	1:40
3×10^4	1:400
2×10^4	1:400
3×10^4	1:400
2×10^4	1:400

It may be concluded that the sorption coefficient for these two nuclides is 10^4ml.g^{-1}, and it is almost independent of particle size, solid liquid ratio and of the cement formulations studied.

Some preliminary batch sorption studies of lead and 14–carbon on crushed cement grouts containing limestone aggregate (250–500μm) have been started. These have shown that the Rd values for 14–carbon increase with increasing 14–carbon concentration from 10^{-9} to 10^{-7}M. It appears that not all of the carbon in the solid phase is available for exchange with the 14–carbon. After some 250 days, differences in values of Rd of an order of magnitude are observed between SRPC and OPC/BFS cement formulations, with the SRPC producing an Rd of 6000ml.g^{-1}. However, the sorption reactions do not appear to have reached equilibrium and the differences may reflect only the relative porosities of the two grouts. The lead experiments show that sorption onto cement particles increases with decreasing lead concentration and values of Rd after 70 days are 500ml.g^{-1} for SRPC and 1300ml.g^{-1} for OPC/BFS at lead solution concentrations of 10^{-3}M [12]

Effect of naturally occurring organic material

Some of the UK clay formations contain significant quantities of organic material which could give rise to groundwaters containing 1–10mg/l of dissolved organic carbon. These organics are high molecular weight acids, some of which could be regarded as colloids. Because of the acidic functional groups, these molecules may have the capacity to form complexes with some radionuclides which could enhance their solubility or reduce the sorption on components of the near field. It is equally possible that, because these complexes will be physically large, the result of the complex formation may be a net reduction in radionuclide transport because of filtration and entrapment in the pore structure of the concrete or the host clay. Studies have concentrated on the experimental determination of the effect of humic acid (HA) on the solubility and sorption of americium in cement systems, as an example of the behaviour of actinides.

A set of experiments using a commercial (Aldrich Chemicals) humic acid and a slag cement equilibrated water showed a thirty fold increase in americium solubility with 5ppm of humic acid. Higher concentrations of humic acid could not be achieved in this water since the acid coagulated, which we attributed to the increased ionic strength of this water. Figure 5 illustrates this effect.

Sorption experiments conducted on a PFA/OPC concrete show a tenfold or greater reduction in sorption coefficient at humic acid concentrations over 10ppm. The solid-liquid phase separation in these experiments was by centrifuging at 1500g for a few minutes.

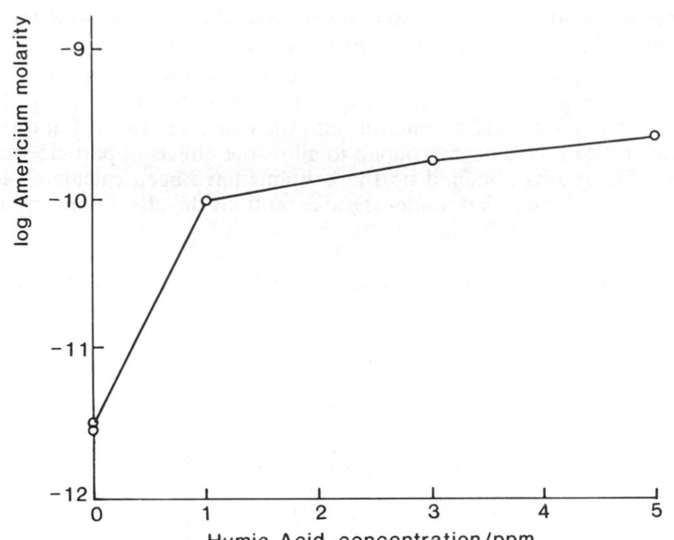

Figure 5. The solubility of americium in the presence of humic acid

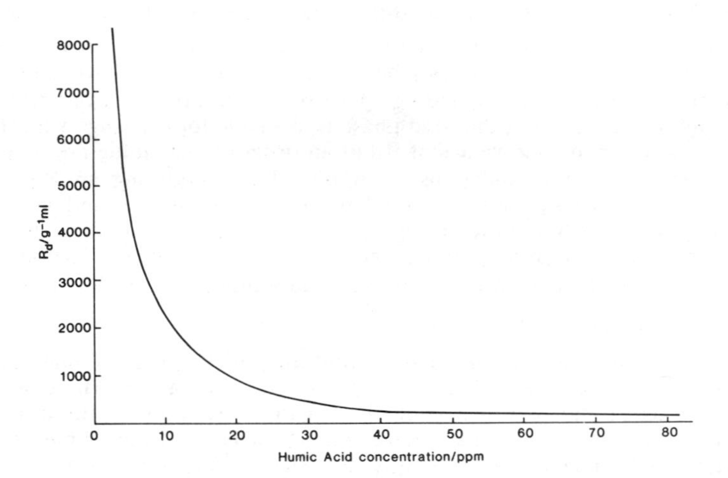

Figure 6. Effect of humic acid on the sorption of americium on cement.

It is thus reasonably clear that some complexing of the americium by the humic acid does occur. Since the molecular weight of the humic acid and the complexes is about 100,000, there is a nice distinction to be made between the definition of a solution and a colloidal suspension. The apparent solubility and sorption subsequent to complex formation will depend strongly on the method of the solid-liquid phase separation used in the experiment. In order to apply the results of such experiments to the actual repository, a more rigorous definition of a soluble species derived from a knowledge of the pore size of the concrete and clay host geology is desirable.

Effect of waste organics material

Some wastes contain organic materials which decompose under disposal conditions. Unfortunately because of the wide range of organic materials present in wastes, and the complexity of the reaction pathways, it is difficult to predict the products which will be formed. The most importance is attached to water soluble organic species, which might complex with long lived radionuclides increasing solubilities and reducing sorption. These organic materials can decompose by several mechanisms; amongst these, the important reactions are those involving the alkaline chemistry of cement pore water, and those induced by the α,β,γ radiation field and by microbial action. The waste products that are of principal interest in this work are those which contain a number of major organic components (polyvinylchloride, polyethylene, neoprene rubber, Hypalon rubber and cellulosic materials) and a host of other materials which will include at least traces of a wide variety of commercially available plastics, rubbers and other materials [13].

There are three classes of organic material which may decompose by different chemical mechanisms in a repository.

(i) Polymers produced by free radical polymerisation (e.g. polyethylene, neoprene, Hypalon) These materials are not particularly susceptible to decomposition under alkaline conditions but random decomposition of the polymer chains particularly in the presence of oxygen could take place.

(ii) Condensation polymers (e.g. polyesters, nylon, polyurethane). These will be particularly susceptible to depolymerisation under alkaline conditions giving small monomeric units.

(iii) Cellulose containing materials (e.g. wood, paper). These are a special type of condensation polymer being particularly susceptible to alkaline hydrolysis under mild conditions to give a complex series of products.

Experimental studies in which the chemical decomposition reactions are being accelerated are underway. In these experiments organic materials are immobilised in cement and held under water saturated conditions at above ambient temperature. In order to check for the presence of water soluble complexing agents, the solubility of plutonium is measured in the water in equilibrium with the solids (the leachate). The effects of radiation are being studied using high specific activity α–emitters. Some high irradiation rate experiments are being carried out using a cyclotron. These data are shown in Table III together with those from chemical degradation. It is clear that there is a modest enhancement in plutonium solubility following chemical degradation for the typical waste material (major mix) and some other materials. The highest solubility has been obtained under the unrealistic conditions of a waste containing cellulose alone. The pattern of results following radiation degradation is rather similar, with cellulose and minor mix giving the greatest enhancement in solubility. It is interesting that the enhancement in solubility for a mixture which contains 10% cellulose is about a factor of 30 while that for cellulose alone is ca 10^4.

Table III

Plutonium solubility– Effect of Organic Degradation Products

Organic Material	Solubility /M	
	Chemical degradation	Radiation degradation
None	5×10^{-10}	5×10^{-10}
PVC	6×10^{-8}	7×10^{-9}
Neoprene	1×10^{-7}	2×10^{-8}
Hypalon	4×10^{-8}	1×10^{-8}
Polyethylene	3×10^{-8}	6×10^{-9}
Cellulose	7×10^{-6}	5×10^{-6}
Major mix	1×10^{-7}	2×10^{-8}
Minor mix	7×10^{-7}	2×10^{-6}

Note

Major mix:–

50% PVC, 10% cellulose, 10% white Hypalon, 10% black Hypalon, 10% polyethylene, 10% neoprene

Minor mix:–

12.5% polymethylmethacrylate, 12.5% polytetrafluorethylene, 12.5% polystyrene, 12.5% nylon, 12.5% bakelite, 12.5% latex, 12.5% polypropylene, 12.5% polyethylpentene

A few intermediate level waste streams contain complexing agents (e.g. EDTA, citric acid) which arise from decontamination operations. The potential for these reagents to change near field performance must also be considered. Initial results suggest that trace levels of complexing agents have little effect on plutonium solubility but large quantities can have a significant effect. Calculations of solubilities in the presence of some complexing agents have been carried out and generally lower solubilities than those found experimentally are predicted.

Microbiological Effects

Micro-organisms have been shown to be inevitable contaminants of typical radioactive wastes [14] and undoubtedly some degree of microbial decomposition will occur in a post-closure repository situation. Microbes, and the products they make, may influence radionuclide sorption and solubility and also repository integrity in terms of voidage formation, gas production and enhanced deterioration of concrete and metal surfaces. The rate of microbial decomposition of radioactive wastes is determined by the availability of the essential intermediates required for microbial proliferation. These can be broadly divided into the major biomass constituents:

(i) carbon, nitrogen, oxygen, hydrogen, phosphorus and sulphur.

(ii) the minor biomass components, trace metals and chlorine.

(iii) energy sources required for the combination of these intermediates into biomass [15].

When all the nutritional intermediates are present in the waste, then the environment imposed by the repository will moderate microbial activity.

The microbial research programme investigates the extent of microbial activity in relation to the possible intermediates present in the waste and the imposed environmental factors. Predictive computer models are being developed which describe the extent and influences of microbial activity and are tested by their ability to mimic experimental model systems.

Experimental microbial degradation is carried out in column reactors containing representative waste and in the presence or absence of crushed OPC. Columns which contained OPC attained a pH of 11.5 within 24 hours and showed no microbial growth. Columns containing unsterilised waste gave considerable leachate growth, $10^3 - 10^9$ microbes/ml, as detected on an aerobic nutrient agar at pH 7, and in the presence of additional nitrogen in the form of ammonium nitrate the microbial numbers are enhanced. The leachates from the columns were tested for the effect of microbially derived products on plutonium solubility following the same methods used for the solubility studies above. Sterile leachate gave a solubility of 7×10^{-8}M, compared with 5×10^{-10}M in the absence of organic material, and leachate known to initially contain microbial activity gave solubilities up to 9×10^{-7}M.

The pH of the water in the repository is likely to be the major factor in determining the growth of microbial activity. To investigate this effect, 176 microbial isolates obtained from experimental systems, novel alkaline environments and clay sites have been tested for their pH tolerance when growing in a complex glucose medium [16]. Of these, 63% of the isolates had growth optima in the pH range 9–12 and 20% in the range 11–12, the latter isolates were bacterial *Bacillus* species.

The growth of a type strain *Bacillus subtilis* NCIB 3610 has been studied on a glucose mineral salts medium at pH 7. The culture medium from this experiment, when all the glucose had been consumed, was tested for its effect on plutonium solubility at pH 11; the value found was 20nM, an increase by a factor of 40 above the normal.

The data derived from the experimental programme has been used to construct a list of isolates likely to be important in a repository. This list constitutes the worst-case mixture and is the base case condition for the modelling work. The computer model attempts to describe the microbial processes which could occur in the lifetime of a repository. The approach has been to consider the repository as a closed system within which microbial activity develops. Microbial decomposition is considered as a series of linked metabolic pathways, each isolate providing at most two major chemical routes for waste degradation. The model identifies that pathway which predominates at a particular imposed environmental condition and allows calculation of the rates of production of endproducts by use of experimental and literature derived values. The performance of the model has been tested by reproducing the experimentally determined growth kinetics of *B. subtilis*.

Equilibrium Leach Tests

In order to evaluate the combined effects of the repository components, not including microbial activity, a set of experiments have been devised in which the componemts of the repository are brought into contact with water under static conditions. These are known as equilibrium leach tests (ELT) [17]. In the test, the water is sampled and analysed at regular intervals until the concentration of the radionuclide of interest levels off. The solids are crushed where possible to speed the rate of equilibration.

The ELT has been applied to representative intermediate level wastes:

(i) Combustible plutonium-contaminated material (PCM), a major alpha waste.

(ii) A mixture of Magnox swarf and the sludge that forms on corrosion of the alloy. This waste arises from the decanning of Magnox fuel elements, and contains traces of irradiated fuel.

(iii) Stainless steel hulls that will result from dissolving irradiated fuel from sheared sub-assemblies from advanced gas-cooled reactors (AGRs).

(iv) Ferric/aluminium hydroxide flocs, which are used to decontaminate certain waste streams from plutonium and americium.

The leach liquors were analysed for Pu–239, Am–241, Np–237, U–238, I– 129, Sr–90 and Cs–137. The experimental results were also calculated from solubility and sorption data obtained separately, under chemical conditions relevant to these tests [18]. The experimental results and the calculated data are shown in Table IV. If the initial inventory of the radionuclide was sufficiently high, the residual concentration of radionuclide in the aqueous phase following sorption would exceed its solubility limit under the conditions of the ELT. Under these circumstances solubility, rather than sorption, would be expected to determine equilibrium levels. The plutonium inventory of the tests with combustible PCM were sufficiently high and, in general, the concentrations found were, within reasonable experimental limits, those which would be expected. None of the ELT results for the radionuclides in the remaining three wastes came up to their solubility limits, although for the ferric/aluminium hydroxide floc and the AGR hulls, some of the results for uranium came close when the cement backfill was used.

An inspection of the results show that for about 70% of the experiments, the ratios between the experimental and the calculated values lie in the range 1–100. The range arises for two main reasons: (i) problems in analysing extremely low concentrations of radionuclides, both in the ELT itself and in the supporting sorption measurements; and (ii) the differences in the experimental conditions between the equilibrium experiments and the supporting measurements. The trend for calculated values to exceed those observed is not unexpected, particularly given the substantial differences in the timescales for the sorption measurements (10 days) and the equilibrium determinations (up to a year). Other sorption studies (see above) with actinides and cement have suggested that between 10 and 30 days, sorption coefficients as measured could increase by up to an order of magnitude. In general, the experimental results for U, Np, Pu and Cs and some of those with Sr and Am are consistent with equilibrium having been achieved.

Under certain conditions of the ELT, the equilibrium concentrations of some radioelements tended to be much lower than anticipated:

(a) Sr, the behaviour of Sr could be anomalous due to isomorphic substitution for Ca in the structure of the cement or isotopic dilution by the appreciable quantities of inactive Sr in the cement.

(b) Am, where the experimental results may reflect coprecipitation or sorption on the hydrated oxides of U(VI) present.

(c) Pu, with the Magnox waste. The magnesium hydroxide in the waste may form magnesium silicate hydrate with silica-containing additives such as PFA and BFS to supplement the calcium silicate hydrate present in the original cement.

Table IV
Equilibrium Concentrations in Leach Tests

Waste	Matrix	Element	Equilibrium Concentration /M	
			observed	calculated
PCM	PFA/OPC	Pu	8×10^{-9}	8×10^{-8}
Magnox decladding waste	BFS/OPC	U Pu Np Am Cs Sr	$<3\times10^{-8}$ 1×10^{-12} $<1\times10^{-11}$ $<2\times10^{-14}$ 5×10^{-9} 5×10^{-10}	8×10^{-8} 8×10^{-10} 2×10^{-12} 2×10^{-11} 2×10^{-8} 6×10^{-8}
AGR hulls	BFS/OPC	U Pu Np Am Cs Sr I	$<5\times10^{-8}$ 9×10^{-12} 8×10^{-12} 6×10^{-14} 2×10^{-7} 1×10^{-9} 6×10^{-7}	1×10^{-6} 3×10^{-10} 3×10^{-12} 3×10^{-10} 1×10^{-6} 3×10^{-7} 9×10^{-7}
Ferric/ aluminium hydroxide flocs	BFS/OPC	U Pu Np Am Cs Sr	3×10^{-8} 6×10^{-12} 1×10^{-11} 4×10^{-13} 1×10^{-12} 1×10^{-11}	9×10^{-8} 5×10^{-10} 5×10^{-11} 5×10^{-10} 6×10^{-12} 6×10^{-9}

Note

All experiments contained 10:1 PFA/OPC to represent a backfill and were carried out under oxidising conditions, Eh ca. 200 mV.

Conclusions

The near field research programme described considers aspects of radioactive waste disposal which may have a direct bearing on the release of radionuclides to the geosphere. Because of the probable design of the repository, the near field chemistry will be dominated by the soluble constituents of the concrete and by the corrosion of the iron components: the chemistry is thus almost independent of the host geology in which the repository is emplaced.

The research programme has enabled an understanding of the chemistry of cement and of the corrosion of iron to be developed and used as a basis for models which have been satisfactorily validated by experiment. A coupled transport and chemical modelling code has been written which will predict the development of the repository chemistry in time and space. The chemical conditions thus defined have been used to design further experimental programmes which are studying the solubility and sorption of the important radioelements. These results have been used to refine and validate the thermodynamic database used for the chemical modelling. The research has studied not only these primary controls of radionuclide release, but also such secondary complicating effects such as organic degradation, microbial effects and the effect of natural organics. The long term equilibrium experiments, although conducted under different redox conditions, have, to a large extent, confirmed the predictive capability of the single nuclide studies.

The results of these experiments lead to the conclusion that a cement based repository, (sited in a suitable geology), will provide a large degree of containment, within itself, of the important radionuclides over extended timescales. This retention arises from the fundamental chemical properties of the system, such as solubility and sorption, which depend only on the composition of the repository and not on near field physical barriers.

Acknowledgements

We acknowledge those scientists at Harwell whose contributions have made this paper possible, but are too numerous to list in this paper.

This work is funded by U.K. Nirex Ltd., the U.K Department of the Environment, the Commission of the European Communities and by BNFL., as part of their respective Radioactive Waste Management Research programmes, and we gratefully acknowledge their support and permission to publish this paper. These results will be used in the formulation of Government policy but the views expressed do not necessarily represent Government policy.

References

1. A.Atkinson, UKAEA Report AERE R 11777, (1985)
2. S.A.Greenberg and T.N.Chang, J.Phys. Chem., 69, (1965), 182.
3.D.L.Parkhurst, D.C.Thorstenson and L.N.Plimmer, Report USGS/WRI 80–96, NTIS Tech. Rep. PB81–167801, 1980, Revised 1985
4. A.Haworth, S.M.Sharland, P.W.Tasker and C.J.Tweed, 'The Evolution of the Groundwater around a Repository', to be presented at the Symposium for the Scientific Basis for Nuclear Waste Management at the MRS Fall Meeting, Boston, December 1987.
5. S.J.Wisbey, N.L.Jefferies, C.J.Tweed, 'The Chemical Effects of a Cement Based Repository on Nuclide Migration', to be presented at the Symposium for the Scientific Basis for Nuclear Waste Management at the MRS Fall Meeting, Boston, December 1987.
6. S.M.Sharland, P.W.Tasker and C.J.Tweed, UKAEA Report AERE R 12442, (1986)
7. F.T.Ewart, R.M.Howse, H.P.Thomason, S.J.Williams and J.E.Cross, in IX Symp. on Scientific Basis for Nuclear Waste Management , Ed. L.Werme, Publ. Elsevier, New York, (1986)
8. F.T.Ewart, S.J.M.Gore and S.J.Williams, UKAEA Report AERE R 11975, (1985)
9. P.Vitorge, T.Berthoud, A.Billon, N.Delorme, I.Grenthe, P.Roubouch, C.Dautel, O.Michau and M.Thyssier, Final Report 1983/4, Contract 83/361/7, CEA Fontenay-aux-Roses 92265, Cedex, France, (1985)
10. J.E.Cross, F.T.Ewart and C.J.Tweed, UKAEA Report AERE R 12324, (1987)
11. A.Avogadro, A.Billon, A.Cremers, P.Henrion, J.I.Kim, B.Skytte-Jensen, P.Venet and P.Hooker in Radioactive Waste Management and Disposal , ed. R.Simon, publ. Cambridge University Press (1985)
12. S.Bayliss, F.T.Ewart, R.M.Howse, J.L.Smith–Briggs, H.P.Thomason and H.A.Willmott, 'The Solubility and Sorption of Lead–210 and Carbon–14 in a Near–Field Environment', to be presented at the Symposium for the Scientific Basis for Nuclear Waste Management at the MRS Fall Meeeting, Boston, December 1987
13. J.D.Wilkins, UKAEA Report AERE R 12719 (1987)
14. A.J.Francis, S.Dobbs and B.J.Nine, Appl. Environ. Microbiol. 40 (1980) 108–113.
15. R.Y.Stanier, E.A.Adelberg and J.L.Ingraham, General Microbiology, 4 Edition(1978), Macmillan Press, London.
16. M.A.D.Collins and S.Y.R.Pugh, European Congress on Biotechnology, 3 (1987).
17. P.Biddle and J.H.Rees, UKAEA Report AERE R 12646 (1987)
18. D.C.Pryke and J.H.Rees, J. Less Common Metals, 122, (1986), 519

CALCULATIONS OF THE DEGRADATION OF CONCRETE IN A FINAL REPOSITORY FOR NUCLEAR WASTE

A Study of Multicomponent Chemical Transport with the Computer Code CHEMTRN

Anders Rasmuson
Ivars Neretnieks
Ming Zhu

Department of Chemical Engineering
Royal Institute of Technology (KTH)
S-100 44 Stockholm (Sweden)

ABSTRACT

Concrete is to be used as a barrier for enclosing low and intermediate level radioactive waste in SFR, a Final Repository for Reactor Waste in Sweden. The concrete will be degraded due to various chemical reactions between the concrete and degrading agents in the environment. The degrading agents in the near-field environment include components in the bentonite clay backfill and in the groundwater in the bedrock. The rate of supply of these agents is governed by transport processes in the concrete and bentonite/groundwater.

Following preliminary estimations with a simplified Shrinking Core Model, the degradation of the concrete walls in the SFR was simulated with the solute transport code CHEMTRN. Three cases, calcium hydroxide depletion, sulphate intrusion, and the full problem accounting for "all" the possible reactions, were calculated with CHEMTRN on the computers CDCyber 170/730 and CRAY-1.

The results of the calculations show that calcium hydroxide in the concrete is depleted to a depth of about 9 cm after 1000 years and the concrete is intruded by sulphate to form ettringite to a depth of some 37 cm at the same time. It was also found that calcium carbonate will precipitate to a depth of about 11 cm, while no precipitate of calcium sulphate forms. The pH in the porewater of the concrete is decreased from 13.4 to about 11 when the transport processes considered have ceased.

The results from the calculations with CHEMTRN are found to be comparable to those obtained from the Shrinking Core Model. When the same boundary conditions are used, the shrinking core model provides an estimate of 9.6 cm for the depletion depth of $Ca(OH)_2$ and 38.2 cm for the penetration depth of sulphate into the concrete.

CALCULS DE LA DEGRADATION DU BETON DANS UNE INSTALLATION DE STOCKAGE DEFINITIF DES DECHETS NUCLEAIRES

Etude du transport chimique à composants multiples à l'aide du programme de calcul CHEMTRN

RESUME

On utilisera du béton comme barrière pour le confinement des déchets de faible et de moyenne activité dans le SFR, installation de stockage définitif des déchets provenant des réacteurs de Suède. Ce béton subira une dégradation due à diverses réactions chimiques entre le béton et les agents de dégradation présents dans l'environnement. Parmi ceux-ci, dans le champ proche, figurent des composants de l'argile bentonitique utilisée comme matériau de remblayage, et de l'eau phréatique contenue dans le soubassement rocheux. Le taux d'apport de ces agents est régit par les processus de transport dans le béton et la bentonite/eau phréatique.

A la suite d'estimations préliminaires effectuées à l'aide d'un Modèle simplifié de réduction de la partie centrale (Shrinking Core Model), la dégradation des parois en béton dans le SFR a été simulée à l'aide du programme de calcul CHEMTRN de transport du soluté. Trois cas, à savoir la baisse de la concentration en hydroxyde de calcium, l'intrusion de sulfate et le problème global prenant en compte "toutes" les réactions possibles, ont été calculées à l'aide du programme CHEMTRN sur des ordinateurs CDCyber 170/730 et CRAY-1.

Les résultats de ces calculs montrent que l'hydroxyde de calcium dans le béton présente une baisse de concentration jusqu'à une profondeur d'environ 9 cm après 1000 ans et que le béton subit une intrusion par du sulfate entraînant la formation d'ettringite jusqu'à une profondeur de 37 cm pendant la même durée. On a également constaté que du carbonate de calcium précipite jusqu'à une profondeur d'environ 11 cm, alors qu'aucun précipité de sulfate de calcium ne se forme. Le pH de l'eau interstitielle du béton subit une décroissance passant de 13,4 à environ 11 lorsque les processus de transport considérés ont cessé.

On a constaté que les résultats des calculs effectués à l'aide du programme CHEMTRN sont comparables à ceux obtenus à l'aide du Modèle de réduction de la partie centrale. Lorsque l'on utilise les mêmes conditions aux limites, le modèle de réduction de la partie centrale fournit une valeur estimée de 9,6 cm pour la profondeur de baisse de la concentration de $Ca(OH)_2$ et de 38,2 cm pour la profondeur de pénétration du sulfate dans le béton.

1. INTRODUCTION

A final repository for low and intermediate level reactor waste is under construction near the nuclear power station at Forsmark in Sweden [1]. The repository is constructed in the bed rock about 50 m under the Baltic Sea. The host rock is a crystalline gneiss-granite with dykes of pegmatite and amphibolitic compositions. The SFR has various storage chambers with different barriers. It consists of a concrete silo and excavated rock caverns.

The silo contains 57 cells where intermediate level radioactive waste is to be disposed. The silo will be surrounded by a bentonite clay barrier with a low permeability, thus the transport will be dominated by diffusion and the transport due to flow is negligible.

Another part of the SFR is the 160 m long rock caverns that are prepared for the radwaste with lower activities. The caverns are in different sizes and shapes. No clay barrier is designed for this part of the repository, thus the release of waste from these caverns will mainly be governed by the interactions with the groundwater in the host rock.

The major transport process of the radionuclides from the SFR is diffusion through the concrete canister and, if present, through the bentonite clay backfill to the flowing water in the fissures in the bedrock. Degrading agents in the bentonite backfill and/or in the groundwater in the rock will react with the concrete and eventually degrade it. A number of reactions may take place simultaneously in the concrete/bentonite/groundwater system including hydroxide (e.g. NaOH, KOH, and $Ca(OH)_2$) diffusion from the concrete to the bentonite, sulphate intrusion from the bentonite to the concrete to form ettringite or calcium sulphate, magnesium intrusion to form magnesium hydroxide, and carbonate intrusion to form calcium carbonate, etc. The reactions most severe to the concrete, according to reported experiences of concrete, are calcium hydroxide depletion and sulphate intrusion.

2. DEGRADATION OF CONCRETE

2.1 Concrete

Concrete can be described as a homogeneous porous solid with water filled pores. The solid part of it consists of two main phases [2,3]:

1. A continuous phase - the cement paste - to which the porewater of concrete is bound. The main components of the cement are calcium-, aluminate-, silicate-, and iron-containing compounds.
2. A particle phase - the ballast consisting of such minerals as sand, gravel, and stone. The ballast particles are greater than 0.125 mm. Concrete usually contains a large amount (65-70% by volume) of ballast. Ballast is usually inert in chemical reactions and is not considered in the calculations.

The concrete chosen to be used in the SFR is Standard Portland Concrete. The pH of the porewater of new concrete is about 13.4. When hydrated, the concrete forms a gel phase with some crystalline constituents like

$Ca(OH)_2$. The chemical composition of the concrete is CaO(64.5%), SiO_2(22%), Al_2O_3(5%), Fe_2O_3(2.5%), MgO(3.2%), SO_3(1.8%), K_2O(0.8%), and Na_2O(0.2%) [3].

2.2 Bentonite

The bentonite clay to be used in the SFR as backfill is in the form of Na-montmorillonite, which was converted earlier from a Ca-montmorillonite. The highly compacted bentonite has a high porosity value of 55%. It also has a high swelling pressure when saturated with water, ensuring a low permeability for the backfill. Due to the high resistance to water flow, the chemical transport through the bentonite is completely governed by diffusion.

The bentonite contains sulphate which is partly dissolved in the porewater and partly bound in the solid phase. It also contains carbonate which is some 2-3% by weight of the sodium carbonate added to the bentonite to transfer it from the Ca-type to the Na-type. The pH of the bentonite porewater is in the range of 8.5-9.0 [4]. The bentonite backfill slit between the silo and the rock is accounted for in the calculations. As the chemistry of the bentonite porewater has not been fully studied yet, a few assumptions necessary for the calculations are made on the basis of the data available:

1. There are no calcium-aluminates in the bentonite.

2. There is a vast amount of bentonite compared to the concrete.

2.3 Groundwater

The chemistry of the groundwater in the host rock at Forsmark was analyzed [5]. The groundwater contains calcium, sodium, potassium, magnesium, sulphate, and carbonate. It has a pH in the range of 6.9 to 7.7.

2.4 Degradation Mechanisms

A full description of the problem illustrating the concrete slab in contact with the bentonite backfill (or the groundwater which can be treated similarly) is shown in Figure 1. The degradation reactions that may occur simultaneously in the concrete of the repository are as follows:

1. Hydroxyl neutralization - Outward diffusion of NaOH and KOH from the concrete through the bentonite to the groundwater.

2. $Ca(OH)_2$ dissolution and calcium diffusion from the concrete to the bentonite.

3. Intrusion of carbonate - Calcium mainly dissolved from $Ca(OH)_2$ will react with intruding CO_3^{2-} from the bentonite to precipitate as $CaCO_3$ in the concrete or in the bentonite depending on where the solubility is exceeded. This is modelled as if the precipitation takes place in the concrete.

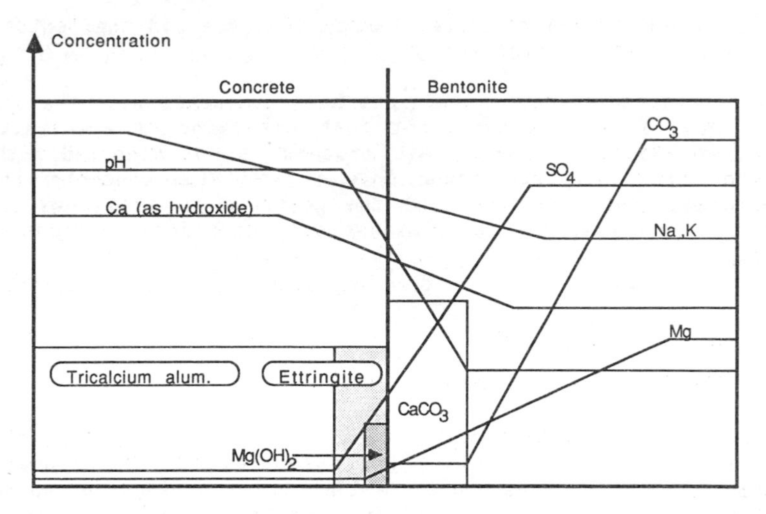

Figure 1. A conceivable distribution of some major ions and solid phases near the interface between concrete and bentonite. Concentration not to scale.

4. Intrusion of sulphate - SO_4^{2-} of the bentonite will diffuse into the concrete and react with tri-calcium-aluminate to form ettringite thus releasing OH^-. The tri-calcium aluminate is modelled as $X(OH)_2$ and the reaction is:

$$\underset{\text{Tri-calcium-aluminate}}{X(OH)_2\ (s)} + SO_4^{2-} = \underset{\text{Ettringite}}{XSO_4\ (s)} + 2\ OH^- \quad (1)$$

   Calcium may also react with intruding SO_4^{2-} to form $CaSO_4$ (s) in the concrete or in the bentonite.

5. Mg intrusion from the bentonite into the concrete to form $Mg(OH)_2$.

These reactions will either destroy the formation of the reinforced concrete or produce a considerable increase in the volume of solid. As a result, this will cause the concrete to loose its structural strength and a lowering of its pH. Reported experience of concrete structures indicate that the reactions of calcium hydroxide depletion and sulphate intrusion are likely to be the most severe.

The degradation of the concrete will probably take place in several steps. First, the alkali hydroxides, NaOH and KOH, will diffuse away from the concrete thus decreasing the pH of concrete porewater from 13.4 to 12.4. Second, the outward diffusion of $Ca(OH)_2$ will become dominant. This sequence of transport mechanisms in the concrete has already been verified by a number of investigators [2,6]. At the same time, CO_3^{2-} and SO_4^{2-} may intrude to react with the concrete. The pH in the porewater will now decrease to about 11. Finally, other components in the cement of concrete like SiO_2 may also diffuse out. The transport processes will probably continue for tens of thousands of years until the pH in the concrete reaches the level in the bentonite and/or in the groundwater of the host rock.

The diffusion of NaOH and KOH has been evaluated with the instationary diffusion equation and it turns out that this kind of alkalinity will be depleted after about 100 years. This process is not modelled with CHEMTRN. Because magnesium has been found only in very low concentrations in the bentonite clay, its intrusion into the concrete is also neglected in the calculations. Therefore, the calculations only consider reactions 2-4 in the list above.

3. CALCULATIONS WITH CHEMTRN

3.1 The Computer Code CHEMTRN

CHEMTRN is a FORTRAN computer code that was developed at the Lawrence Berkeley Laboratories, California [7]. It couples one dimensional advection and dispersion/diffusion to a number of chemical reactions including precipitation/dissolution, aqueous complexation, water dissociation, ion exchange, and surface complexation. The chemical reaction, transport, and boundary condition equations are expressed in a differential/algebraic form and solved simultaneously. CHEMTRN can handle the transport of a large number of chemical species at the same time.

Three cases are studied with CHEMTRN. The cases of calcium depletion in the concrete and of sulphate intrusion from the bentonite to the concrete have been calculated separately. Then the full problem accounting for "all" the possible reactions are studied. These cases consider the chemical transport at either the concrete/bentonite or the concrete/groundwater interface.

The movements of the calcium depletion and sulphate intrusion fronts are compared with a simple Shrinking Core Model. In this model the accumulation in the water is neglected and the reaction with the solid is assumed to be instantaneous leading to penetration depth $\alpha(\text{time})^{\frac{1}{2}}$.

3.2 Case 1: Calcium Diffusion from Concrete to Groundwater

Because of the high concentration gradients between the concrete and the groundwater, NaOH and KOH and other forms of hydroxides will dissolve and diffuse out to the groundwater. It is estimated that the main alkalinity will be depleted after about 100 years. It should be pointed out that in all the calculations of this report the time is set to be zero when the alkali hydroxides are consumed. This case considers the dissolution and diffusion of $Ca(OH)_2$ (portlandite) from concrete after that time period. Corresponding to the transport in the caverns in the SFR, the concrete is in direct contact with the groundwater and there is no bentonite clay surrounding the concrete.

A concrete slab is modelled with groundwater of known composition on one side of the slab. In a volume of 1 m^3, the concrete contains 210 kg CaO in its cement paste. The amount of portlandite formed during hydration is 3.74 $kmol/m^3$. Because CHEMTRN requires concentration in moles per liter solution, this is divided by the porosity of degraded concrete (20%) which is equivalent to 18.7 mol/l porewater. Then the total concentration of calcium, if it were dissolved, is 18.7 mol/l. The actual concentration of free calcium ions in the porewater is $6.52*10^{-3}$ mol/l. The basis species chosen are Ca^{2+}, OH^-, and H^+ and the chemical reactions in association with Ca and the water dissociation are considered. The pH in the initially portlandite saturated pore water is 12.4. As long as there is portlandite present, the pH will remain at this level. But when the portlandite is gone, the pH will decrease.

The calculation with the chemical equilibrium code, MICROQL [8], gives a first estimate of the initial species concentrations. CHEMTRN may also be used as an equilibrium code (static mode). It turns out that the results are identical to those obtained from MICROQL when the activity coefficients are set equal to 1 in CHEMTRN.

The pore diffusivity of the concrete is chosen as $5*10^{-10}$ m^2/s or $1.58*10^{-3}$ m^2/year. Assuming that advection and flow will not occur in the concrete, both the dispersion coefficient and the velocity of flow are set equal to 0.

At the concrete/groundwater interface, it is assumed that there is a large volume of groundwater containing 10^{-4} mol/l Ca^{2+} at pH 7.5. The groundwater has the capacity to wash out the species released from the concrete or to supply the concrete with the species present in it without changing its own composition. Therefore, a constant concentration boundary condition is imposed for the groundwater. In the center of the concrete slab, a zero flux condition is imposed because of the symmetry. However, it is anticipated that the concrete will not be fully penetrated, so that the conditions at the inner side of the concrete are of no importance.

3.3 Case 2: Sulphate Intrusion from Bentonite to Concrete

Sulphate intrusion from the bentonite into the concrete is considered in this case. Sulphate in the bentonite will migrate towards and eventually into the concrete, where it reacts with the calcium-aluminates to form ettringite. If the aluminates are represented as X, the reactions can be modelled as the precipitation of sulphate and X as shown below:

$$SO_4^{2-} + X^{2+} = XSO_4\ (s)\ \text{(ettringite)} \qquad (2)$$

Because of the limitations of CHEMTRN, the problem of sulphate migration towards the concrete is treated as a hypothetic XSO_4 outward diffusion from the concrete. These cases are identical to each other if the penetration front only is of interest.

This case is now similar to that of the calcium hydroxide depletion if only one basis species, XSO_4, is considered. The total amount of XSO_4 is 0.135 $kmol/m^3$ concrete, which is taken as the capacity of the concrete to react with sulphate from bentonite. This is again transformed to 0.900 mol/l porewater using the porosity of 15% for aged but not degraded concrete. The solubility product of XSO_4 (s) is chosen as $1.04*10^{-2}$ mol/l, so that the porewater which is initially saturated with respect to this solid has $1.04*10^{-2}$ mol/l XSO_4 (aq). This corresponds to the solubility of sulphate in the bentonite porewater. Again, a no-flux condition is imposed at the center of the concrete slab. At the boundary where the concrete is in contact with bentonite, it is assumed that the water has a constant concentration and XSO_4 is equal to zero.

The concrete and the bentonite have different transport properties, the diffusivity in bentonite is higher than that in concrete. When sulphate intrudes into the concrete, the bentonite itself will also be depleted. This is taken into account by using a weighted effective diffusivity according to

$$D_{e2}^* = \frac{D_{e2}}{1 + \frac{D_{e2}\ q_2}{D_{e1}\ q_1}} \quad . \tag{3}$$

We get $D_{e2} = 1.30*10^{-11}$ m^2/s corresponding to a weighted pore diffusivity of $8.67*10^{-11}$ m^2/s or $2.73*10^{-3}$ m^2/yr.

3.4 Case 3: Full Problem

3.4.1 Description of the case to be considered

This case consider reactions 2-4 as given in Section 2.4.

In the calculations, the presence of groundwater in the rock has not been taken into consideration. Therefore the system to be studied is at the concrete/bentonite interface where the exchange of species between these materials occur.

Chemical reactions which probably will take place in the considered system are obtained from Stumm and Morgan [9] and Freeze [10]. Preliminary equilibrium studies with the code MICROQL have been performed. The basis species chosen are Ca^{2+}, X^{2+} (denotes tri-calcium aluminate), CO_3^{2-}, SO_4^{2-}, H^+, and OH^-. The five precipitates to be considered are $Ca(OH)_2$, $CaCO_3$, $CASO_4$, $X(OH)_2$, and XSO_4. The aqueous complexes include $CaOH^+$, $CaHCO_3^+$, $CaCO_3$ (aq), $CaSO_4$ (aq), HCO_3^-, and H_2CO_3.

The complex HSO_4^- was originally accounted for but its concentration was found to be very low in both the bentonite and the concrete. Therefore this species is neglected in the full run. Magnesium is not accounted for because of its low concentration in the bentonite. Sodium is also dropped from the list, although it has a high content in both the concrete and the bentonite, because it exists mainly in the form of Na^+ in the solution which only influences the ionic strength. This is of no importance since the ion effect is negligible in comparison to the chemical reactions. Silica is excluded because it is present essentially in the solid phase and because the silicate will probably be the last to dissolve after thousands of years.

3.4.2 Assumptions and input parameters

a. Concrete

It is assumed that the concrete initially contains $Ca(OH)_2$, tri-calcium-aluminate in the cement paste, and the six basis species along with the six aqueous complexes in the porewater.

In 1 m^3 of concrete there is 210 kg CaO in the cement paste. The amount of portlandite transformed from CaO is 3.74 $kmol/m^3$. This is divided by the porosity of concrete (15%) and is equivalent to 24.9 mol/l porewater. The content of Ca^{2+} in the porewater, $6.52*10^{-3}$ mol/l, is negligible. Thus the total concentration of calcium which is not dissolved is 24.9 mol/l. The pH in the initially saturated portlandite porewater is 12.4. An initial guess of the free concentrations is obtained from previous CHEMTRN runs.

The sulphate intrusion from the bentonite to the concrete is modelled by Equation (1). The sulphate diffuses into the concrete and reacts with the tri-calcium-aluminate which is assumed to exist in its hydroxide form $X(OH)_2$. The reaction forms a precipitate which is probably ettringite (XSO_4). The total amount of $X(OH)_2$ available in the concrete to react with the intruding sulphate is 0.135 kmol/m^3 concrete. This is transformed to the equivalent concentration in the porewater of 0.900 mol/l. The solubility products of $X(OH)_2$ (s) and XSO_4 (s) are assumed to be 10^{-8} and $10^{-11.5}$ mol/l, respectively. These are somewhat arbitrary figures and are chosen such that when SO_4^{2-} intrudes the concrete, XSO_4 (s) will form thus releasing OH^-. The results are not sensitive to the choice of these numbers as long as the more or less quantitative expulsion of OH^- is ensured.

The pore diffusivity, D_p, through the concrete porewater is taken to be $2*10^{-10}$ m^2/s or $6.31*10^{-3}$ m^2/yr. Because of symmetry, a zero flux boundary condition is assumed for the center of the concrete.

b. Bentonite

As mentioned above, a number of assumptions are made for the properties of the bentonite in the computation.

1. The pH of the bentonite porewater is 8.5.

2. There is no tri-calcium-aluminate in the bentonite (a very low concentration for X is used) and species other than those listed for the concrete are not considered.

3. There is a vast amount of bentonite compared to the concrete and the transport will not affect the bentonite composition, so that a constant concentration condition is imposed for the bentonite.

The contents of calcium and carbonate in the bentonite are not well known but are determined by the following analysis. When 1 m^3 of dry bentonite (1050 kg) is put into water and compacted to 1 m^3, the weight of the wetted bentonite is 1600 kg. Therefore the water which has come into the pores has a weight of 550 kg and the volume of the pores is 0.550 m^3. The porosity of the bentonite is then 55%.

It is known that some 2.5% by weight of sodium-carbonate is added to the dry bentonite to change the bentonite from the Ca-type to the Na-type. The amount of Na_2CO_3 used to transform 1 m^3 bentonite is equivalent to 0.248 kmol Na_2CO_3. Calcium is exchanged for sodium according to the ion-exchange relation:

$$Na_2CO_3 + \overline{Ca} = 2\ \overline{Na} + CaCO_3 \qquad (4)$$

Assuming a complete transformation, this will release 0.248 kmol Ca^{2+} and introduce an equal amount of CO_3^{2-} to the bentonite, and will in part precipitate as $CaCO_3$. Therefore the total amount of Ca and CO_3 is specified as either 0.248 mol/l bentonite or 0.450 mol/l porewater.

The concentration of dissolved sulphate in the bentonite porewater was determined experimentally [11]. The concentration of sulphate in the porewater was found to be $1.04*10^{-2}$ mol/l.

4. RESULTS

The following results are from the calculations for the three cases studied using CHEMTRN.

4.1 Case 1: Calcium Diffusion

This case was run on the CRAY for up to 1000 years. This required 200 seconds of CPU time. The results where compared to the Shrinking Core Model results. Based on the calculation of CHEMTRN, the concrete is depleted more than 16.0 cm due to the $Ca(OH)_2$ (portlandite) diffusion after 1000 years. This is consistent with the results obtained from the Shrinking Core Model, which yields an estimation of 18.2 cm for the depletion depth.

4.2 Case 2: Sulphate Intrusion

In order to save CPU time, CHEMTRN was run by setting the activity coefficients equal to 1. This required only 60 seconds CPU on the CDC to run up to 1000 years. The penetration depth of sulphate into the concrete is 20.3 cm at 785 years and less than 24.7 cm after 1000 years as given by CHEMTRN. This again compares reasonably well with the value of 25.1 cm at 1000 years obtained from the Shrinking Core Model.

4.3 Case 3: Full Problem (Depletion of Bentonite not Considered)

The full problem was run up to 1000 years which required 693 seconds CPU. The deterioration depths in the concrete due to outward calcium diffusion and sulphate intrusion are given in Table I. The concentration profiles for 100 years are illustrated in Figure 2a-c. The calcium hydroxide in the concrete depletes 8.8 cm in 906 years, and the tri-calcium-aluminate is intruded by the sulphate from the bentonite to a depth of 30 cm within 530 years. Therefore, the intrusion of sulphate from the bentonite has a more severe effect on the concrete than does the outward calcium diffusion. An extrapolation predicts 37.4 cm penetration at 1000 years for the sulphate intrusion and 9.4 cm depletion for the calcium migration. The ettringite formation depth is reduced to 24.7 cm if the diffusion resistance in the bentonite is taken into consideration.

The computation results from CHEMTRN are found to be comparable to those obtained from the Shrinking Core Model, according to the penetration of sulphate from the bentonite and the depletion of $Ca(OH)_2$ from the concrete which are 38.2 and 9.6 cm, respectively, after 1000 years. It is noticed that CHEMTRN provides a slightly more optimistic estimate than the Shrinking Core Model for both the sulphate intrusion and the calcium migration.

Table I. Deterioration Depth in the Concrete (Case 3).

Node	Depth (cm)	Penetrated Time Due to Sulphate Intrusion (yr)		Depleted Time Due to Calcium Depletion (yr)	
		CHEMTRN	Shrink. Core*	CHEMTRN	Shrink. Core*
2	0.338	0.169	0.078	2.31	1.23
3	0.744	0.611	0.378	8.46	5.98
4	1.23	1.50	1.03	20.6	16.3
5	1.82	3.06	2.26	42.3	35.8
6	2.53	5.90	4.37	78.3	69.1
7	3.37	9.86	7.76	136.	122.
8	4.39	16.4	13.2	228.	208.
9	5.61	26.5	21.5	366.	340.
10	7.08	41.8	34.2	577.	541.
11	8.85	64.7	53.5	906.	846.
12	11.0	99.2	82.6		1307.
13	13.5	150.	124.		1968.
14	16.6	224.	188.		2976.
15	20.3	335.	281.		4450.
16	24.7	494.	416.		6589.
17	30.0	530.	615.		9720.
	9.61				1000.
	38.2		1000.		

Shrinking Core Model: * $x = 1.21 \cdot t^{0.5}$
** $x = 0.304 \cdot t^{0.5}$

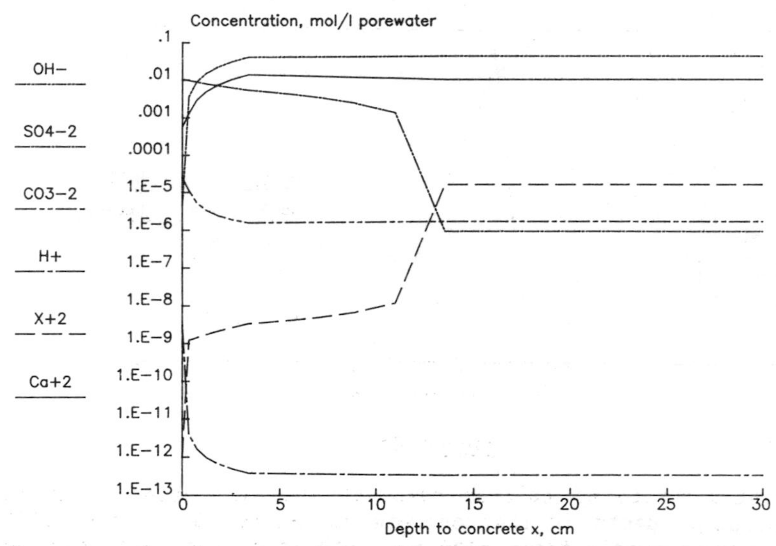

Figure 2a. Concentration distributions of various elements present within the concrete barrier for the time of 100 years. The results are calculated for case 3.

a) Concentration profiles of basis species
b) Concentration profiles of aqueous complexes
c) Solid phase profiles

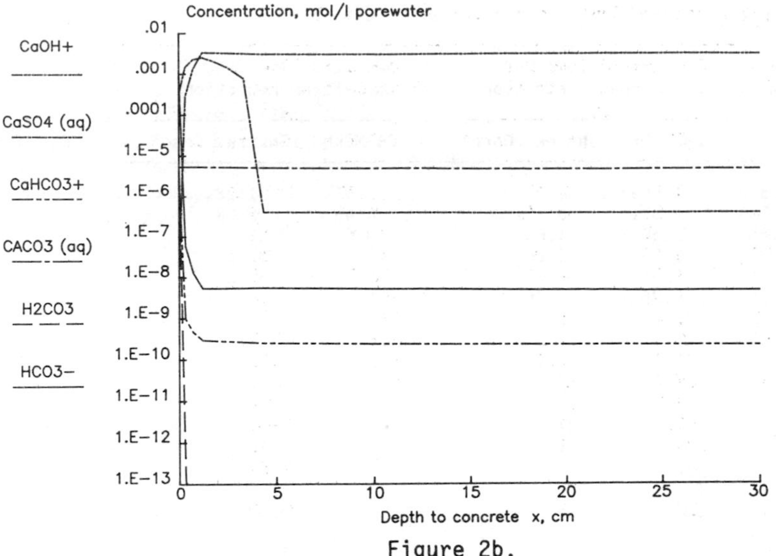

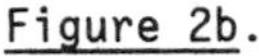

Figure 2b.

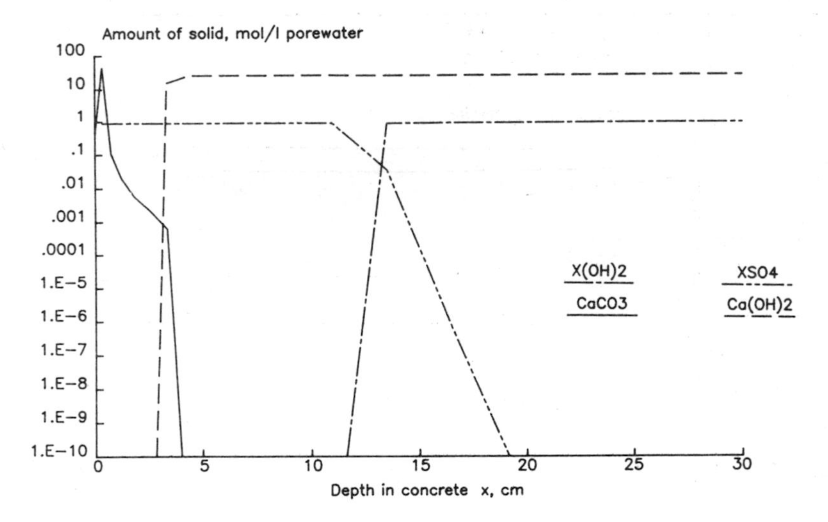

Figure 2c.

The precipitation of calcium carbonate is also observed at the same time. Its migration depth in the concrete is around 11 cm after 1000 years with a large amount precipitated near the interface. The same phenomenon was observed by Atkinsons's group when they performed an analysis of the concrete which had been buried in a clay for 43 years [6].

No precipitate of gypsum ($CaSO_4$) is formed in the concrete or in the bentonite. The pH of the concrete porewater is decreased when SO_4^{2-} has intruded forming ettringite and the $Ca(OH)_2$ has migrated out.

5. DISCUSSION AND CONCLUSIONS

The two simple test cases where the results of CHEMTRN are in reasonably good agreement with those of the Shrinking Core Model indicate that CHEMTRN can be used to compute the movement of precipitation and dissolution fronts. The front of sulphate intrusion into the concrete moves faster than that of calcium depletion from the concrete. The full problem was then calculated for several species and fronts moving simultaneously. The results agree well with the simplified calculations and in other respects seem to be reasonable.

It is noticed that the degradation depths of the concrete obtained from CHEMTRN and from the Shrinking Core Model are rather close in value, but are not identical. The values differ especially at earlier times. The latter is due to space discretization errors. In the test cases, the Shrinking Core Model provides the more conservative estimate for the concrete degradation.

Calculation with the chemical equilibrium code MICROQL gives a first estimate for the initial species concentrations. CHEMTRN may also be used to make equilibrium calculations. The results are identical to those obtained from MICROQL when activity coefficients are set equal to 1 in CHEMTRN. When activity coefficients are used, CHEMTRN gives species concentrations that can differ by up to a factor of 2.

REFERENCES

1. Hedman, T., and I. Aronsson: "The Swedish Final Repository for Reactor Waste (SFR)", IAEA-SM-289/16 (1986).
2. Andersson, K. and B. Allard: "The Chemical Conditions Within a Cement Containing Radioactive Waste Repository", Preliminary report, Stockholm (1986).
3. Bergström, S.G., G. Fagerlund, and L. Rombén: "Bedömning av Egenskaper och Funktion hos Betong i Samband med Slutlig Förvaring av Kärnbränsleavfall i Berg", KBS Technical Report No. 12, Stockholm (1977). (In Swedish)
4. Pusch, R.: "Buffertar av Bentonitbaserade Material i Siloförvaret", SFR Preliminary report, Stockholm (1985). (In Swedish)
5. Wikberg, P.: "Grundvattenkemiska Undersökningar i SFR", Department of Inorganic Chemistry, Royal Institute of Technology (KTH), Stockholm (1985). (In Swedish)
6. Atkinson, A., D.J. Goult, and J.A. Hearne: "An Assessment of the Long-Lived Durability of Concrete in Radioactive Waste Repositories", Mat. Res. Soc. Symp. Proc., 50, p. 239-246 (1985).
7. Miller, C.W., CHEMTRN User's Manual, LBL-16152, Lawrence Berkeley Laboratory, California (1983).
8. Westall, J.: "MICROQL I. A Chemical Equilibrium Program in BASIC", EAWAG, Swiss Federal Institute of Technology (1979).
9. Stumm, W., and J.J. Morgan, Aquatic Chemistry, 2nd ed, John Wiley & Sons (1981).
10. Freeze, R.A., and J.A. Cherry: Groundwater, Prentice-Hall (1979).
11. Neretnieks, I., K. Axelsson, P. Wikberg, E. Bergman, and L. Rombén: "Measurements of the Sulphate in the Pore Waters of Bentonite Type GEKO/QI", Department of Chemical Engineering, Royal Institute of Technology, Stockholm (1986).

α AND LONG-LIVED β γ WASTE SOURCE TERM. A FIRST GENERATION MODEL FOR A DEEP CEMENTED WASTE REPOSITORY

P. LOVERA (*), J.P. MANGIN (*), M. JORDA (*), J. LEWI (**)

(*) Commissariat à l'Energie Atomique - Institut de Recherche Technologique et de Développement Industriel - Département de Recherche et Développement Déchets - Service d'Etudes de Stockage des Déchets
(**) Commissariat à l'Energie Atomique - Institut de Protection et de Sûreté Nucléaire - Département d'Analyse de Sûreté - Section d'Analyse et d'Evaluation de la Sûreté des Déchets
B.P. n° 6 - F - 92265 FONTENAY AUX ROSES CEDEX

ABSTRACT

According to the normal scenario of radioactivity release to the biosphere, only long-lived nuclides are able to migrate significantly to the surface. A first generation model, concerning a cemented waste of hulls and ends deeply disposed of in a granitic medium is in progress at CEA. Two nuclides have been selected : 237-Neptunium (as a reference of α emitters) and 135-Caesium (as a reference of long-lived β emitters). Attributing the long term activity to these both nuclides leads to a model which is conservative beyond ca. 150000 years. Principal difficulties arise from physico-chemical behaviour of Neptunium in aqueous phase, and from non-linear Caesium adsorption on various media. CONDIMENT code (versions 2 and 3), which is developed parallely to the present model is conceived to take account for these phenomena.

RESUME

Dans un scenario normal de relâchement d'activité à la biosphère, seuls les nucléides à très longue période pourront migrer de façon significative jusqu'à la surface. Un modèle de première génération, concernant un déchet de coques et embouts cimentés et stockés en profondeur en milieu granitique est en cours de développement au CEA. Deux nucléides ont été sélectionnés : le Neptunium 237 (référence des émetteurs α) et le Césium 135 (référence des émetteurs β à longue vie). Un regroupement de l'activité présente à long terme sur ces deux radionucléides conduit à un modèle conservatif au-delà de 150000 ans environ. Les principales difficultés proviennent du comportement physico-chimique du Neptunium en phase aqueuse, et de l'adsorption non linéaire du Césium sur certains milieux. Le code CONDIMENT (versions 2 et 3) développé en parallèle du modèle est conçu pour en tenir compte.

INTRODUCTION

According to one of the schemata studied in France, packages of cemented hulls ans ends (issued from the reprocessing of PWR fuel) are disposed of in a hard rock. In the case of a granitic repository, the free room between package and host rock is filled with an "engineered barrier" (compacted clay or concrete). Before solubilizing radionuclides, located at the surface (α emitters or fission products) or in the bulk (activation products) of metallic waste, water has to diffuse through host rock, engineered barrier and a part of the cement matrix. Solubilized radionuclides have then to diffuse back in the interstitial solution through cement and engineered barrier towards the host rock.

Consequently, it is important to assess the source term generated by such a package, taking into account the physico-chemical properties of nuclides, water and solid media (cement and clay). A first generation model, intending to assess the activity released into the host rock is in progress at CEA.

SOURCE TERM MODELLING

A - CONDIMENT COMPUTATION CODE

The "CONDIMENT" code (developed at CEA and CISI - Compagnie Internationale de Services Informatiques) solves the diffusion-convection equation with radioactivity and source term expressed as :

$$\epsilon \tau \left(\frac{\partial C}{\partial t} + \lambda C\right) = \text{div} \left(D \,\mathbf{grad}\, C - U C \right) + S$$

in which :

ϵ is porosity

τ is retardation coefficient

D is diffusion coefficient

U is DARCY velocity

T is radioactive period

$\lambda = \ln 2 / T$ is radioactive constant

S is a source term

The resolution is carried out with the finite differences discretization method, using LAASONEN implicit schema which is unconditionaly stable. This reduces interface problems with far field computation codes (METIS /1/), inside the global risk assessment model MELODIE /2/, which is developed by the IPSN.

Two versions of CONDIMENT are presented : CONDIMENT 2 and CONDIMENT 3. Initial and boundary conditions are different.

CONDIMENT 2

This code deals with transfer of only one ion. Room is divided in regions where physical parameters (diffusion coefficient, porosity) are constant. No time variation of these parameters is foreseen. Source term is constituted eventually by radioactivity of a parent radionuclide. Different kinds of boundary conditions are envisaged :

a) DIRICHLET and NEUMAN boundary conditions :

They are the most classical ones in this kind of problem. A typical application is evolution of alkaline concentration in a deep repository cement. These ions (Sodium, Potassium) diffuse through cement, engineered barrier and host rock, towards an outlet where concentration remains zero (DIRICHLET conditions). At the center, concentration gradient remains zero (NEUMAN conditions).

b) CAUCHY conditions :

CAUCHY conditions (prescribed value of a linear combination of concentration and of its normal gradient) are used for coupling to METIS code. Thus continuity of concentration and of its normal gradient at the interface of both codes may be obtained.

c) Weakly soluble hydroxydes :

An example is free lime in cement, which exists in presence of an interstitial water saturated in Ca^{++} ions. In the region where there is free lime, concentration is determined, no diffusion may occur. However, there may be diffusion outside this region. The limit of solid lime region moves, velocity of this retreat is computed by writing that Ca^{++} flux results from dissolution and ionization of solid lime initially contained in cement. Thus there is calcium diffusion in a zone of variable thickness, with DIRICHLET conditions at boundary. Indeed, concentration on this boundary equals lime saturation concentration. After some mathematical processings, this DIRICHLET condition on a moving boundary is transformed into a CAUCHY condition on a fixed boundary, with a periodical re-meshing of cement medium. pH evolution in cement after alkaline departure may thus be assessed.

CONDIMENT 3

In this code, one computes diffusion of a cation and an anion which may precipitate. This formation of precipitate is the source term. There is no disintegration term in this code, as the behaviour of a radioactive precipitate is unknown. The presence of a precipitate does not modify conditions of numerical stability, unconditional stability of LAASONEN schema remains. The precipitate may change porosity, permeability and diffusion coefficients. Therefore, U and D may vary at each mesh.

A similar formulation allows processing of adsorption isotherms, such as LANGMUIR or FREUNDLICH ones, by a formalism of apparent porosity. In this case, anion (and consequently precipitate) do not play any role. Solubility products may vary mesh by mesh, its variation as a function of temperature (heat emission by radioactive decay), pH (computed by CONDIMENT 2) or redox potential Eh (water radiolysis) may be taken into account.

In the case of an actual precipitate, the only conditions that may be considered without further complication are DIRICHLET or NEUMAN conditions with a zero flux. The case of weakly soluble hydroxyde such as lime then becomes a simple particular case in this new code, without needing further boundary conditions.

B - REFLEXIONS ABOUT THE CHOICE OF REFERENCE RADIONUCLIDES :

The aim of this first generation model (coupled, inside a global risk assessment model - MELODIE -, to other sub-models describing the migration of nuclides in the host rock, and assessing the release into the biosphere and the health effects as a function of time) is to evaluate the release of nuclides escaping from a cemented waste deep repository in a normal scenario, i.e. excluding every geological abnormal event (such as volcanism or earthquake) or accidental (badly sealed shaft) phenomenon. In a first approach, only the behaviour of the repository at periods longer than 10^{5} years is considered. Thus, most of radionuclides have decayed enough to reach the surface at concentrations lower than maximal admissible ones.

Hence, only radionuclides having a very long period (more than 10000 years) have to be considered, and among them ^{237}Np (period 2.1 10^{6} y.) as a α reference emitter, and ^{135}Cs (period 2.3 10^{6}y.) as a β reference emitter have been chosen. It appeared necessary to differenciate α emitters (transuranians) from β emitters (in general fission or activation products) for their migration is not obviously governed by the same phenomena.

237-Neptunium :

Beyond 2 10^{5} years, the only significant activity is due to ^{237}Np, ^{242}Pu (period 3.8 10^{5} y.) and to the filiation of ^{238}Pu (particularly ^{234}U period 2.4 10^{5} y.), and the activity of ^{237}Np and of its filiation becomes preponderant beyond ca. 5 10^{5}years (Fig. 1). Although the maximal activity of these radionuclides is not reached at the initial instant of disposal - and in the case of ^{238}Pu filiation, for instance, the maximal activity arises at ca. 2 10^{5} years - one attributes conservatively to ^{237}Np the maximal total activity of ^{237}Np, ^{242}Pu and ^{238}Pu filiation. Taking into account the activity increase due to the decay of Neptunium, the proposed model supplies a total activity that is higher than the real one beyond ca. 150000 years. Hence, this modelling seems to be adapted for studiing the long term behaviour of deep repositories.

135-Caesium :

135-Caesium is chosen in preference to other long-lived β emitters present in the repository because of its long period (compared to ^{99}Tc), and of its high migration ability (compared to ^{93}Zr). One attributes to ^{135}Cs the total initial activity of long-lived fission products -FP- (^{93}Zr, ^{99}Tc, ^{135}Cs). Whether the activity of activation products -AP- (principally ^{93}Zr) must be taken into account or not is being studied, for the initial activity of AP is widely higher than that of FP, but on one hand, they are less easily available (because they are located in the bulk of metallic waste and not on its surface), and on the other hand, Zirconium is far less mobile than Caesium due to lower solubility of its hydroxyde.

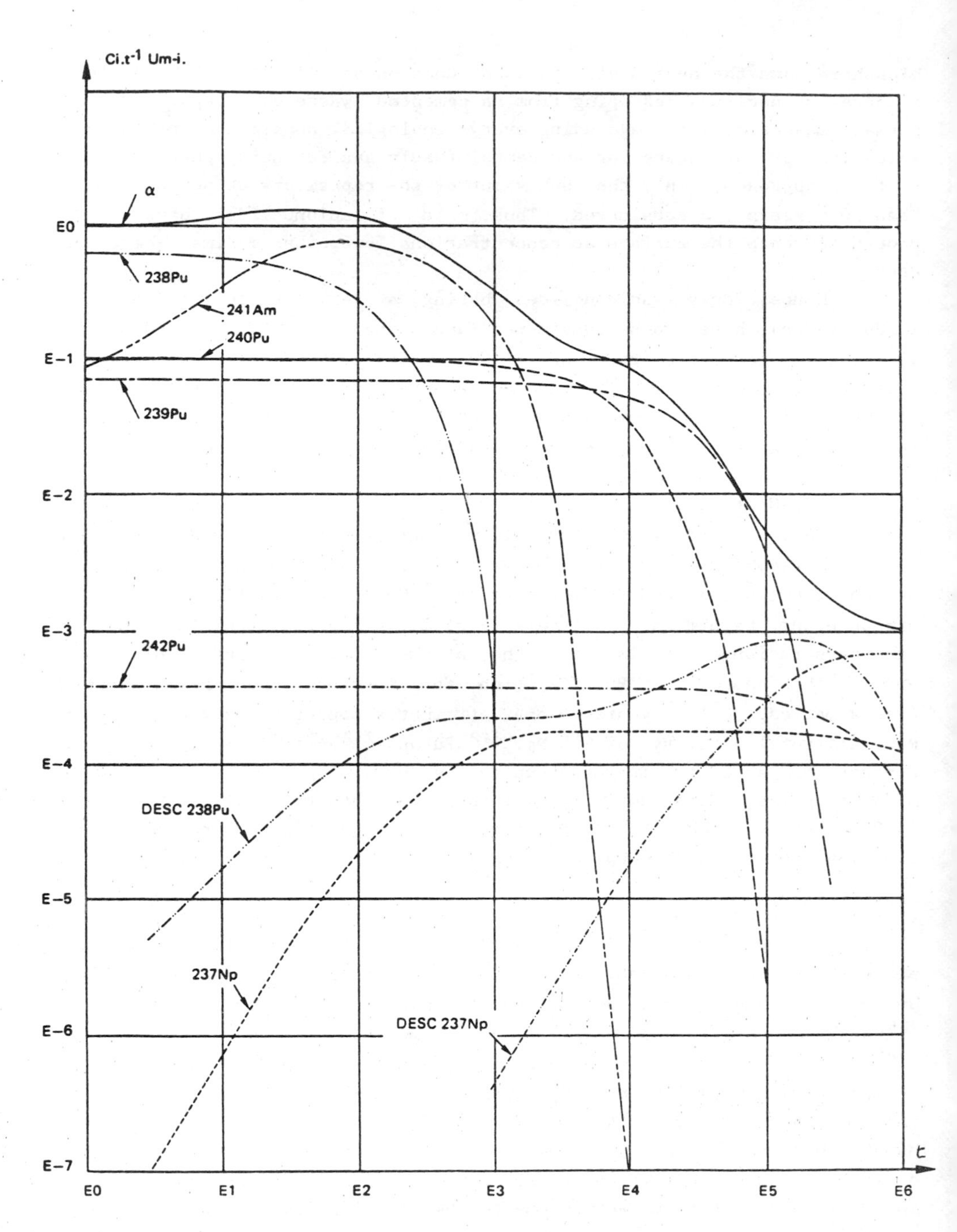

Fig. 1 : Evolution of α Activity vs. Time (Years after Reprocessing)

In conclusion :

The long term activity of a package of hulls and ends, cemented and deeply reposited is modelled (in a first generation of source term model) by taking into account two reference radionuclides :

^{237}Np as α emitter

^{135}Cs as β emitter.

C - BEHAVIOUR OF THE REFERENCE RADIONUCLIDES

Neptunium may exist, according to redox conditions, at several valences (particularly Np IV, Np V and Np VI). Moreover, Neptunium is able to form anionic species in a basic medium. Thus Neptunium solubility depends on two important parameters, redox potential Eh and pH of the bulk solution.

Caesium solubility is insensitive to redox potential and bulk pH, but it appeared a non-linear behaviour during experimental studies concerning its migration in porous media.

Source term modelling thus implies :

- an assessment of physico-chemical characteristics of concrete interstitial solutions as a function of time (study of concrete degradation, under influence of groundwaters),
- an assessment of Np solubility in local (Eh, pH) conditions,
- an assessment of Cs adsorption in various media.

MODELLING OF NEPTUNIUM SOLUBILITY

According to local physico-chemical conditions (Eh, pH), Np solubility may vary up to several orders of magnitude, and although standard potentials of (NpO_2^{++}/NpO_2^{+}) and (Np^{4+}/Np^{3+}) equilibria are relatively well known /3/, it is not the same for (NpO_2^{+}/Np^{4+}) equilibrium /4/.

At a basic pH, the existence of anionic species would have as a consequence an increase of Np solubility when pH increases. Though the existence of these compounds is not proved absolutely /5/, they are considered conservatively.

Fig. 2 shows a proposal of Np (Eh, pH) solubility diagram, in a zone of Eh from -0.6 to +1.45 V, and of pH from 4 to 14 (granitic groundwaters often have a redox potential between -0.45 and 0 V, and a pH between 7 and 9). The existence of oxidizing solutions may be a consequence of water radiolysis by α emitters.

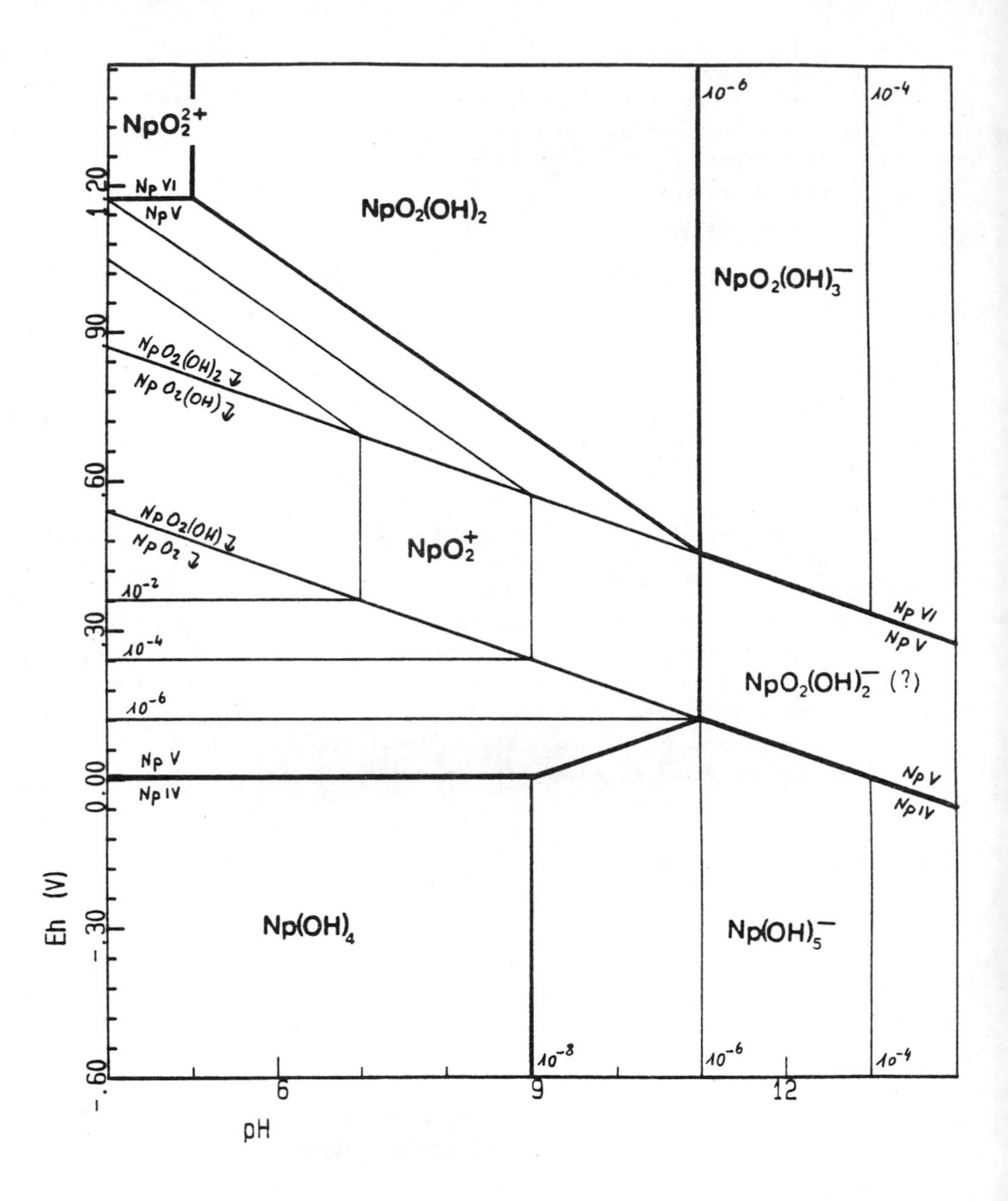

Fig. 2 : Proposal of Neptunium (Eh, pH) Solubility Diagram

Experimental studies /6,7/ reinforce the hypothesis of a NpO_2OH solid phase in equilibrium with NpO_2^+, in contradiction with some previous theoretical studies only considering the existence of NpO_2 and $NpO_2(OH)_2$ as solid phases /4/. Assessed solubilities of Neptunium are fitted to experimental results, but low Np solubilities in basic and weakly oxidizing water /6/ are not explained completely (precipitation of poorly identified compounds ?).

Taking into account the likely presence of colloids which dissolve slowly, the low measured solubility of Np V /6/ is not considered conservatively. An approached formula for Np solubility, independant of redox potential, sufficient for a first generation model and in good agreement with experimental data /7/ may be :

$pH < 11 : \log s = 5 - pH$ (s = Np solubility, mol l^{-1})

$pH > 11 : \log s = pH - 17$

In a first approach, this study does not take into account the possibility of complex formation with other anions present in groundwaters (carbonates, organic materials).

RADIONUCLIDE MIGRATION IN COATING CEMENT

Radionuclides are located initially at the surface of metallic waste (at least α emitters and FP). Now, considering on the one hand the present technology of package filling (hulls and ends are poured at first, then concrete), and on the other hand the imperfect adherence of cement on metal, it therefore may appear a porosity along metallic waste, ca. 100 μm thick. Cement only plays a role of rubble and of alkaline and calcium container, and some more or less discontinuous preferential ways of nuclide migration along metallic waste likely occur. Consequently, classical diffusion and leaching laboratory experiments might underestimate diffusion coefficients and overestimate adsorption capacities of cement.

RADIONUCLIDE MIGRATION IN ENGINEERED BARRIER

The nature of engineered barrier is not settled definitively (concrete and/or clay). However, it is likely that a clay will be chosen in the case of a deep repository in granite. Hereafter, we principally consider Np and Cs migration in clay. As neither its nature not its apparent density are known with certitude, only a parametric study may be envisaged, based on experimental studies performed in closely related conditions.

Migration of a radionuclide may be characterized by :

- an effective diffusion coefficient De (expresing migration in permanent flow)

- a retardation coefficient τ (expressing retention of radioelement by adsorption on porous material).

In a permanent flow, effective diffusion coefficient of all nuclides should be nearly identical and close to the one determined with tritiated water -HTO- for instance. (In practice, experimental diffusion coefficients measured for various radionuclides are lower than the one measured for HTO, perhaps because the other nuclides are heavier and larger than HTO. Particularly if porosity is very fine, the "available porosity" is lower than "HTO porosity"). Very conservatively, it may be assumed that diffusion coefficient of every nuclide in permanent flow is assimilated to that of HTO in the same medium.

Retardation coefficient is often expressed as a "Kd", representing the equilibrium ratio between adsorbed concentration Ca (moles/g of dry material) and dissolved concentration Cd (moles/cm^3 of aqueous solution), and one often writes :

$\tau = 1 + \rho\ Kd\ (1-\theta)/\theta$ (ρ is bulk density of material, θ is porosity).

Such a formulation is correct, provided that some assumptions on adsorption and desorption of radionuclide in material are fulfilled :

- Equilibrium must be reversible,
- Adsorption and desorption must be very fast, compared to diffusion,
- Concentration of adsorbed element Ca must be proportional to the concentration of dissolved element Cd,
- no saturation of material must occur.

In the case of Caesium diffusion, desorption may seem slow at laboratory scale, but it is likely that it may be considered as instantaneous at geological scale (which is conservative in any case). On the other hand, one of initial assumptions is not validated experimentally, and power laws such as :

$Ca = Cd^n$ (with n = 0.5 to 0.7)

have been pointed out in various media /8,9/.

- if n=0.5, adsorption-desorption equilibrium could be interpreted as a cationic exchange :

$Ca^{++}ads + (Cs^{+}+K^{+})sol === Ca^{++}sol + (Cs^{+}+K^{+})ads$

in which "ads" and "sol" represent resp. adsorbed species on the material, and species in aqueous solution. This equilibrium is probably available in fresh cement with dilute Cs solutions.

- A linear isotherm (n=1) might be interpreted by an exchange between a Ca^{++} adsorbed ion and 2 Cs^{+} ions in solution, or between an alkaline adsorbed ion and a Cs^{+} ion in solution.

- an intermediate value of n could express a competition between both reactions.

If it is assumed (in opposition to concrete) that clay is not damaged, an equilibrium relation determined according to laboratory results may be likely extrapolated to the duration of storage (provided that margins on coefficients are taken into account)

A non linear adsorption formula may be introduced into CONDIMENT, instead of a linear formulation with a retardation coefficient. A coefficient Kd^{*} (C) = Ca/Cd (as a function of Cd) is computed at each time step.

Data concerning Neptunium adsorption are rare, and do not allow to assess influence of concentration. However, clays are principally cationic exchangers, it is thus likely that Np will not be very strongly adsorbed if present as an anionic or neutral species. It is conservative, in this case, to consider existence of anionic Np species, therefore Np adsorption in clay may be neglected.

D - EVOLUTION OF SOURCE TERM MODEL

The first generation model of α and long life $\beta\gamma$ source term of cemented waste only will take into account two nuclides, with very conservative assumptions about their migration. In a second generation model, it is foreseen to model the behaviour of Plutonium, which is, with Americium, the principal α emitter before 10^{5} years. In the same way, study of $\beta\gamma$ emitters will be improved by considering migration of 99-Technetium (shorter period than Caesium, but retardation coefficient likely lower than that of Cs, especially as anionic species TcO_4^-), and that of Zirconium. The model has to be improved according to assumptions for Np and Cs migration less conservative.

CONCLUSIONS

The source term model of a cemented waste, currently in progress at CEA, will be able to describe the behaviour of a repository. In a first generation of this model, in which ^{135}Cs and ^{237}Np are considered, the behaviour is conservative beyond only ca. 150000 years. Consequently, assessment of abnormal scenarios is not possible with it. A more complete study of the migration of other nuclides, such as Technetium or Plutonium, will provide a model conservative sooner. The interactions between reference nuclides and bulk (Np solubility as a function of redox conditions and pH, Cs adsorption on various barriers,..) have been studied more particularly. These phenomena are taken into account in an improved version of CONDIMENT code.

ACKNOWLEDGEMENTS

We would like to thank Mr. P. VITORGE for fruitful discussions about Neptunium behaviour.

REFERENCES

/1/ METIS - Notice d'utilisation - Rapport CIG - Ecole des Mines LHM/RD/85/41

/2/ J. LEWI, M. ASSOULINE, J. BAREAU, P. RAIMBAULT, "MELODIE : "A Global Risk Assessment Model for Radioactive Waste Repositories", Congrès de Modélisation Mathématique pour les Sites de Déchets Radioactifs, Madrid, 1987

/3/ C. RIGLET, P. ROBOUCH, P. VITORGE, to be published in Radiochim. Acta

/4/ M. SCHWEINGRUBER - "Löslichkeits- und Speziations-berechnungen für U, Pu, und Th in natürlichen Grundwässern. Theorie, thermodynamische Dateien und erste Anwendungen", EIR Bericht Nr 449 - Würenlingen 1981

/5/ P. VITORGE, "Comportement physico-chimique du Plutonium et du Neptunium en formations granitiques et dans les barrières artificielles" - Rapport CCE WAS 210 81 F

/6/ F.T. EWART, R.M. HOWSE, H.P. THOMASON, S.J. WILLIAMS, J.E. CROSS - "The solubility of Actinides in the Near-Field", Mat. Res. Soc. Symp. Proc., Vol 50, 701-8, (1985)

/7/ C. LIERSE, W. TREIBER, J.I. KIM, "Hydrolysis Reactions of Neptunium V", Radiochim. Acta, Vol 38, 27-8, (1985)

/8/ F.P. GLASSER, A.A. RAHMAN, R.W. CRAWFORD, C.E.C. McCULLOCH, M.J. ANGUS "Radioactive Waste Management - Immobilization of Radioactive Waste in Cement Based Matrices", DOE RW 84050, (1984)

/9/ D. RANçON, J. ROCHON, "Rétention des Radionucléides à Vie Longue par Divers Matériaux Naturels", Proc. of the Workshop on "The Migration of Long-lived Radionuclides in the Geosphere", 301-22, Brussels (1979)

INTERACTION OF Am(III) WITH ORGANIC MATERIALS ; SOME RESULTS

V. Moulin, P. Robouch, D. Stammose and B. Allard
Commissariat à l'Energie Atomique, CEN-FAR
DRDD/SESD/SCPCS/LCh
BP 6, 92265 Fontenay-aux-Roses, FRANCE

ABSTRACT

The study of the influence of natural organics as humic substances in the near-field chemistry of radioactive wastes is investigated in two directions : formation of soluble complexes with actinides and their effect on retention properties of actinides on solid surfaces. Considering the stable complexes they form with trivalent elements, some results obtained with americium are presented : formation constants and distribution coefficients obtained with different humic materials.

RESUME

L'étude de l'influence des composés organiques naturels, tels que les substances humiques dans la chimie du champ proche des stockages nucléaires est abordée dans deux directions : la formation de complexes solubles avec les actinides et le rôle des matières organiques sur les propriétés rétentrices des actinides vis-à-vis de surfaces solides. Connaissant la stabilité des complexes formés avec les éléments trivalents, les résultats obtenus avec l'américium III sont présentés : constantes de formation et coefficients de distribution pour différents acides humiques.

INTRODUCTION

The knowledge of chemical behaviour of actinides in the near-field environment of radioactive waste repositories is of primary importance for their safety conception and the risk analysis.

In low and intermediate level waste disposals organic coumpounds may occur both naturally and be introduced in or with the waste form [1]. The naturally occuring organic compounds as humic and fulvic acids constitute the major part of organic carbon of most natural waters (surface - and ground waters), soils and sediments [2,3], and present very strong complexing properties towards cations, particularly actinides [4,5]. So they may enhance the migration of disposed radionuclides by increasing their solubilities in a disposal system. Moreover humic substances have high affinities for various surfaces (oxides, clays, ...) [6,7] and therefore can modifythe transport of radionuclides. However, these aspects of humic substances are not fully understood, and further researches are required in this area . This study attempts to illustrate the role of humic substances on americium behaviour : the present paper deals with the homogeneous (soluble complex formation) and heterogeneous (sorption studies) interactions of americium III with humic substances from various origins. Speciation calculations are also presented in natural water conditions and in interstlcial concrete solutions.

EXPERIMENTAL

Materials

Humic (HA) and fulvic (FA) acids studied are from different origins : surface water (HASW, FASW), ground water (FAGW),soil (HA-A) and lake sediment (HALB). Alumina (crystalline α-Al_2O_3) used for the sorption studies was perchased from Ventron.

Analytical Procedure

The spectrophotometric study was performed on a Cary 17D instrument by recording the displacement of the absorbance peak of Am (III) at 503.1 nn im the presence of increasing humic concentrations.

The sorption study was performed by a batch procedure where the alumina is prequilibrated with humic materials before adding americium.

Precise descriptions of the procedure for these two studies are given in [8] for the spectrophotometric technique, in [9] for the sorption study.

Speciation calculations

All computations for Am speciation in different media were made according to the commonly used equation % species = [species]/Σ[species]. No colloid or pseudocolloids formation was considered. The formation constants β for the hydroxyde and carbonate systems (Table I) were recalculated at a ionic strength of 0.1 M using the Specific Interaction Theory [10].

RESULTS AND DISCUSSION

Complex Formation

From the absorbance spectra (Figure 1a) obtained and the curve giving the molar extinction coefficient, ε (the ratio : absorbance on total americium concentration) at the wavelength corresponding to the complex formed (between 504.8 and 505.5 nm according to the ligand studied, figure 1b) as a function of humic/fulvic concentration, conditional formation constants β have been calculated assuming the formation of a 1:1 complex. A complete description of β calculation is given elsewhere [8].

Table II summarizes the β values determined by spectrophotometry at pH 4.65, I = 0.1 M $NaClO_4$, and β values given in the literature for Am-humate complexes. The data are also reported in the same unit system (for the definition of ligand capacity) due to the rather different systems used by the authors. No important differences between β values for the humic and fulvic acids of various origins exist, and these values confirm the fact that humic substances form rather stable complexes with americium (III). A fair agreement is observed between the data reported in Table II. However the problem of the possible existence of 1:1 or 1:2 complexes, and of the nature of the bondings has not yet been resolved, these concerns are discussed in [8].

Despite the fact that more studies have to be performed on americium-humics system, in particular the determination of formation contants by other techniques (to compare and validate) and the pH dependence of these formation constants (due to the polyelectrolyte properties of humic substances), the americium behaviour in solution can be approached by considering the ligands (hydroxyde (OH^-), carbonate ($CO_3^{=}$) and organic ligands) playing an important role towards the Am speciation. Using the model developed previously for the Am-organic system with the reasons and limitations here explained [11], species distribution plots are presented Figure 2 (I = 0.1 M $NaClO_4$, CO_2 partial pressure pCO_2 = $10^{-3.5}$ atm) for different humic acid concentrations.

The particular case of interstícial concrete solutions (pCO_2 = 10^{-13}atm imposed by $CaCO_3$ and $Ca(OH)_2$ solids ; pH of 12 $\mp$ 1) has been also modelized (Figure 3). In these conditions americium is entirely present as $Am(OH)_3$.

Sorption studies

From the precedent study, americium behaviour in presence of naturally occuring ligands becomes more comprehensible. So the effect of humic acids on americium sorption on solid surface will be more easily understood. The results obtained with a particular surface, crystalline alumina, are presented Figure 4 where the influence of humic concentration, ionic strength, pH and contact time is represented. Same kinds of results have been obtained for fulvic acids [9].

At low pH range (pH < 7) the presence of humic acids at low concentration (1 mg/l) enhances americium adsorption, while at higher concentration (10 mg/l) there is a general decrease of americium retention in all pH range studied. At high pH-range an important decrease in the americium adsorption occurs in the presence of 1 mg/l of humic acids.

Sorption affinities of humic substances for mineral surfaces (oxides, clays, hydroxides, ...) [6,7] are known, and, Table III gives the percent of humic acids fixed on alumina as a function of pH and humic concentration.

Humic acids are sorbed on alumina at pH below the point of zero charge (around 8 for this material) forming a surface layer (Table III). Americium is thus uptaked by this organic coating, at pH below 7 and for a concentration of 1 mg/l, inducing an enhanced adsorption in the system.

At high humic concentration (10 mg/l) most of the organics remain in solution (Table III) and thus americium is present in the aqueous phase as a non-sorbing organic complex. The aqueous composition could be predicted from speciation plots (Figure 2) : americium exists as a 1:1 complex with humic acids.

This study shows that the behaviour of americium in presence of humic materials and a solid phase (for which humics have some affinities) would be controlled by the behaviour of humic acids. In the case of humic sorption on the solid phase, there would be an enhanced americium adsorption (immobilization), whereas when humic acids remain in solution, americium would be present as an organic complex (migration). The formation of pseudo colloids with humic acids is not disregarded [14,15] and will be more studied.

More extensive research needs to be conducted on humic substances field related to actinides behaviour in a nuclear waste disposal in order to have more precise idea on their role in the near-field as well in the far-field environment : retention or mobility enhancement of actinides.

ACKNOWLEDGMENTS

Mr M.THEYSSIER and Mr. M.T. TRAN are gratefully aknowledged for their assistance.

This work received financial support from C.E.C., contract FI 1W/0068. It has been performed within the framework of a cooperation between C.E.A. - France and S.K.B. - Sweden.

REFERENCES

[1] TOSTE, A.P. and MYERS, R.B. : "The relative contributions of natural and waste-derived organics to the subsurface transport of radionuclides", Proceedings of an NEA Workshop, OECD/NEA, Paris, 1985, 57-76.

[2] STUMM, W and MORGAN, J.J. : Aquatic Chemistry, 2nd ed., Wiley Interscience, New-York, 1981.

[3] CHRISTMAN, R.F. and GJESSING, E.T. : Aquatic and Terrestrial Humic Materials, Ann Arbor Science, 1983.

[4] CHOPPIN, G.R. and ALLARD, B. : "Complexes of Actinides with naturally occuring organic compounds", Handbook on the Physics and Chemistry of the Actinides, Eds. FREMAN, A.J. and KELLER, C., 1985, Vol. 3, 407-429.

[5] KIM, J.I. : "Chemical Behaviour of Transuranic Elements in Natural Aquatic Systems", Handbook on the Physics and Chemistry of the Actinides, Eds. FREEMAN, A.J. and KELLER, C., 1986, Vol. 4, 413-455.

[6] GREENLAND, D.J. and HAYES, M.H.B. : The Chemistry of Soil Consti tuants, Wiley, 1981.

[7] DAVIES, J.A. : "Adsorption of natural dissolved organic matter at the oxide/water interface", Geochim. Cosmochim. Acta, 46 (1982), 2381-2393.

[8] MOULIN, V., ROBOUCH, P., VITORGE, P. and ALLARD, B. : "Spectrophotometric Study of the Interaction between americium III and Humic materials", Inorg. Chim. Acta, 140, (1987), in press

[9] ALLARD, B., MOULIN, V., BASSO, L., TRAN, M.T. and STAMMOSE, D. : "Americium Adsorption on Alumina in presence of humic materials", Geoderma, submitted for publication.

[10] ROBOUCH, P. : "Contribution à la prévision du comportement de l'américium, du plutonium et du neptunium dans la géosphère ; données chimiques", Thesis, Novembre 1987, Université Louis Pasteur de Strasbourg.

[11] BERTHA, E.L. and CHOPPIN, G.R. : "Interaction of humic and fulvic acids with Eu(III) and Am (III)", J. Inorg. Nucl. Chem., 40 (1978), 655-658.

[12] TORRES, R.A. and CHOPPIN, G.R. : "Europium (III) and Americium (III), Stability constants with Humic Acid", Radiochim. Acta, 35 (1984), 143-148.

[13] YAMAMOTO, M. and SAKANOUE, M. : "Interaction of Humic Acid and Am (III) in Aqueous Solution", J. Radiat. Res., 23 (1982), 261-271.

[14] MOULIN, V., ROBOUCH, P., VITORGE, P. and ALLARD, B. : "Environmental Behaviour of Americium in natural waters", Radiochim. Acta, Submitted for publication

[15] OLOFSSON, U., BENGTSSON, M. and ALLARD, B. : "Formation and Transport of Americium Pseudocolloïds inaqueous systems", Mat. Res. Symp. Proc., 50 (1985), 729-736

[16] KIM, J.I., BUCKAU, G. and ZHUANG, W. : "Humic colloid generation of transuranic elements in groundwater and their migration behaviour", Mat. Res. Symp. Proc., 10 (1987) in press.

Table I - Formation constants for the americium-hydroxyde system, and americium-carbonate system from [10] at I = 0.1 M $NaClO_4$

Species	log β_1	log β_2	log β_3
$Am(OH)_i^{3-i}$	- 7.0	- 15.3	- 24.3
$Am(CO_3)_i^{3-2i}$	6.3	10.1	11.3

Table II - Formation constants for the americium-humate or fulvate system (1:1 complex)

Ligand	C^a meq/g	log β_{11} (l/eq)	log β_{11}	log β^e_{11} (l/g)	Ref
FA-GW	0.88	6.2 0.2	-	3.1	[8]
FA-SW	1.22	6.0 0.2	-	3.1	"
HA-SW	1.20	7.0 0.2	-	4.1	"
HA-LB	1.03	7.0 0.3	-	4.0	"
HA-A	(1.4)	(7.5)		(4.7)	"
HA-LB			6.83^b	4.44	[11]
HA-LB			9.26^c	6.56	[12]
HA(soil)			6.40^d	3.14	[13]

a. Capacity corresponding to a 1.1 complex ; pH 4.65, I = 0.10 M $NaClO_4$, [Am] = 10^{-5} M, [HA] = 0 to 60 mg/l

b. determined by Ion-Exchange ; pH 4.5, I = 0.10 $NaClO_4$, tracer Am, [HA] = 0 to 1 g/l ; β expressed in l/eq (H^+)

c. determined by Solvent-Extraction ; β calculated at pH 4.65, I = 0.10 M $NaClO_4$, from log β = 10.58 α + 3.84, tracer Am, [HA] = 0 to 100 mg/l ; β expressed in l/eq (COOH)

d. determined by Ion-Exchange ; pH 6.5, I = 0.10 M $NaClO_4$, tracer Am, [HA] = 0 to 27 mg/l ; β expressed in l/mole (MW)

e. β expressed in l/g (common units for comparison)

Table III - Retention of humic acid on α-alumina I = 0.1 M $NaClO_4$

[HA] mg/l initial concentration	pH 5	pH 7	pH 9
1	94	66	68
10	21	9	9

(results expressed in % of fixed humics)

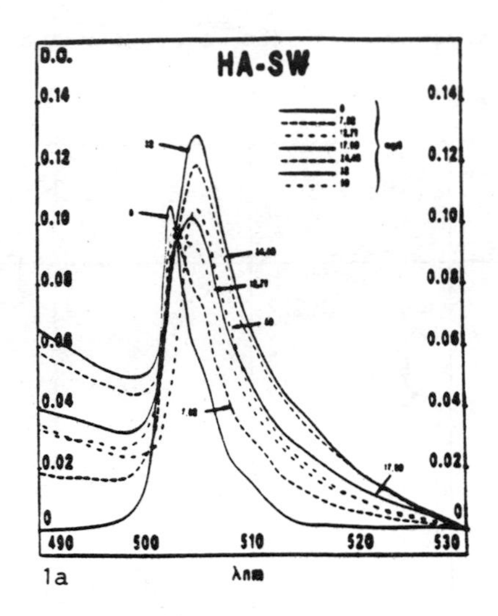

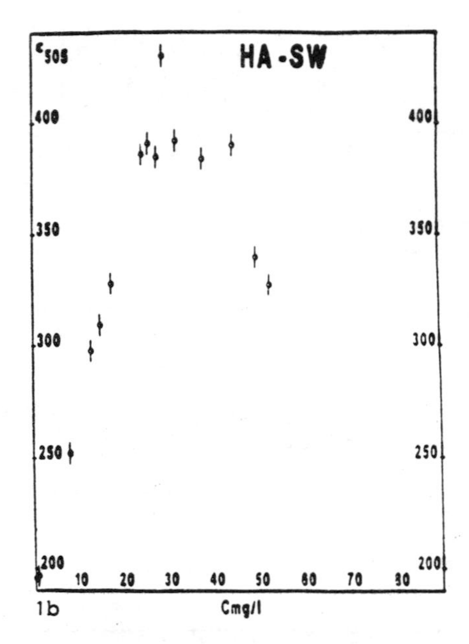

<u>Figure 1</u> - Spectrophotometric study of Am - Humate interactions
1a: Absorbance spectra obtained for increasing HASW concentrations $[Am]_T = 3 10^{-5} M$; pH = 4.65 ; I = 0.10 M $NaClO_4$

1b: Molar extinction coefficient ε (calculated at 505 nm) as a function of total humic concentration (C mg/l)

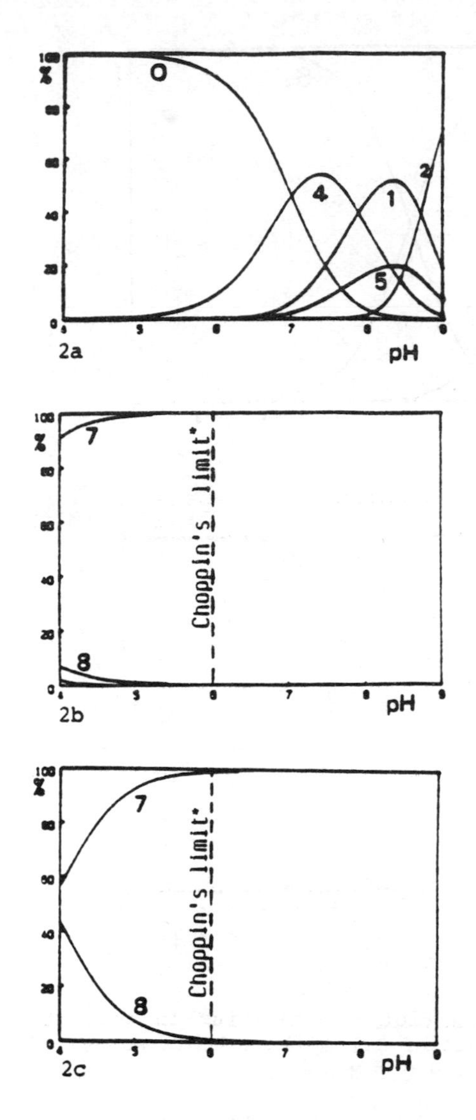

Figure 2 - Species Distribution Plot of americium in the system Am - OH - CO_3 - Humate (HALB) (*from the model used in [12,14])
I = 0.1 M $NaClO_4$; $pCO_2 = 10^{-3.5}$ atm

2a: [HA] = 0 mg/l 2b: [HA] = 1 mg/l 2c: [HA] = 10 mg/l

0 = Am^{3+} 4 = $Am(OH)^{2+}$ 5 = $Am(OH)_2^{+}$
1 = $Am(CO_3)^{+}$ 2 = $Am(CO_3)_2^{-}$
7 = Am A 8 = Am A_2 (where A is the humate ligand)

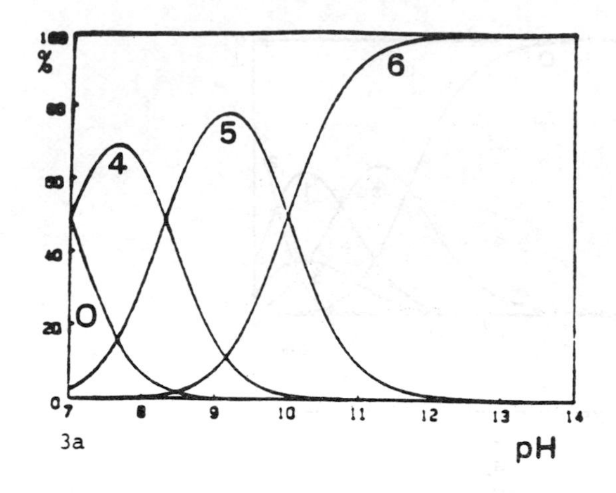

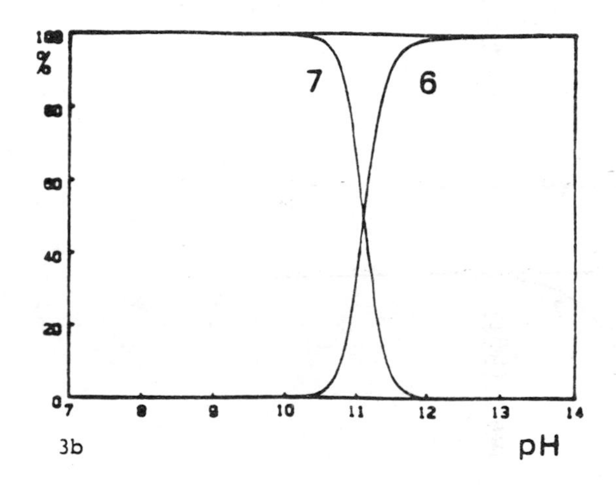

Figure 3 - Species Distribution Plot of americium in concrete intersticial solutions in presence of 0.1 mg/l of humic acids
$pCO_2 = 10^{-13}$ atm ; I = 0.1 M $NaClO_4$

3a: without humic acids - 3b : with humic acids
0 = Am^{3+} 4 = $Am(OH)^{2+}$ 5 = $Am(OH)_2^+$ 6 = $Am(OH)_3$ 7 = AmA

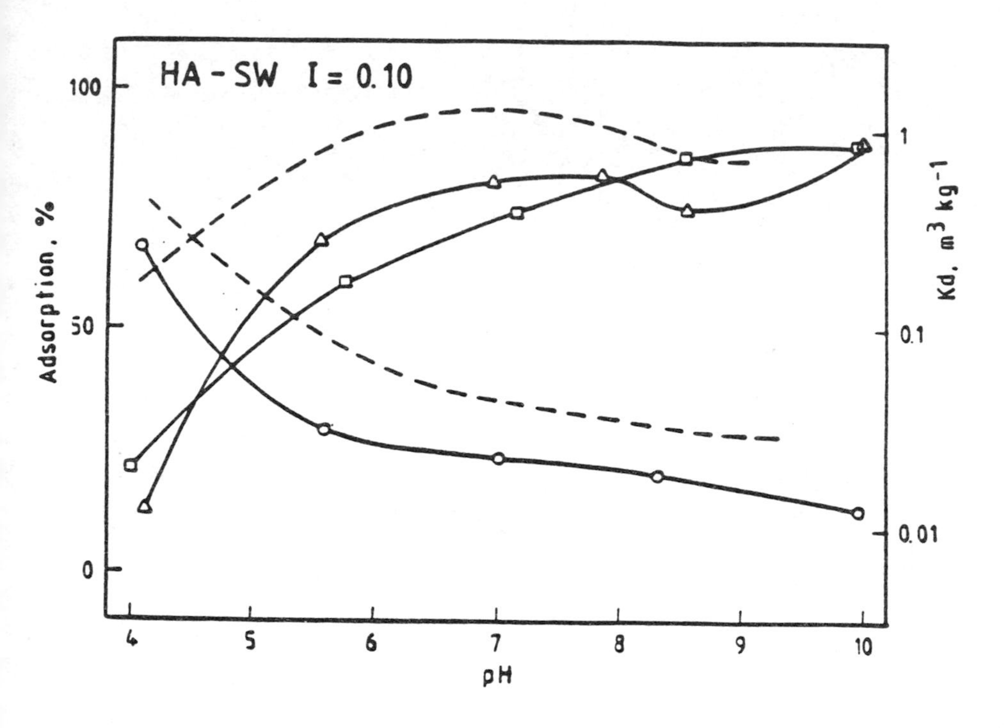

Figure 4 - Adsorption of americium on alumina as a function of pH at various concentrations of humic acids (HASW) at I = 0.1 M $NaClO_4$

□ no humics   Δ 1 mg/l humic   0 10 mg/l humic
solid line = 1 day of contact time
dotted line = 1 week of contact time

THE NEAR-FIELD CHEMISTRY OF A SWISS L/ILW REPOSITORY

Urs Berner, Janet Jacobsen and Ian G. McKinley*
EIR, Würenlingen, Switzerland
*Nagra, Baden, Switzerland

ABSTRACT

Understanding near-field chemistry is an essential component of a repository safety assessment programme as such background information is required in order to specify important parameters such as radionuclide solubilities and sorptive properties. Analysing how the near-field chemistry evolves with time requires integrating models of chemical degradation of engineered barriers and waste forms, reactive solute transport and far-field hydrology and geochemistry.

This paper concentrates on a description of one part of the chemical model - the evaluation of cement/pore water interaction. Recently a model has been developed to describe the incongruent dissolution of hydrated calcium silicates. Based on this model, the degradation of two cement modifications in pure water and in a carbonate-rich groundwater has been modelled using a static mixing tank approach. Work has also begun on coupling the dissolution model with a reactive chemical transport code, in order to investigate temporal and spatial variations in near-field chemistry. Such a development not only allows better evaluation of the temporal variation in near-field chemistry but also identifies regions of strong chemical gradients which are likely locations for perturbing processes such as colloid formation or microbial activity.

Although not discussed in detail, additional Swiss work on cement/bitumen leaching, bentonite/cement interaction, radionuclide speciation modelling and microbial perturbations will be briefly summarised.

CHIMIE DU CHAMP PROCHE D'UN DEPOT SUISSE DE DECHETS DE FAIBLE ET DE MOYENNE ACTIVITE

RESUME

La connaissance de la chimie du champ proche est un élément essentiel d'un programme d'évaluation de la sûreté d'un dépôt, car de telles informations de base sont nécessaires pour définir des paramètres importants tels que les solubilités et les propriétés de sorption des radionucléides. L'analyse de la manière dont la chimie du champ proche évolue avec le temps, nécessite l'intégration de modèles de la dégradation chimique des barrières ouvragées et des formes de déchets, du transport des solutés réactifs, ainsi que de l'hydrologie et de la géochimie du champ lointain.

La présente communication est axée sur une description d'une partie du modèle chimique -- à savoir l'évaluation de l'interaction entre le ciment et l'eau interstitielle. Récemment, un modèle a été mis au point en vue de décrire la dissolution inhomogène des silicates de calcium hydratés. Sur la base de ce modèle, on a modélisé la dégradation de deux formes modifiées de ciment dans de l'eau pure et dans de l'eau phréatique riche en carbonate en utilisant la méthode d'une cuve de mélange statique. On a également entrepris des travaux sur le couplage du modèle de dissolution avec un programme de calcul du transport chimique réactif afin d'étudier les variations dans le temps et dans l'espace de la chimie du champ proche. Un tel développement permet non seulement de mieux évaluer la variation dans le temps de la chimie du champ proche, mais également de définir des zones à forts gradients chimiques qui sont les lieux probables de phénomènes perturbateurs, tels que la formation de colloïdes ou l'activité microbienne.

Bien qu'ils ne soient pas examinés en détail, des travaux complémentaires menés en Suisse sur la lixiviation du ciment/bitume, l'interaction bentonite/ciment, la modélisation de la formation d'espèces de radionucléides et les perturbations microbiennes, seront brièvement récapitulés.

1. INTRODUCTION

From the safety assessment viewpoint, the desired output of a near-field model are profiles of the release of each significant radionuclide (with chemical speciation specified) as a function of position and time. Two main factors, however, limit development of such a model - the chemical complexity of the L/ILW near-field and its inherent heterogeneity.

In the base case of the Project Gewähr 1985 (PG '85) safety analysis /1/, the L/ILW near-field was treated in an extremely simplistic manner - containment is assumed to fail immediately on repository closure and the total inventory of waste radionuclides is assumed to be instantaneously leached. The only factor retarding transport of radionuclides from the near-field into the geosphere was sorption. Sorption was simply modelled by a 'Kd' value which was conservatively taken to be 0.1 m^3/kg for elements expected to be strongly sorbed (C, Zr, Tc, Ra, Th, Pa, U, Np, Pu, Am, Cm) and 0 for all other less-sorbed elements.

Since 1985, the main aim of the near-field chemistry studies was to develop a more realistic approach, which takes credit for the very low solubilities of some elements in such an environment and includes more reasonable sorption data. Given the requirement that the overall model must be shown to be conservative, however, this requires an extensive programme of work.

2. OVERVIEW

Swiss work related to near-field chemistry can be subdivided into a number of areas:-

a) Definition of the major element chemistry which, for the planned Swiss repository, is primarily determined by cement hydration reactions. Recent modelling work in this field is described in more detail in the following section but support laboratory studies are also planned (at the Eidgenössisches Institut für Reaktorforschung (EIR), Würenlingen). Further experimental studies of direct interaction between cement and bentonite are in progress at the Institut für Grundbau und Bodenmechanik ETH (IGB), in Zürich.

b) Specification of redox conditions. Although trapped air in the repository is expected to be quickly consumed by corroding metal, a profile of redox conditions as a function of time has yet to be modelled. It is hoped to build on existing literature studies (e.g. /2/) as part of a more general study of natural redox buffering and artificial redox control.

c) Empirical leaching studies of various waste types (real or simulated) in their appropriate immobilisation matrices (resins, bitumen and cement). Earlier studies concentrated on simply measuring radionuclide release rates by standard methods but, in the future, there will be more effort on leachate characterisation in terms of radionuclide speciation, concentration of complexants, colloids, etc.

d) Sorption studies of single radionuclides or leachates on cement/ concrete. Data available to date are limited to Cs and Sr but a more extensive programme is planned with focus on elements identified as important in the conservative PG '85 analysis.

e) Modelling of radionuclide solubility and speciation using standard chemical thermodynamic codes (MINEQL/EIR, PHREEQE). A major limitation to the applicability of such models is the poor thermodynamic database for elements of interest, particularly for the hyperalkaline conditions involved. Database expansion is underway at EIR with initial emphasis on Se, Pd and Ni /3/ and I /4/.

f) Problems areas - organic complexants, colloids and microorganisms. Although overlapping with many of the areas above, it is generally easier to consider the factors which complicate a simple "inorganic pure solution" system separately. Various literature studies of the possible production or rôle of organics in the Swiss L/ILW near-field have been produced (e.g. /5/,/6/) and further work to examine the best approach to modelling this factor is underway. This more theoretical study will be supported by laboratory work. On the colloid side, extensive development work on separation and characterisation techniques has taken place at EIR /7/ which will feed into some of the other laboratory studies described above. An extensive overview of the potential rôle of microbes in a Swiss L/ILW repository /8/ together with a long-running study of bitumen degradation at the Institut für Pflanzenbiologie at the ETH in Zürich has resulted in an integrated programme of microbiological lab studies and model development which should commence at the beginning of 1988.

g) Finally, natural analogues are used to test the applicability of some of the safety assessment models to real systems. Chemically, natural groundwaters from Oman are very similar to the porewaters expected in an aged repository. An initial study of Oman springs has shown that the existence of significant microbiological groups in hyperalkaline conditions cannot be discounted and indicated some possible problems with the existing radionuclide thermodynamic database /9/.

As insufficient space is available to describe all this work in detail, only the example of the cement degradation modelling will be considered in the next section. This choice is based not only on the intrinsic importance of such work to provide a base for the other studies, but also because this work has recently advanced considerably.

3. MODELLING THE CHEMISTRY OF CEMENT POREWATER

3.1 Degradation of hydrated cement in natural groundwaters

A normal hydrated Portland cement consists of several hydrated phases which are formed essentially from the cement components: CaO, SiO_2, Al_2O_3, Fe_2O_3, SO_3, MgO and H_2O. A general distribution for the hydrated phases in the cement is shown in Table I.

For blended cements, containing for example silica fume, slag, trass etc., the percentage content of the "main" phases may be shifted such that the $Ca(OH)_2$ - percentage is decreased and those of CSH and AFm, AFt are increased.

Recently, a model has been developed which describes the incongruent dissolution of hydrated calcium silicates /10/. Within this model the CSH-phase is represented by two independent components. The nature of these components and their solubilities are determined by one single parameter; the calcium/silicon ratio (C/S) of the CSH-phase. A short compilation of the results of this dissolution model is given in Table II.

In the present work the results given in Table II are used to model the degradation of actual cements in real groundwaters.

3.2 Description of model system

A standard Swedish Portland cement (SPP) and a blend of this cement and silica fume (80% SPP + 20% silica fume; ⟶ SIP) were chosen as the model materials. The two cement modifications (SPP, SIP) are described in detail elsewhere /11/,/12/, their compositions are given in Table III.

The cement compositions given in Table III show a simplified picture of the cements. AFm, AFt-phases and also the ferritic phases are not considered. At present, there is no dissolution model available for these phases. However, a comparison with the percentages given in Table I shows that the simplified compositions given in Table III cover 70-95% of the components of the original cement. Additionally Al and Fe concentrations are generally low in pore solutions /11/. It is therefore assumed that ignoring the Al and Fe containing phases does not essentially affect the quality of the results.

MgO is likely to be bound in hydrated Mg-silicates or Mg-aluminates, more detailed information about Mg-containing phases is not available at present. Therefore, due to the low Mg concentrations found in pore solutions, Mg-containing phases are represented by $Mg(OH)_2$. Compared with CSH-phases and with $Ca(OH)_2$, the alkali hydroxides are only present in small quantities. Nevertheless, the composition of the pore solutions may be strongly affected by the alkali hydroxides, at least during the initial stages of degradation, since they do not exhibit solubility limits.

A modified Valanginian marl groundwater was used as a model groundwater /13/ and its composition is given in Table IV. An important characteristic of this groundwater is the high HCO_3^- and F^- ion concentrations. Both of these anions can react with calcium from the cement to form new solid phases ($CaCO_3$, CaF_2). Calculations were also performed using pure water as the contacting liquid (base case calculation).

3.3 The mixing tank model

The successive degradation of a normalised quantity (1 kg) of hydrated cement is described using the following scheme:

i) Equilibrate the cement phases with a volume of groundwater equivalent to the existing porosity.

ii) Calculate the quantities of dissolved/precipitated phases. Calculate new cement compostion.

iii) Calculate the new porosity of the cement, using densities and quantities of dissolved/precipitated phases.

iv) Calculate the new solubility of the CSH-phase (according to the dissolution model; Table II) depending on the new composition. Go to step (i).

In the following, steps i) to iv) are called a cycle. The time period of a cycle is not known. It depends on several parameters and has to be determined separately, from transport calculations (see below). However, the concept of time-normalised cycles enables the influence of chemical parameters on the degradation of cement systems to be studied relative to this standard.

The above scheme (in particular step i) is applicable only for solids exhibiting a solubility limit. This means that an additional assumption is necessary for the very soluble alkali hydroxides. From pore solution analysis, it is clear that only a portion of the available alkali hydroxides is dissolved in the pore solution of freshly hydrated cement (cycles = 0). This portion depends on the cement composition, the water/cement ratio and the cation in question (Na, K) and lies between 1 and 70 % /14,15/.

It is supposed that the dissolved portion of the alkali hydroxides decreases to between 1 and 10 % during subsequent degradation (cycles $>$ 0) /16/. The present model describes the dissolution of the alkali hydroxides in the following manner: Before step i), a percentage of the remaining solid alkali hydroxide is dissolved in the groundwater. Quantities less than 0.1 % of the initial quantity are neglected.

To perform the calculations, the code DISSOLVE was developed. This code contains the geochemical speciation code MINEQL/EIR as a main subroutine. For this reason, the speciation calculations are based on the thermodynamic database MINEQL/EIR. To calculate a full chemical speciation at every cycle would be very time-consuming (several hours of CPU-time). In order to save CPU-time, a new speciation calculation is performed only if the water chemistry changes significantly during the cycle. Thereby the CPU-time is decreased to several minutes.

3.4 Results and discussion

In principle, a complete chemical speciation of the pore solution and an inventory of the remaining quantities of all solid phases is available at the end of each cycle. However, to give details of all the results of the different calculations would only serve to confuse the important issues. Therefore, only the parameters relevant for the main solid (CSH-phase), namely pH, Ca_{tot} and SiO_{2tot}, are compared in this section. Figures 1 and 2 show the evolution of these parameters during the degradation of the standard SPP cement in pure water and the Valanginian marl groundwater. Four distingishable domains of pore solution composition can be identified.

Domain 1 covers the first 30 cycles. Within this domain the pore solution is mainly composed of KOH and NaOH. The pH is high (12.5 - 13.5) and Ca_{tot} is held at a low level by the solubility of $Ca(OH)_2$ in this strongly alkaline medium. The SiO_{2tot} concentration is generally low (10^{-6}-10^{-5} M).

In domain 2 (ca. 30 to ca. 1,000 cycles) the alkali hydroxides have dissolved completely and the composition of the solution is determined by the dissolution of $Ca(OH)_2$, whose solubility constant is determined by the C/S ratio of the CSH-phase. The actual Ca_{tot} concentration also depends on the groundwater composition. For the particular groundwater example used here (Table IV), the maximum Ca_{tot} concentration is fixed at ca. 18 mmoles/l (saturation with respect to calcite; fig. 2) whereas in pure water a concentration of ca. 22 mmoles/l is calculated (fig. 1). The pH is relatively constant at approximately 12.5 ($Ca(OH)_2$ saturation), the concentration of silica remains generally low (10^{-6} - 10^{-5} M).

Within domain 3 (more than ca. 1000 cycles) the C/S ratio of the CSH-phase has decreased to C/S $\lesssim$ 1. For this reason pH and Ca_{tot}-concentration also decrease (rapidly, on the log scale) and the silica concentration strongly increases.

A peculiarity in the behaviour of the system should be noted at this point: it may happen that Ca and SiO_2 are dissolved from the solid in a ratio which is equal to C/S ratio. This results in a subsequently congruent dissolution of the remaining CSH-phase. In pure water this situation arises at C/S ratio of $\sim$ 0.79. The congruent dissolution state is never reached in the groundwater, due to its composition (HCO_3^- concentration). This also explains the significant increase of SiO_2 concentration in the case of the groundwater as compared with the pure water.

Domain 4 starts after complete dissolution of the CSH-phase. In pure water, only $Mg(OH)_2$ has to be considered and thus the pH is fixed at 10.4 during the next 10,000 cycles. The degradation in the groundwater shows a different situation. After the dissolution of the CSH-phase, the solids SiO_2(am), $Mg(OH)_2$, $CaCO_3$ and CaF_2 have to be considered, whereby $CaCO_3$ and CaF_2 have been precipitated from cement-groundwater reactions during degradation. Complete dissolution of SiO_2(am), $Mg(OH)_2$ and CaF_2 requires about 640 cycles, leaving ca. 440 g of $CaCO_3$ per kg of originally hydrated cement. This remaining $CaCO_3$ is dissolved very "slowly" in the nearly calcite-saturated groundwater.

The different solid phases as a function of the number of cycles are given in figures 3 and 4.

The parameter which best illustrates the influence of cement composition and groundwater composition on the "lifetime" (number of cycles to reach full dissolution) of the cement is pH (fig. 5).

The "lifetime" of the standard SPP cement in pure water is ca. 16,000 cycles whereas the lifetime of the blended SIP cement is ca. 24,300 cycles. There is no doubt that the increased lifetime of the SIP cement is due to its initial composition. The SIP cement contains twice the quantity of silica and ca. 10 % less $Ca(OH)_2$ than the SIP cement (Table III). The lifetime of the CSH-phase, for C/S ratios greater than unity, essentially depends on the total amount of $Ca(OH)_2$ in the hydrated cement. It takes ca. 1,000 cycles to reach $C/S \sim 1$ for both cements. The lifetime of the low calcium CSH-phase ($C/S \lesssim 1$) depends mainly on the quantity of silica available and thus the SIP cement needs more cycles to dissolve completely.

The influence of groundwater composition can be illustrated using both cements as examples. The lifetime of the SPP cement in pure water is ca. 16,000 cycles compared with the groundwater case, where pH has already decreased to the background level after only 2,200 cycles. The SIP cement however, shows a less pronounced relative decrease in lifetime (24,300 cycles in H_2O compared to 5,900 cycles in groundwater).

The reason for the shorter lifetime in groundwater is the formation of new insoluble solid phases ($CaCO_3$, CaF_2) during degradation. With each cycle an additional amount of calcium is removed from the CSH-phase through precipitation with $CO_3{}^{2-}$ and F^-. On one hand, the original CSH-phase is dissolved as it is in the pure water case and, on the other hand, an additional portion of the CSH-phase is converted to $CaCO_3$ and CaF_2 (fig. 4). The evolution of Ca_{tot}-concentration for all cases considered is given in figure 6.

3.5 Results of coupled-code calculations

The mixing tank model may be used to demonstrate **relative** differences in the evolution of different cement systems. It is not possible to model the real-time behaviour of a cement system with the mixing tank model since the cycle-time is unknown. This cycle-time would, amongst other parameters, depend on the penetration depth of the degradation front.

For this reason, the real temporal behaviour of the system was modelled using a coupled code. The reactive chemical transport code THCC /17/, a derivative of CHEMTRN /18,19/, was modified by incorporating the dissolution model previously described. The modified code is called THCCDM.

This modified code is not yet fully developed, e.g. numerical problems arise below C/S=1. Therefore, results are available only for C/S ratios above unity. These partial results, describing the degradation of the CSH-phase in pure water, are given in figure 7. The calculations are based on the following conditions:

- one dimensional calculation with constant boundary conditions
- no advection (diffusion only)
- a single solid phase (CSH)
- a single diffusion coefficient valid for all species ($5 \cdot 10^{-10} m^2/sec$)
- species considered: Ca^{2+}, $Ca(OH)^+$, $H_2SiO_4{}^{2-}$, $H_3SiO_4{}^-$, H_4SiO_4, H^+, OH^-

In figure 7, the C/S ratio of CSH-phase within the solid after 10, 50 and 100 years is given as a function of the distance from the cement surface. As already mentioned, C/S ratios less than unity are not available at present. To get the results given in figure 7, ca. **3 hours** of CPU-time were needed, and it may be estimated that the calculation of the whole profile (C/S = 0 at 1.0 cm) would take another **50 hours** of computer-time!

From figure 7, the velocity of the reaction front was estimated to be $\sim$ 0.3 mm/year, based on the position of the medium C/S ratio at different times. This is considered to be a maximum value since it should be noted that the reaction front velocity will decrease (to a constant value) with increasing penetration depth.

A second estimation is based on the corresponding mixing tank model using a mean C/S ratio integrated over the whole front and the number of cycles needed for complete dissolution. A reaction front velocity of $\sim$0.05 mm/year was estimated. For the same reasons as for the first estimation, this is also a maximum value.

3.6 Conclusions

The results show that the influence of different chemical parameters on the degradation of cement can be demonstrated using the simple mixing tank model. An important disadvantage inherent in this model is the absence of temporal scaling. To calculate temporal scaling, coupled codes have to be used. The coupled code used in the present study is too slow for use as a "production code". The need for better solutions in this field is clearly indicated.

The chemical lifetime of a cement is increased by blending the cement with stabilising silicates (i.e. silica fume). The requirement for a high pH in the near field to limit the solubilities of certain important radionuclides is only slightly affected by these blending agents. The composition of the groundwater is much more important. Groundwaters containing reactive species (especially $CO_3{}^{2-}$) may drastically decrease the lifetime of the cement. Calcium and carbonate are of major importance. High calcium concentrations increase and high carbonate concentrations decrease lifetimes. The levels of silica concentrations usually found in groundwaters do not have a significant influence.

Assuming exclusively diffusive transport of solvents, maximum dissolution rates of 0.3 and 0.05 mm/year were estimated for the SPP cement in pure water using coupled and mixing tank models respectively. However, it is supposed that a realistic dissolution rate will be even lower. A value of 0.005 to 0.01 mm/year would be comparable with results calculated in /20/.

4. FUTURE DEVELOPMENTS

As has been shown, modelling of the basic cement porewater chemistry is proceeding well. Parallel studies of redox conditions in the near-field and the rate of production and rôle of organics are at a much earlier stage but, hopefully, may eventually be integrated to form a more complete model of near-field chemistry. Given the problems highlighted above, however, coupled models of chemistry and transport seem to be prohibitively expensive in computer time without dramatic improvements in calculational efficiency.

Radionuclide speciation and retardation can, to some extent, be decoupled from the basic chemistry modelling and work is in progress to fill the current hiatus in laboratory data. Finally, the less well defined problems caused by possible microbial activity or colloids are being studied - initially with the aim of determining if they are likely to cause significant perturbations to the "inorganic, true solution" base case.

5. REFERENCES

/1/ Nagra, Project Gewähr 1985, Reports NGB 85-07, 85-08 (in German), NGB 85-09 (English summary). Nagra, Switzerland, (1985).

/2/ K.E. Heusler: "Ueberlegungen zur Thermodynamik und Kinetik der Reaktion zwischen Eisen und Wasser." Report NTB 85-22, Nagra, Switzerland, (1985).

/3/ B. Baeyens, I.G. McKinley: "A thermodynamic database for Ni, Pd and Se". Nagra report in preparation.

/4/ L. Yuan-Fang: "Migration chemistry on behaviour of iodine relevant to geological disposal of radioactive waste - a literature review with a compilation of iodine sorption data." Nagra Report in preparation.

/5/ D. Lüscher, R. Bachofen: "Mögliche Mikrobiologische Vorgänge in unterirdischen Kavernen im Hinblick auf die Endlager radioaktiver Abfälle (Literaturstudie)". Report NTB 84-07, Nagra, Switzerland.

REFERENCES (cont.)

/6/ B. Allard: "Organic complexing agents in low and medium level radioactive waste". Report NTB 85-19, Nagra Switzerland.

/7/ C.A. Degueldre, B. Wernli: "Characterization of the natural inorganic colloids from a reference granitic groundwater." Analytica Chimica Acta, 195, (1987), 211-223.

/8/ H.A. Grogan (ed.): "The significance of microbial activity in a deep repository for L/ILW." Internal report NIB 87-05, Nagra, Switzerland.

/9/ A.H. Bath et al.: "Trace element and microbiological studies of alkaline groundwaters in Oman, Arabian Gulf - a natural analogue for cement pore water." BGS Report FLPU 87-2/Nagra Report NTB 87-16.

/10/ U. Berner: "Modelling the incongruent dissolution of hydrated cement minerals". Radiochimica Acta, Proc. Int. Conf. "Migration 87", Munich Sept. 1987, in press.

/11/ K. Andersson: "Transport of radionuclides in water/minerals systems". Chalmers University of Technology, Goeteborg, Thesis (1983).

/12/ U. Berner: "Modelling porewater chemistry in hydrated portland cement." Mat. Res. Soc. Symp. Proc., 84, (1987), p. 319.

/13/ Nagra, Switzerland: unpublished data from Nagra-databank (1987).

/14/ F.P. Glasser et al.: "The chemical environment in cements", Mat. Res. Soc. Symp. Proc., 44, (1985), p. 849.

/15/ U. Berner: "Radionuclide speciation in the porewater of hydrated cement; I. The hydration model." EIR Internal Report TM 45-86-28, Würenlingen (1986).

/16/ F.P. Glasser, University of Aberdeen, personal communication, (1987).

/17/ C.L. Carnahan: "Simulation of uranium transport with variable temperature and oxidation potential; The computer programm THCC." Mat. Res. Soc. Symp. Proc, 84, (1987), p. 713.

/18/ C.W. Miller: Lawrence Berkeley Laboratory Report No. LBL-16152, (1983).

/19/ C.W. Miller, L.V. Benson: Water Resour. Res., 19, (1983), p. 381.

/20/ L.O. Höglund: "Degradation of concrete in a LLW/ILW repository." Nagra Technical Report NTB 86-15, (1987).

Table I Simplified distribution of phases in hydrated Portland cement (w/w).

Hydrated calcium silicates (CSH-phase)	40 - 50 %
Portlandite ($Ca(OH)_2$)	20 - 25 %
AFm, AFt-type phases, ferritic phases	10 - 20 %
Pore solution	10 - 20 %
Minor components (KOH, NaOH, $Mg(OH)_2$ ...)	0 - 5 %

Table II The incongruent dissolution behaviour of hydrated calcium silicates. The table gives the independent model components and their solubilities as a function of C/S ratio as described in /10/.

C/S range	model component	solubility product of model component as a function of C/S ratio
$0 < C/S \leq 1$	SiO_2	$\log K_{SO} = - 2.04 + 0.792/ (C/S - 1.2)$
	CaH_2SiO_4	$\log K_{SO} = - 8.16 - (1-C/S) \cdot (0.78 +0.792/ (C/S-1.2))/(C/S)$
$1 < C/S \leq 2.5$	$Ca(OH)_2$	$\log K_{SO} = - 4.945 - 0.338/ (C/S-0.85)$
	CaH_2SiO_4	$\log K_{SO} = - 8.16$ (constant)
$C/S > 2.5$	$Ca(OH)_2$	$\log K_{SO} = - 5.15$ (constant)
	CaH_2SiO_4	$\log K_{SO} = - 8.16$ (constant)

Table III Composition of cements used as model materials (from /11,12/). Values are given as mmoles solid/kg of hydrated cement.
SPP: Standard Swedish Portland Cement
SIP: 80% SPP + 20% silica fume

	SPP	SIP
$Ca(OH)_2$	470	-
CSH-phase	2277 (C/S = 2.5)	4391 (C/S = 1.25)
$Mg(OH)_2$	562	505
KOH	156	150
NaOH	43	48
Density	1940 kg/m^3	2100 kg/m^3
Porosity	27.5 %	18.6 %

Table IV Composition of model groundwater. The composition is that of a modified Valanginian marl groundwater /13/. Ion concentrations are given in mmoles/litre.

Component	total concentration	Component	total concentration
Ca^{2+}	0.024	CO_3^{2-}	14.76
Mg^{2+}	0.045	SO_4^{2-}	1.19
Sr^{2+}	0.0043	Cl^-	0.43
K^+	0.18	F^-	1.05
Na^+	19.21	Br^-	0.01
H^+	13.96	NO_3^-	0.1
		$H_2SiO_4^{2-}$	0.00015

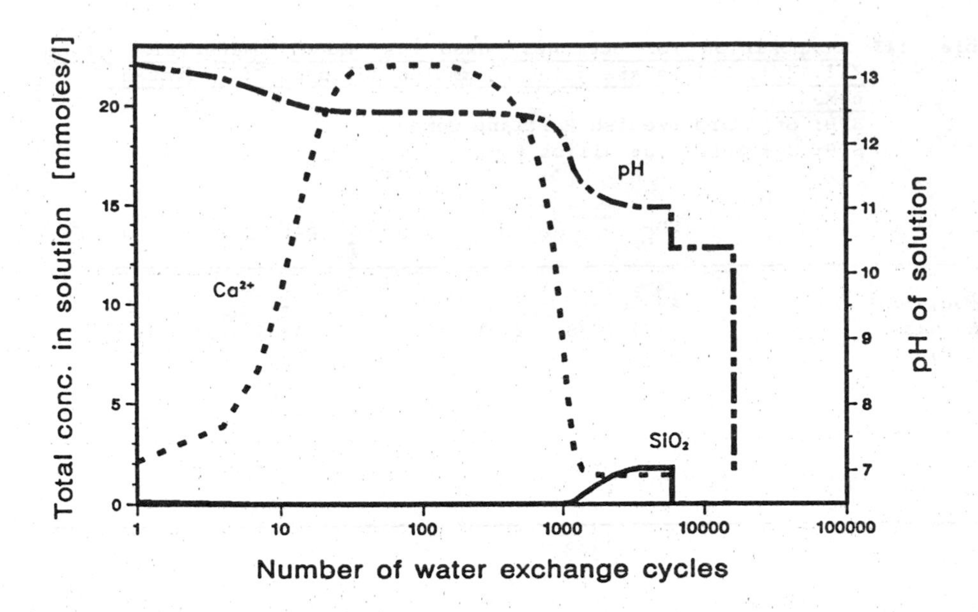

Figure 1: Evolution of pH, Ca_{tot} and SiO_{2tot} during degradation of standard Portland cement in pure water.

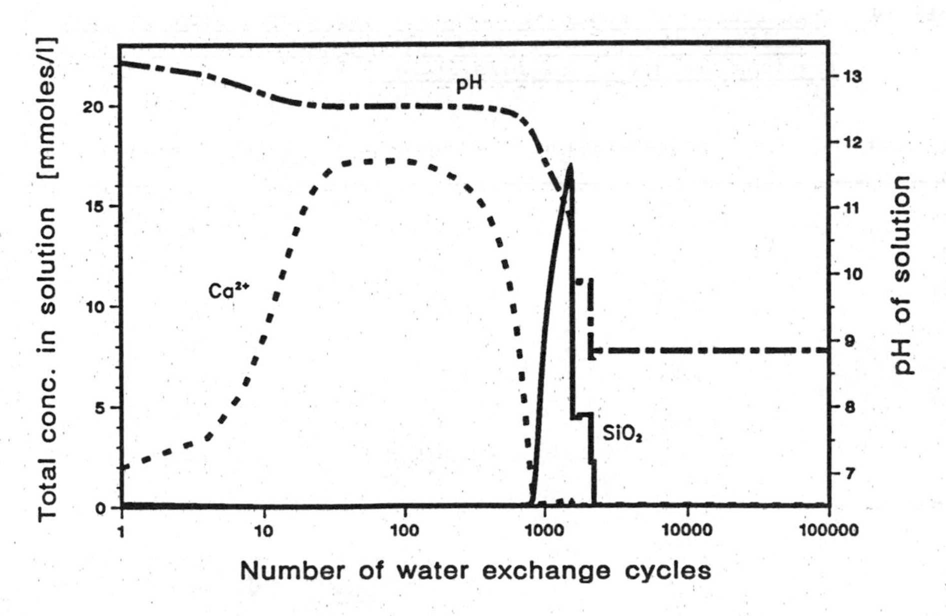

Figure 2: Evolution of pH, Ca_{tot} and SiO_2 during degradation of standard Portland cement in a modified Valanginian marl groundwater.

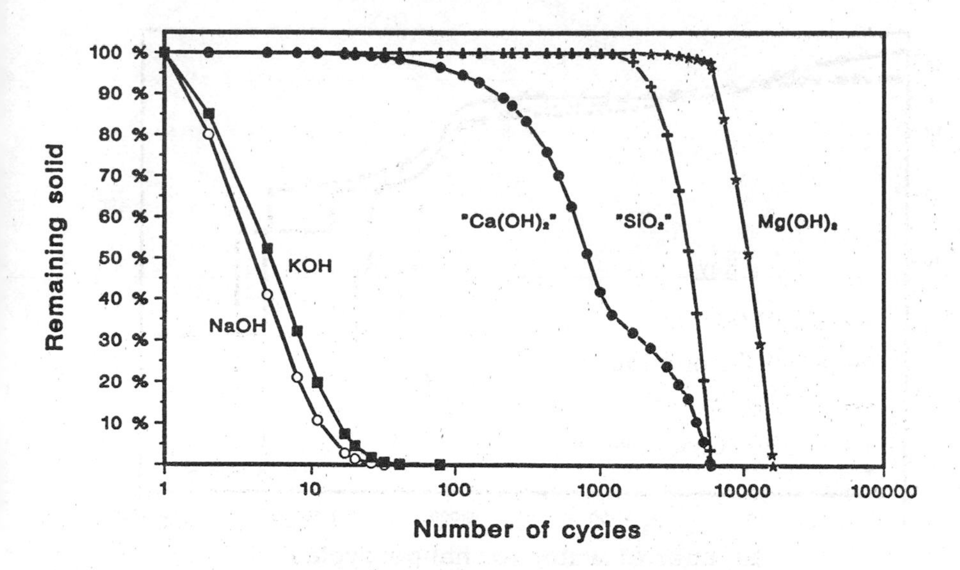

Figure 3: Distribution of solid phases during degradation of standard Portland cement in pure water. Percentages are based on initial quantities.

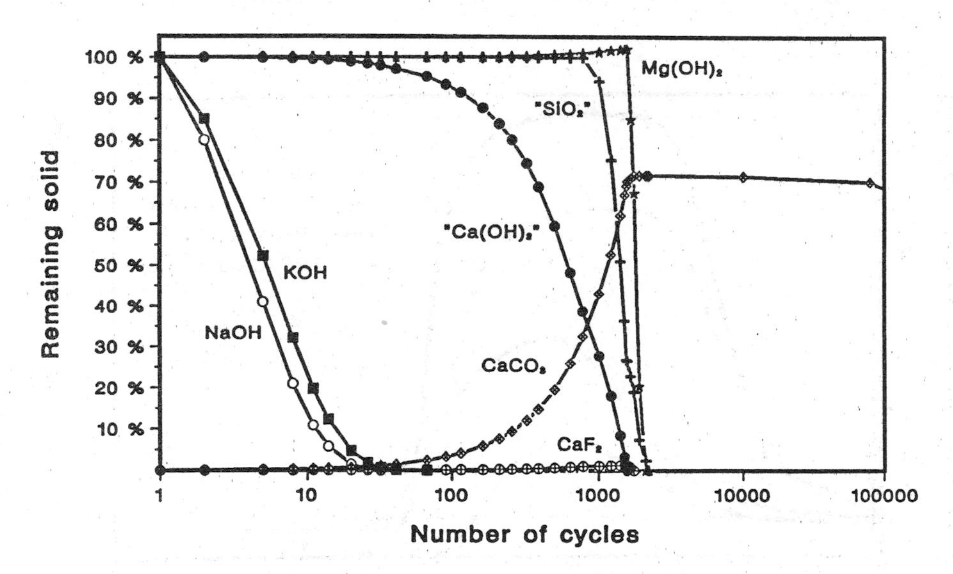

Figure 4: Distribution of solid phases during degradation of standard Portland cement in a modified Valanginian marl groundwater. Percentages are based on initial quantities. Percentages of precipitated solids ($CaCO_3$, CaF_2) are based on initial $Ca(OH)_2$ content.

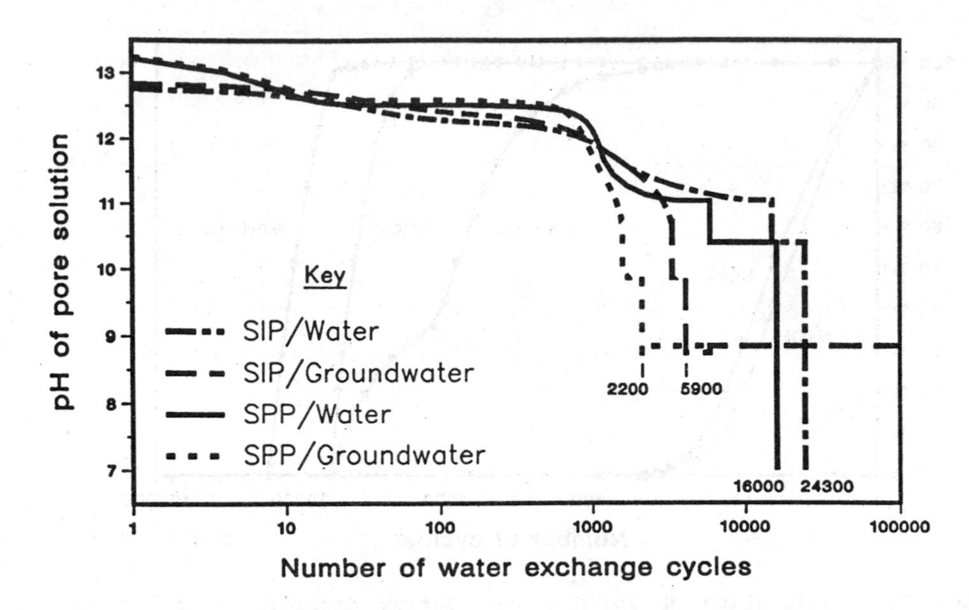

Figure 5 Influence of cement composition and groundwater composition on the "lifetime" of the hydrated cement. The pH of the pore solution is given as a function of the numbers of exchange cycles.

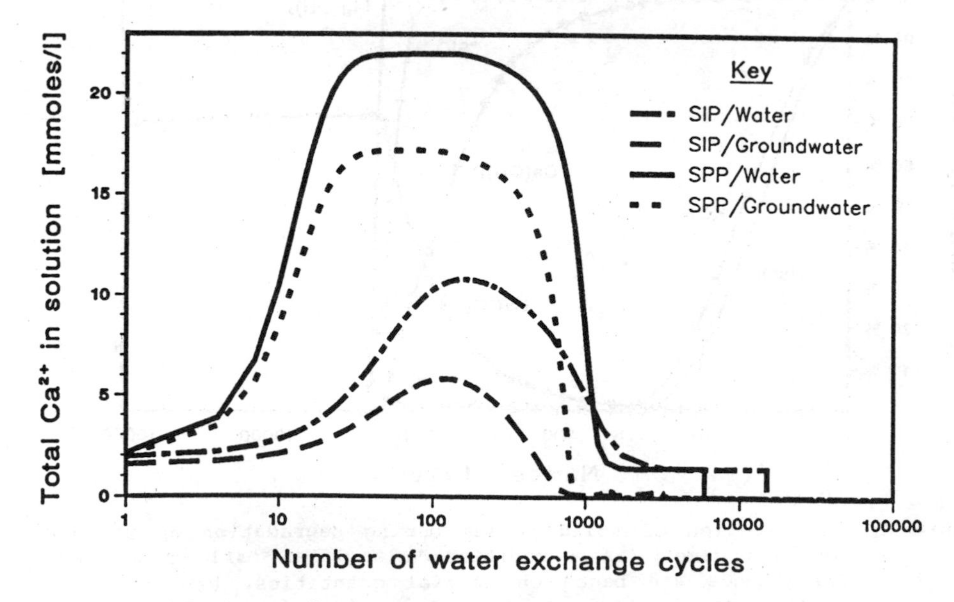

Figure 6: Influence of cement composition and groundwater composition on the Ca_{tot} concentration of the pore solution.

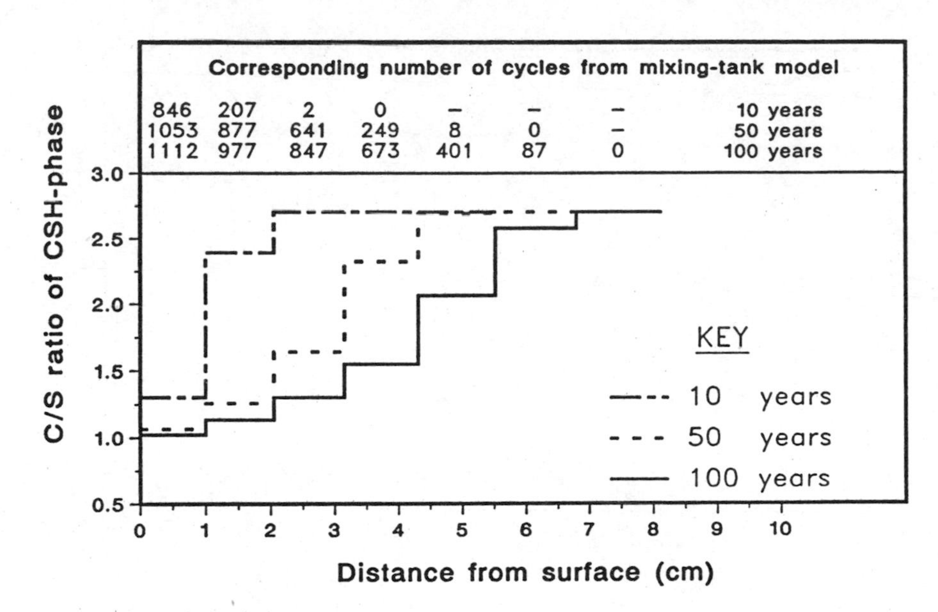

Figure 7: The degradation of CSH-phase in pure water. Results from coupled-code calculations. The calculations are based on a single diffusion process (diffusion coefficient $5 \cdot 10^{-10} m^2/sec$).

SESSION IV

NEAR-FIELD TRANSPORT MODELLING

SEANCE IV

MODELISATION DU TRANSPORT DANS LE CHAMP PROCHE

Chairman - Président

D. HODGKINSON

(United Kingdom)

MODELLING OF PHYSICAL AND CHEMICAL EFFECTS ESSENTIAL TO THE RELEASE OF RADIONUCLIDES FROM A REPOSITORY IN A SALT DOME

R. P. Hirsekorn
Gesellschaft für Strahlen- und Umweltforschung mbH München
Institut für Tieflagerung
Theodor-Heuss-Strasse 4
D-3300 Braunschweig, Germany F.R.

ABSTRACT

The modelling of effects which determine the release of radionuclides from a LLW-repository in a salt dome is presented and simplifying assumptions which have to be made to obtain results within an acceptable computation time are discussed. Physical and chemical effects are covered which are essential to the transport of radionuclides within the repository. The following three groups of effects are discussed in more detail:
Mobilization of radionuclides due to corrosion of the waste canisters and degradation of the waste matrix, retention of radionuclides in the repository due to solubility limits, transport of radionuclides by convective exchange processes and expulsion of contaminated brine by the creep of the rock. Results from performed assessments for a reference site are given to illustrate the effectiveness of the described effects.

MODELISATION DES EFFETS PHYSIQUES ET CHIMIQUES ESSENTIELS POUR LE REJET DE RADIONUCLEIDES A PARTIR D'UN DEPOT DANS UN DOME DE SEL

RESUME

L'auteur présente la modélisation des effets qui déterminent le rejet de radionucléides à partir d'un dépôt de déchets de faible activité dans un dôme de sel et examine les hypothèses simplificatrices qui doivent être formulées pour obtenir des résultats dans un temps de calcul acceptable. L'étude couvre les effets physiques et chimiques qui revêtent une importance cruciale pour le transport de radionucléides à l'intérieur du dépôt. Les trois groupes d'effets suivants sont examinés plus en détail : mobilisation des radionucléides due à la corrosion des conteneurs de déchets et à la dégradation de la matrice des déchets, rétention des radionucléides dans le dépôt due aux limites de solubilité, transport des radionucléides par des processus d'échanges convectifs et expulsion par le gonflement de la roche de la saumure contaminée. Les résultats des évaluations effectuées dans le cas d'un site de référence sont présentés afin d'illustrer l'efficacité des phénomènes décrits.

Low level radioactive waste (LLW) disposal concept in a salt dome

In the LLW disposal concept [Sto85] the wastes are conditioned by different methods (cementation, bitumisation, pressing) and sealed in containers. These containers are normally thin-walled, generally their lifetimes are expected to be below 100 years. The containers are disposed of in large chambers constructed in salt mines. After backfilling the chambers, the whole mine is backfilled and various sealings and dams are built to hinder a flow of brine accidentally intruding (cf. altered evolution accident scenario). In Fig. 1 the principle of the construction of a repository (including MLW and HLW disposal boreholes on the right) is given. Here the repository is reduced to representatives for the different types of waste locations. On the left side of the repository disposal chambers contain LLW and MLW without heat production, while on the right side heat producing MLW and HLW are stored in boreholes of 300 m depth measured from the drift level.

After disposal and backfilling the convergence process of the salt rock reduces the remaining void volume (cf. modelling of the convergence process). If a porosity of $n_f = 0.001$ is reached we consider a corresponding part of the mine as closed by convergence, which means that there is no more permeability for intruding brine. The value of $n_f = 0.001$ is chosen, for it is the average porosity of natural German halite. It is checked that using smaller values has no significant influence on the resulting dose rates.

Normal evolution scenario

In the normal evolution of a repository in a salt dome the residual voids in drifts are closed by convergence within a time period of 550 years at latest (cf. Table I). From this time on brine intrusion through these drifts is impossible. On the other hand, diapirism may lead to an uplift of the repository and the disposed waste canisters may be laid open due to subrosion. Since the repository is built deep inside of the salt dome, this, in general, does not lead to serious consequences for future generations.

Table I: Typical time ranges [a] for volume reduction by convergence.

Segments	Porosity at disposal time	f_r	Time [a] to reach a porosity of 0.100	0.010	0.001
Drifts	0.30	1	40	300	550
Sealings	0.05	1	--	200	450
Chambers	0.27	0.1	310	2800	5300
Chambers	0.32	0.1	480	3200	5800
Chambers	0.33	0.1	520	3300	5900

Altered evolution accident scenario

An accident scenario which can lead to the transport of radionuclides into the biosphere has been selected to demonstrate the methodology developed so far. In this scenario it is expected that a pathway for groundwater from the aquifer system of the overlying strata down to the mine is opened and brine can intrude. The pathway is supposed to lead to the central field of the mine (cf. Fig. 1). Here brine enters the mine and slowly tries to fill the whole residual void volume. On the way to do so the brine is hindered and delayed by dams, sealings and the compacted backfill. Finally, the brine reaches the waste packages. The canisters can be destroyed, the radionuclides can be mobilized and the brine gets contaminated.

Another possible mechanism for brine intrusion is from a brine pocket which may be in the neighbourhood of the repository. But only a very large brine pocket would be able to inundate the rather large chambers. If this brine inclusion is rather compact and near to the repository, it will be detected before diposal of wastes. Intrusion from a flat and extended brine pocket into the LLW wing has not been considered up to now.

After the residual void volume of the mine is inundated, the continuing convergence process, which has a much lower rate now (cf. modelling of convergence rate), squeezes out the contaminated brine. With this flow of brine radionuclides are transported to the overlying strata. Other mechanisms of transport like diffusion and convection are in general small compared to the flow due to convergence (cf. modelling of convective exchange processes).

In Table II some typical time ranges of flooding of the different parts of the repository are given.

Through the overlying strata the radionuclides can be transported via the deep groundwater to the biosphere and may affect humans living in that area.

Table II: Time ranges [a] for flooding different parts of the repository.

Segments	Intrusion time [a]	
	Begin	End
Central field	40	70
Flank drift	70	100
Disposal drift	100	140
Disposal chambers	140	500

Basic assumptions for the repository model

The numerically necessary simplifications in modelling the repository consist of

a. the time discretisation with parameter values kept temporally constant within one time step,
b. the division of the repository into segments with parameter values kept spatially constant within one time interval.

Nuclide transport through the segments of the repository is described with mathematical models, i.e. segment models. Calculation of physical and chemical effects as convergence of void volume, flow of intruding brine, increase of brine pressure, corrosion of canisters, degradation of waste matrix, precipitation of elements and radionuclide transport mechanisms are described with effect models. For a repository in a salt dome sorption of radionuclides at the surface of the backfill or on degradation products of canisters and conditioning materials has not yet been taken into account.

Procedure in the case of these models relies on discrete time steps. The size of such a time step is determined either by relative changes in physical parameters, such as brine flow or nuclide flux, or by specific time points which should be aimed at. These are, for example, the time of waste emplacement, time of brine intrusion, time of filling-up a segment or reaching final porosity.

Some of the individual physical and chemical effects treated in the segment models are described in detail in the following, where we focus on the LLW wing, i.e. drifts and chambers placed on the left in Fig. 1.

Mobilisation

Break-down of the canisters

The canisters of the waste packages can be destroyed either by mechanical load due to the convergence of the host rock or by corrosion due to intruded brine. Conservatively, we model the break-down with a constant rate, starting when the brine reaches the waste packages and ending after a predefined maximum lifetime of the canisters. Generally, the maximum lifetimes are between 10 and 100 years. After the break-down of a canister the brine starts to affect the conditioned waste.

Degradation of cemented wastes

For cemented wastes a constant reaction rate for degradation is used which can be element specific. The reaction rate is chosen in a way that after a pre-defined (element specific) maximum time the degradation is completed. Generally, these maximum degradation times are for cemented waste packages between 10 and 500 years. In Fig. 2 the relative reaction rates and the cumulated reaction rates are plotted versus time.

Parameter variations have shown that, due to the long period required for the brine to be squeezed out, these reaction rates do not have a serious influence on the resultant dose rates in the overall calculation.

Resulting leaching rates

The nuclides are assumed to be distributed homogeneously in the conditioning materials. Thus, in a first step, they are released parallel to the degradation. Afterwards their concentration may be reduced by sorption (not considered so far) or due to element specific solubility limits. The modelling of this effect is sketched below. The resulting concentration gives the amount of nuclides which is mobile in the sense of the modelling of the whole system.

New leaching model for cemented wastes

In the above described degradation model the reaction rates do not depend on the amount of intruded brine. This seems to be very unrealistic, since in realistic cases the amount of brine is small compared to the amount of conditioning materials. In a new model (which is under preparation and of which numerical results are not yet available) the first step is a degradation of cement again, but this is controlled by the available amount of magnesium in the brine, which is assumed to replace the calcium of the cement. After the magnesium is used up the degradation comes to an end and the release of radionuclides is by diffusion only. Generally, the release by diffusion is about three orders of magnitude smaller than the release by degradation.

Convergence of void volume

Convergence rate

The decrease of the volume of cavities and voids due to convergence of surrounding salt rock is approximated by

$$\frac{d}{dt} V(t) = -K(p,n,T;t)\, V(t). \qquad (1)$$

Here, K is the normalized convergence rate which depends on the hydrostatic pressure p, the porosity of sealing and backfill materials n, and in general on the temperature T. All these parameters vary with time and are functions of space. Therefore, K is an implicit function of space and time. An explicit time dependence has not been considered.

Following the assumptions leading to space discretisation, for each segment one has a convergence rate determined by the hydrostatic pressure, the porosity, and the temperature which are obtained for this segment. In general, one may have different values of K for the sealing, for the backfill and eventually another value for voids in the canisters.

Neglecting a temperature dependence of K for the LLW wing the complex dependence of K on the parameters p, and n is simplified to

$$K(p,n;t) = K_r\, f_r\, f_1(p(t))\, f_2(n(t)) , \qquad (2)$$

where K_r is a reference (initial) value for p = 0 and n = 0.3 (reference value) derived from experiments and numerical calculations, f_r is a reduction factor (cf. Table I) used to modify the convergence of the rock surrounding chambers which contain waste canisters and backfill. It is set to 1 after contact with brine. The factor $f_1(p)$ accounts for the pressure dependence of the convergence rate

$$f_1(p) = (1 - p/p_p)^m \qquad (3)$$

with p_p the petrostatic pressure. The exponent m is obtained from theoretical investigations of the convergence of cylindrical and spherical voids using the constitutive law for secondary creep of rock salt. This formula is supplied by laboratory experiments with macroscopic void spaces.

The supporting capacity of backfill or sealings is given by the factor f_2. This is obtained from theoretical investigation of the convergence of the pores of backfill with large porosity, taking into account the microscopic geometric properties of the backfill grains [Ste85]. Conservative generalization of this result yields f_2 as a function of porosity n

$$f_2(n) = n(1-n/n_r) \left[(1-n/nr)^2 + [n(1-n/n_r)]^{1/m}\right]^{-m} , \qquad (4)$$

where $n_r = 0.3$ is the reference porosity. For drift backfill this is the initial value n_i of the porosity at the time of backfilling. For sealings the initial value $n_i = 0.05$, i.e. a sealing of salt concrete is modelled as if it were made of backfill material which has had enough time to converge to a porosity of $n = 0.05$. Chambers have initial porosities of 0.27 to 0.33.

Reduction of cavities and voids

The convergence of voids is calculated by integrating Eq. (1) from t to t+dt. In general K is a slowly varying function of time, i.e. if the time interval dt is small enough K can be assumed to be a constant. An appreciable change of K in most cases occurs due to a jump of the pressure p(t) (cf. below). Therefore, the assumption of smoothness of K still leads to acceptable results if the end of the time interval is at that point where p(t) jumps, the next following time interval then should be very small. Hence, for a numerical calculation the simplified formula

$$V(t+dt) = V(t) \exp[-K(t)\, dt] \qquad (5)$$

is used. Cross sections A of sealings and drifts are obtained with the assumption that their lengths L are conserved. The reduction of height and width of a drift is calculated within the assumption that the absolute change of both quantities is equal.

Bearing in mind that the volume of the solid component of a converging cavity is conserved, the porosity n(t+dt) can be calculated from n(t) by

$$n(t+dt) = 1 - [1-n(t)] \exp(K\, dt) \; . \qquad (6)$$

Fig. 3 shows the porosity of different locations as a function of time.

Transport of brine

Resistance of segments

Dams, sealings, and backfill of crushed salt offer resistance to the flow of brine within the repository. Their resistances R [MPa a/m^3] are calculated from their permeability k [m^2] according to the relation

$$R = \mu L/(A\, k) \; , \qquad (7)$$

where μ [MPa a] is the dynamic viscosity, L [m] the length and A [m^2] the cross section of the resistance.

Dams are assumed to have a constant permeability k_D. For sealing and backfill material the permeability is calculated from the porosity by the equation

$$k = f_p \; c \; n^q \quad , \tag{8}$$

with c and the exponent q being empirical constants obtained from experimental data.

The variation factor f_p for the permeability makes the sealing and backfill more permeable in the fill-up phase, since a settling of the backfill materials during the first contact with brine cannot be excluded.

If a sealing or backfill has attained a porosity n_f (final porosity), the permeability of the resistance is assumed to be as low as that of solid rock salt. The segment in question no longer allows penetration of brine. Since n_f is not very well determined a priori, it has been considered as a numerical cut-off parameter. Local sensitivity studies show that values below $n_f = 0.001$ have nearly no influence on release of radioactivity. This value has, therefore, been chosen for all applications.

Calculation of brine pressure

In each segment the voids which serve as a reservoir for inflowing brine and the flow resistance are placed in series, with the resistance on the outside and the void space on the inside. Consequently, along a sequence of segments, resistances and voids alternate.

In the case of water intrusion through the central field the void volumes of the segments are successively filled with brine following the tree structure of the drift system (cf. Fig. 1). Whenever a segment is completely filled, the hydrostatic pressure p of the brine jumps from zero (neglecting atmospheric pressure) to several MPa. If a following inner segment is filled up, the pressure rises again by a certain amount. Finally, if the void volume of the drift system is completely filled, brine is squeezed out against the hydrostatic pressure. Now p attains values between the hydrostatic pressure p_H at the outside of the central field and the petrostatic pressure p_P.

The dynamics of the brine intrusion is determined by the resistances of the segments and the pressure gradient. In the present modelling the flux $\dot{V}$ through the resistance R is a quasi steady state flow with a linear dependence of the difference of pressure on both sides of the resistance (Darcy's law),

$$\dot{V} = R^{-1} \; (p - p_{ex}), \tag{9}$$

where p is the brine pressure of the segment under consideration and p_{ex} that of the outer segment.

In addition, in a segment filled with brine the convergence of the voids leads to an extra flux, the convergence flux $\dot{V}_c$. In the present modelling

$$\dot{V}_c = B(V,n,T) \; (1-p/p_P)^m \quad , \tag{10}$$

which is only a different writing of Eq.(1) with

$$\dot{V}_c(t) = - \frac{d}{dt} V(t) \quad . \tag{11}$$

The factor B is a function of the actual volume filled with brine, and the porosity n defined by

$$B(V,n) = K_r \, f_r \, f_2(n) \, V(t) \tag{12}$$

which is the pressure independent part of the convergence rate times the volume. For each segment one has conservation of total flow

$$\dot{V}_{ex} = \dot{V}_{in} + \dot{V}_c \ , \tag{13}$$

where the outgoing flow $\dot{V}_{ex}$ (directed to the outer segment when positive) is given by Eq.(9) replacing $\dot{V}$ by $\dot{V}_{ex}$.

The incoming flux $\dot{V}_{in}$ can be a sum of several fluxes, if the considered segment has several connections to neighbouring inner segments (treelike structure of the drift system). In analogy with Eq.(9) one has

$$\dot{V}_{in} = \sum_i R_i^{-1} (p_i - p) \ , \tag{14}$$

where the sum runs over all neighbouring inner segments i, which are filled with brine.

Writing equation (13) explicitly as a function of p yields a system of nonlinear equations for the brine pressure in the drift system. For a segment with index j one has

$$B_j \, (1-p_j/P_P)^m + R_j^{-1} \, (p_{k(j)} - p_j) + \sum_{i(j)} R_{i(j)}^{-1} \, (p_{i(j)} - p_j) = 0 \tag{15}$$

where k(j) is the index of the neighbouring outer segment and i(j) that of the neighbouring inner segments with respect to j.

Since the convergence flow $\dot{V}_c$ differs from zero only for segments which are completely filled with brine, equations (15) are applicable for these segments only. Because the void volume V, its cross section A, and the porosity n decrease with time due to convergence, the parameters B(V,n) and $R^{-1}(A,n)$ are varying with time. The solutions of system (15), therefore, yield the pressure in all filled segments at the time t. All empty segments have brine pressure p = 0. For segments which are in the fill-up phase the pressure is the hydrostatic pressure of the intruded brine which can be neglected compared to the brine pressure of several MPa in a filled segment.

Due to the fact that successively more and more segments are filled with brine, the dimension of system (15) changes with time accordingly. On the other hand, if sealing and/or backfill of a segment reach their final porosity n_f, the respective segment and all inner segments are locked and thus excluded from Eq. (15).

With the brine pressure calculated, the flows $\dot{V}_{ex}$, $\dot{V}_{in}$, and $\dot{V}_c$ are then obtained by Eqs. (9), (10), and (14), respectively. Examples of $\dot{V}_{ex}$ are given in Fig. 4.

Numerical algorithm

For each time step system (15) is solved by Newton's iteration with the brine pressure $p_j(t-dt)$ of the previous time step as a starting value. Thus the number of iterations depends on the variation of brine pressure with time. Since the time steps are controlled e.g. by the convergence rate which determines the parameters B and R^{-1} in Eq.(15) rapid variations of brine pressure do not occur, except if an additional segment is linked to the system. But then the starting value for the additional segment is calculated seperately and in any case the number of iterations will be small.

Transport via brine expulsion by convergence

The activity flux due to convergence is generally given as the product of the flux of brine $\dot{V}$ and the concentration of radionuclides

$$\dot{N}_c = \dot{V} \sum_i C(i) , \qquad (16)$$

with C(i) the concentration of the ith nuclide in that segment, where the flow comes from.

Hence, if the flow of brine $\dot{V}_{ex}$ (Eqs.(9),(13)) from the segment to the neighbouring outer segment is positive ($\dot{V}_{ex} > 0$) the flux of activity into the same direction is given by

$$\dot{N}_c = \dot{V}_{ex} \sum_i C(i) \qquad (17)$$

with C(i) the concentration of radionuclides in the segment under consideration. If, on the other hand, the flow of brine $\dot{V}_{ex}$ is negative, the flux of activity from the neighbouring outer segment into the segment is

$$\dot{N}_c = \dot{V}_{ex} \sum_i C_{ex}(i) \qquad (18)$$

with $C_{ex}(i)$ the concentration of radionuclides in the outer segment.

Convective exchange processes

There are several processes and effects which may lead to an exchange of brine. Examples are a flow of gas produced by corrosion or radiolysis and a spatial variation of brine density due to a gradient of cement concentration or temperature. If the brine is contaminated these exchange processes lead to transport of radionuclides.

Hence, a gradient of concentration of radionuclides along the drifts is generated leading to an additional transport mechanism, the diffusive transport of radionuclides. It turns out that, although much smaller, the diffusive transport is the next most important transport mechanism of radionuclides besides the forced convection due to convergence.

The modelling of the diffusive transport assumes a quasi-stationary gradient of concentration, i.e. during one time step for each segment the gradient of concentration of radionuclides is assumed to be a constant. With the space discretisation the flux of activity $\dot{N}_D$ forced by diffusion is given by

$$\dot{N}_D = - D(T) \; A \; n \; (C-C_{ex})/L \qquad (19)$$

with L the length and C the nuclide concentration of the segment and C_{ex} the nuclide concentration of the outer segment. Here D is molecular diffusion constant (temperature dependence neglected).

Superposition of transport mechanism and applications

The total transport of radionuclides is obtained by superposing the transport processes mentioned above. Since these transport mechanisms are very different, the real superposition is very difficult to be obtained. In general, as mentioned above, exchange processes are of minor importance for the transport of radionuclides. Therefore, a simplified modelling is used, i.e. a linear superposition of the flux of activity caused by convergence of the voids and those due to convective exchange processes and diffusion. In general, only one of the exchange processes dominates. Thus summation of nuclide fluxes instead of the correct superposition yields acceptable results. Adding the exchange flux and the convergence flux gives a conservative approximation

$$\dot{N} = \dot{N}_C + \dot{N}_D + \text{additional fluxes.} \qquad (20)$$

Here $\dot{N}_C$ is either given by Eqs.(17) or (18).

Precipitation of elements

The flux of activity $\dot{N}$ is reduced if radionuclides exceed solubility limits. Therefore for all segments of the repository element specific solubility limits have been considered. The solubility limits strongly depend on the chemistry of the solution in the respective segment. Since the chemistry and its temporal evolution in the segments is not very well known, constant solubility limits in the whole repository have been applied.

The transport of radionuclides out of a segment is determined by the dissolved nuclides only.

Release of radionuclides from a repository is then obtained calculating the flow of contaminated brine which comes out of the central field of the repository. It turns out that the most important effect is the convergence of the salt rock. Most important parameters are e.g. permeability of dams and sealing, the convergence rate and the time of brine intrusion into the central field.

References

[Ste85] Stelte, N.: In PSE Final Report, Vol. 15, No. 3, HMI, Berlin 1985
[Sto85] Storck, R.: PSE Final Report, Vol. 16, HMI, Berlin 1985

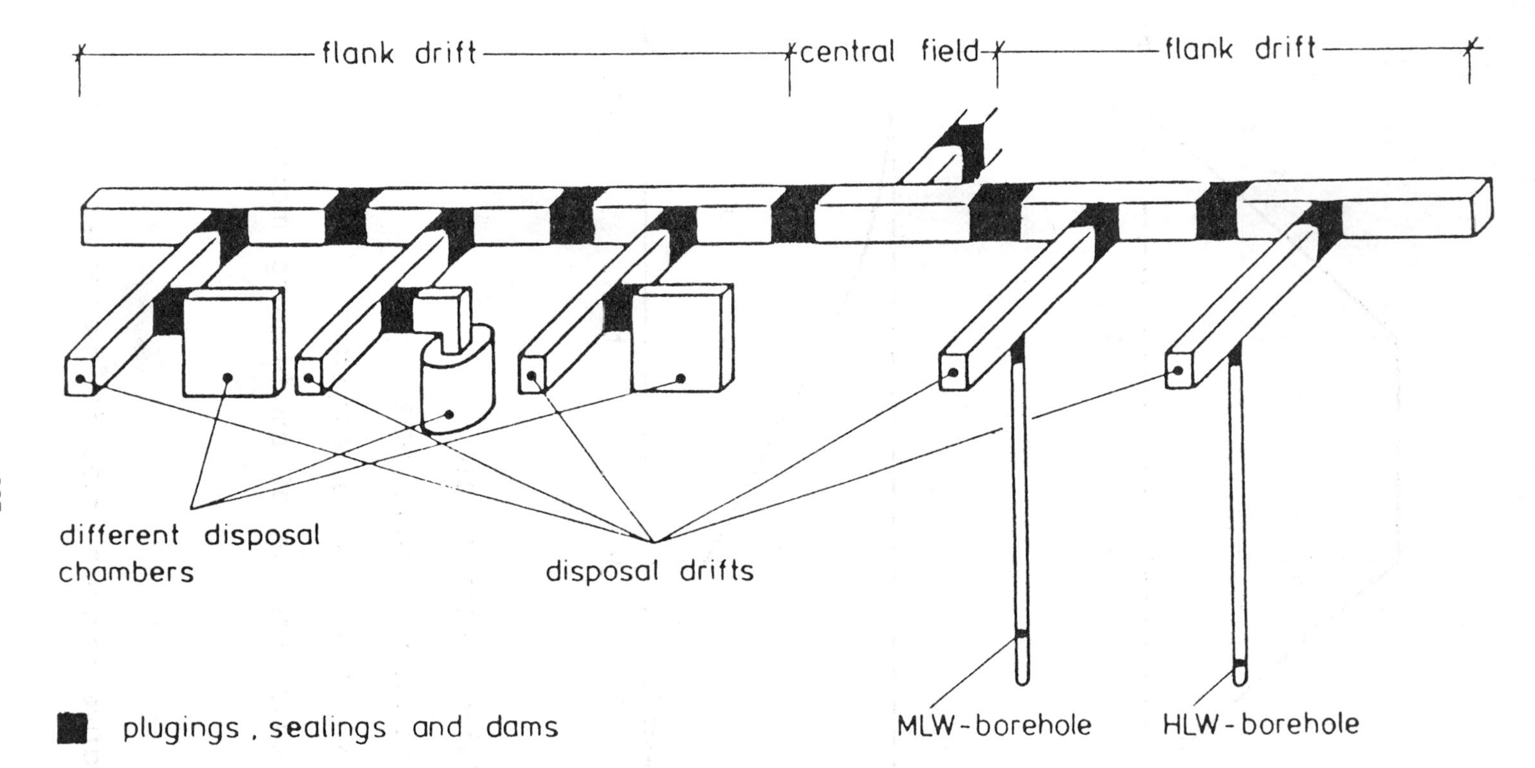

figure 1: principle of the construction of a repository

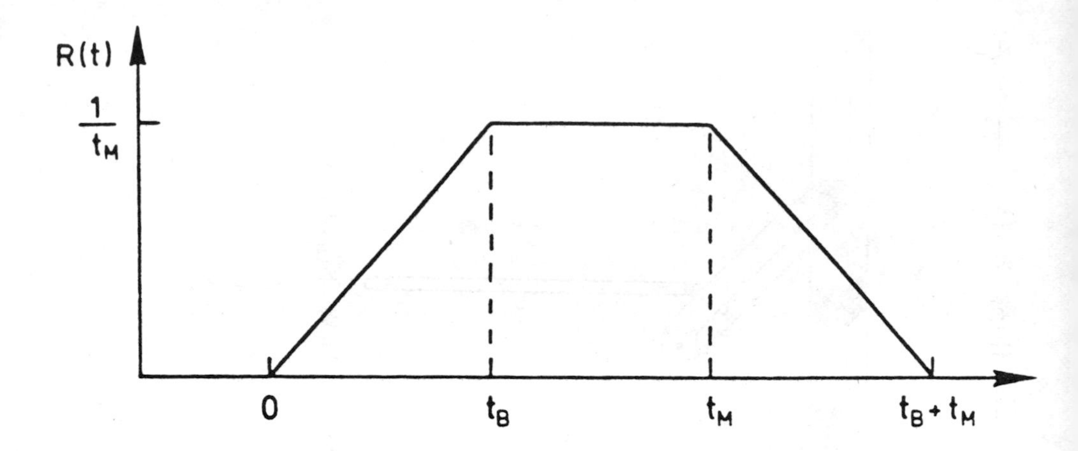

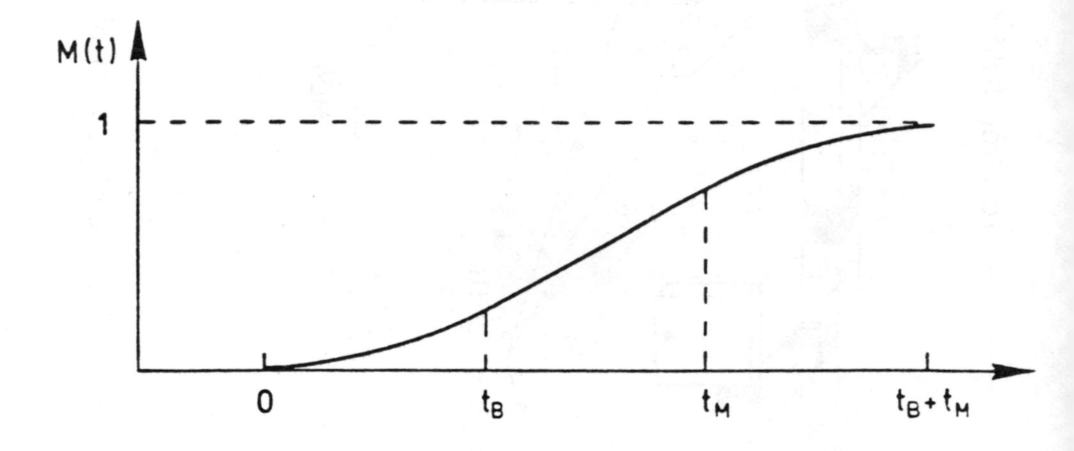

R(t): relative reaction rate

M(t): cumulated reaction rate

t_B = maximal lifetime of canisters

t_M = maximal lifetime of conditioning material

figure 2: relative reaction rates for cemented waste

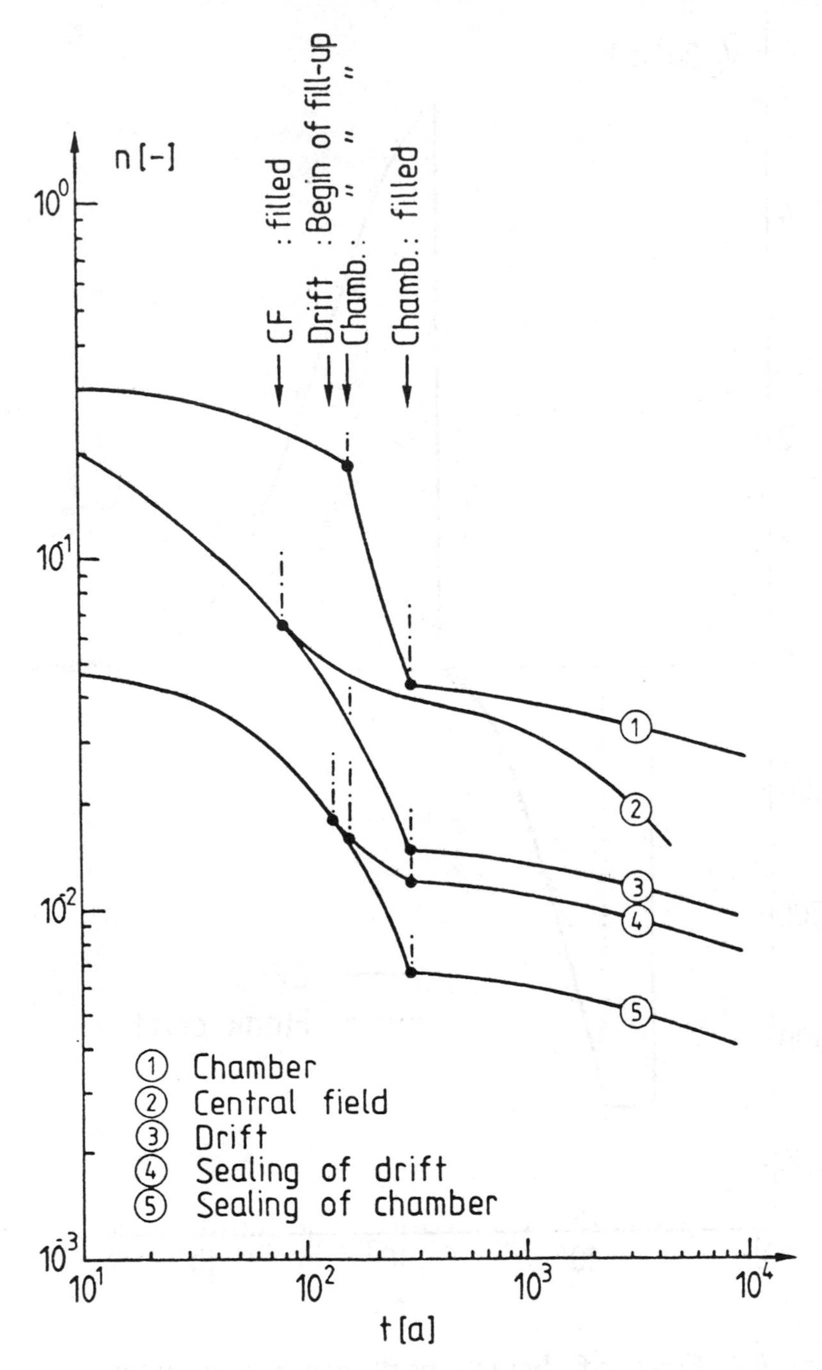

Fig. 3: Temporal evolution of porosity in drifts and chambers.

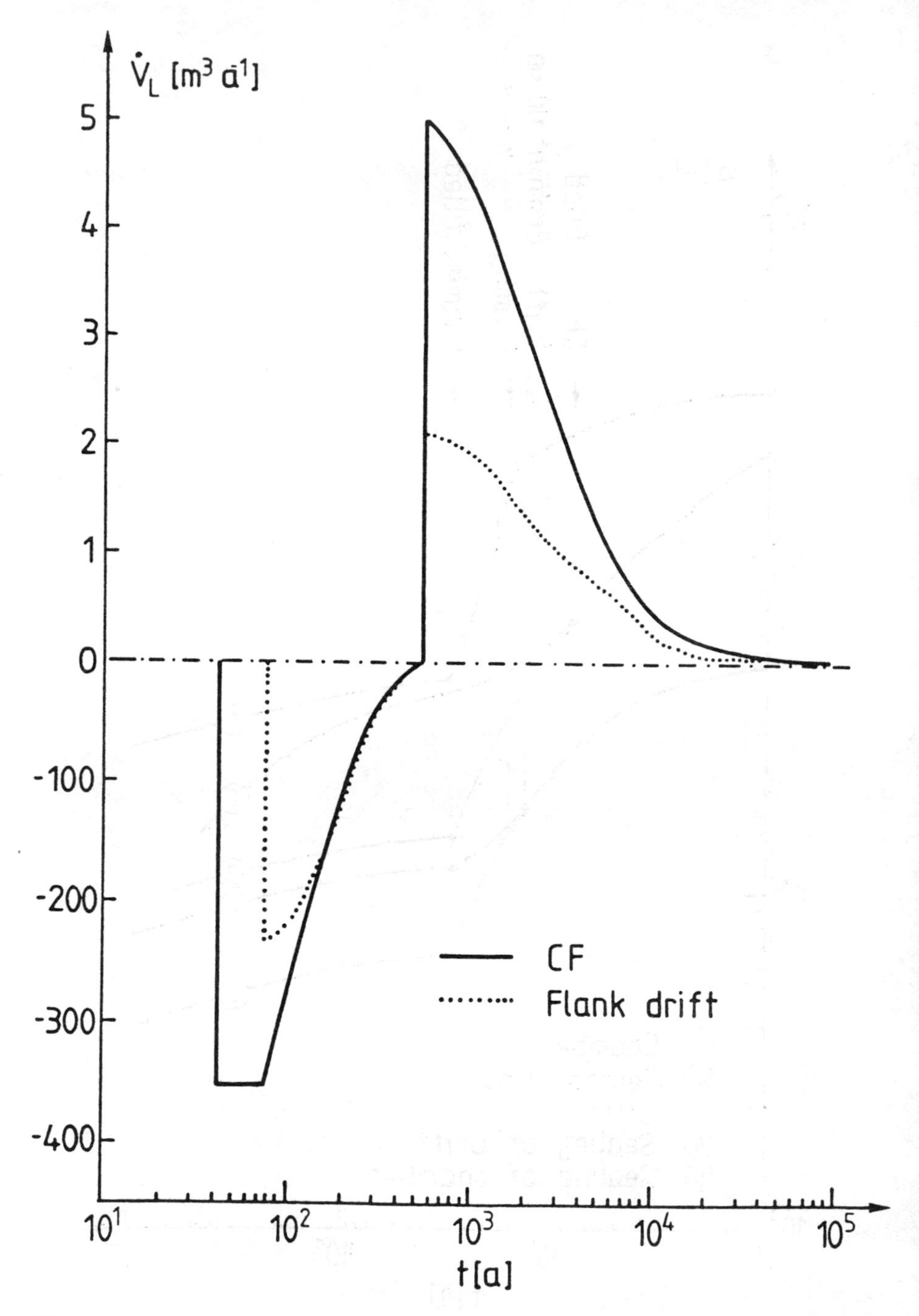

Fig. 4: Flow of brine and brine pressure.

ESTIMATING RADIONUCLIDE RELEASES FROM LOW-LEVEL RADIOACTIVE WASTES

Michael F. Weber
U.S. Nuclear Regulatory Commission
Washington, D.C., U.S.A.

ABSTRACT

Performance assessments of low-level radioactive waste (LLW) disposal facilities are particularly sensitive to assumed radionuclide source terms and release models. The U.S. Nuclear Regulatory Commission (NRC) is currently evaluating alternative approaches for estimating radionuclide releases from LLW. As part of a more comprehensive performance assessment methodology, the NRC is developing source term models to estimate radionuclide releases for safety assessments of LLW disposal facilities. These models range from simple, empirical approaches to more complex models that attempt to simulate the processes of leaching, dissolution, sorption, corrosion, diffusion, and advection. Preliminary results indicate that relatively simple, deterministic approaches are preferable provided that they can be demonstrated to be reasonably conservative. The NRC staff intends to select an approach for estimating radionuclide releases that provides demonstrably conservative estimates and sufficient flexibility to consider releases on a facility-specific basis.

ESTIMATION DES REJETS DE RADIONUCLEIDES A PARTIR DE DECHETS DE FAIBLE ACTIVITE

RESUME

Les évaluations des performances des installations de stockage des déchets de faible activité dépendent tout particulièrement des modèles retenus pour les valeurs du terme source et pour les rejets de radionucléides. La Commission de la réglementation nucléaire des Etats-Unis (NRC) étudie actuellement différentes méthodes permettant d'évaluer les rejets de radionucléides à partir des déchets de faible activité. Dans le cadre d'une méthodologie d'évaluation plus complète des performances, la NRC élabore des modèles de terme source permettant d'estimer les rejets de radionucléides en vue des évaluations de sûreté des installations de stockage de déchets de faible activité. Il s'agit, dans le cas de ces modèles, aussi bien de méthodes empiriques simples que de démarches plus complexes cherchant à simuler les processus de lixiviation, de dissolution, de sorption, de corrosion, de diffusion et d'advection. Les résultats préliminaires montrent que des méthodes déterministes relativement simples sont préférables, à condition que l'on puisse démontrer leur caractère raisonnablement empreint de conservatisme. Le personnel de la NRC a l'intention de choisir une méthode permettant d'estimer les rejets de radionucléides en démontrant leurs faibles niveaux, tout en étant suffisamment souple pour prendre en compte les rejets propres à chaque installation.

INTRODUCTION

The U.S. Nuclear Regulatory Commission (NRC) is responsible for regulating the disposal of commercial low-level radioactive waste (LLW) in the United States. NRC's regulations in 10 CFR Part 61 provide a comprehensive set of licensing requirements for land disposal of LLW. These regulations provide several performance objectives, including one that requires license applicants to demonstrate that concentrations of radioactive materials that may be released to the general environment will not exceed appropriate dose limits [1].

To demonstrate compliance with this performance objective, license applicants need to assess the performance of proposed LLW disposal facilities. Such assessments should consider any significant pathway of radionuclide release, transport, and exposure to humans [2]. The NRC staff considers the groundwater transport pathway as one of the most significant pathways resulting in potential human exposure to radionuclides released from LLW disposal facilities [3]. In addition to evaluating the environmental transport of radionuclides after they reach the groundwater, a realistic performance assessment of a LLW disposal facility should include an evaluation of the release of radionuclides from the waste form as the source term for the transport evaluation.

The processes and conditions that control radionuclide releases from LLW are among the poorest understood components of performance assessments of LLW disposal facilities. Although a substantial amount of source term modeling and experimental studies have been performed, these studies have not significantly reduced uncertainties associated with the estimation of radionuclide releases from LLW [4]. Nevertheless, such estimates are essential to performance assessments of LLW disposal facilities. Previous performance assessments have either assumed radionuclide releases *a priori* or utilized extremely simple models to estimate radionuclide releases. These approaches, however, do not provide demonstrably conservative or best estimates of radionuclide releases. In addition, such approaches do not provide sufficient flexibility to account for enhanced radionuclide containment by engineered barriers and stabilized waste forms. Therefore, additional effort is warranted to improve current performance assessment capabilities to estimate radionuclide releases from LLW disposal facilities.

After a brief review of previous efforts to estimate radionuclide releases from LLW, this paper describes several approaches being considered by NRC and its contractors for estimating radionuclide releases from the waste form into the disposal unit. The paper briefly reviews the diverse characteristics of LLW streams in the United States and suggests that this diversity currently precludes complicated approaches for estimating radionuclide releases. Based on this information, the paper presents an approach and computer code entitled RELEASE that the NRC staff is currently evaluating. The paper concludes by describing planned future work to test and refine the capabilities of RELEASE and alternative approaches.

PREVIOUS WORK

The relative importance of radionuclide release estimates in performance assessment for LLW disposal facilities has been recognized for many years [5]. Without releases of radionuclides into the aqueous phase, groundwater and surface water pathways would not be significant in considering the performance of LLW disposal facilities. In recognition of the importance of radionuclide releases from LLW, numerous attempts have been made to minimize or eliminate the releases from certain types of LLW [6]. However, these attempts have only been partially successful because they have not been able to quantify the small, but finite release rates of the radionuclides over the long periods of concern. In addition, not all types of LLW can be stabilized and treated to minimize the release of radionuclides.

Considerable effort has also been invested in development of standard laboratory leaching tests to screen waste form characteristics and demonstrate their suitability for disposal [7-10]. The validity of extrapolating the results from these tests to estimate radionuclide releases in actual LLW disposal facilities has not been demonstrated. In addition, the technical community has not been able to select a standard leaching test to represent the full range of environmental and waste conditions anticipated at LLW disposal facilities. Thus, the principal utility of the laboratory testing has been to screen waste form characteristics and expand the data base about radionuclide releases rather than to estimate long-term radionuclide release rates. This data base has supported the development and validation of empirical and phenomenological models for estimating radionuclide releases.

Previous investigators have also pursued development of computer models to estimate long-term releases of radionuclides from LLW. For example, NRC developed an empirically-based generic approach in the Impacts Methodology used to support the development of 10 CFR Part 61. The source term algorithm of the comprehensive Impacts Methodology was developed based on the best information about radionuclide releases available when NRC promulgated these regulations in 1982. The algorithm employs an integer-based approach to calculate a waste form and package barrier factor, which is a multiplier in the overall equation used to estimate radionuclide concentrations in the environment. The source term algorithm of the Impacts Methodology accounts for leachability (i.e., waste form), chelating agents, waste segregation, and the accessibility of radionuclides embedded in activated metals [11]. This approach, however, does not attempt to estimate radionuclide releases by explicitly simulating any of the processes or phenomena that may control radionuclide releases (e.g., advection, diffusion, solubility, sorption). Instead, the approach provides release estimates based on simple empirical relationships and radionuclide concentrations observed in leachate at two commercial LLW disposal facilities: West Valley, New York, and Maxey Flats, Kentucky.

More recent approaches have attempted to simulate the processes and phenomena that control radionuclide releases from LLW [12-14]. For example, Reference 14 develops an evaluation methodology that explicitly simulates moisture migration and groundwater flow in and near the LLW, as well as transfer of the radionuclides into the aqueous phase, degradation of waste packages, and transport of radionuclides after they are released. Under contract to the NRC, this effort should continue for the next couple of years and develop an approach to simulate the processes that significantly affect release and near-field transport of radionuclides at LLW disposal facilities. Such complicated analytical approaches, however, may not be compatible with the amounts and types of information that are available for LLW disposal facilities. Therefore, implementation of relatively complicated approaches for source estimation may be precluded by data limitations.

DIVERSITY OF LLW

The release of radionuclides from LLW is controlled at least in part by the characteristics of the waste. Thus, it is important to assess the relative amounts and characteristics of LLW expected to be received at LLW disposal facilities. Figure 1 compares typical radionuclide concentrations in untreated, high-volume waste streams in the United States. This figure illustrates the diversity of waste characteristics in terms of radionuclide types and concentrations. It is important to recognize that the figure provides information on only a few radionuclides, whereas LLW may contain many other radionuclides. It is also important to recognize that radionuclide concentrations are often lower in the large volume LLW streams than in smaller volume waste streams. For example, typical radionuclide concentrations in decontamination resins may be orders of magnitude greater than concentrations in the large volume waste streams depicted in Figure 1. Further, these concentrations reflect average concentrations in "typical" waste streams. Actual radionuclide concentrations may vary considerably as a function of many parameters, including generation process, operational cycle, age of facility, fuel burn-up, housekeeping practices, waste treatment, and age of waste. The diversity of radionuclide concentrations is only one factor that complicates development and use of approaches for estimating radionuclide releases from LLW. Although the diverse characteristics of LLW complicates analysis of radionuclide releases, careful record-keeping of the characteristics of the waste in disposal units is essential to the development of defensible source term models.

In addition to radionuclide concentrations, radionuclide releases from LLW may depend upon many other variables, including composition and characteristics of the waste form, leachant residence times, and the physico-chemical characteristics of the leachant. For example, if the percolation rate of water through the waste is slow, release of radionuclides from the waste may be dominated by diffusion. However, if the flow rates are very slow, radionuclide concentrations in the leachate may be constrained by solubility limits. Alternatively, if percolation rates are relatively rapid, then release of the radionuclides may be controlled by the dissolution rate of the waste form [15]. Table I summarizes the factors that are known or suspected to affect the release of radionuclides from LLW [4]. Given the current limitations of analytical approaches for estimating radionuclide releases from LLW, these complications currently preclude explicit simulation of all of the processes or phenomena that may be important in quantifying the releases. Therefore, approaches to estimate the releases should rely on relatively simple and conservative analyses that attempt to approximate the results of release processes rather than on complicated algorithms derived from fundamental equations (e.g., complicated multi-dimensional flow modeling coupled with geochemical equilibria and speciation modeling).

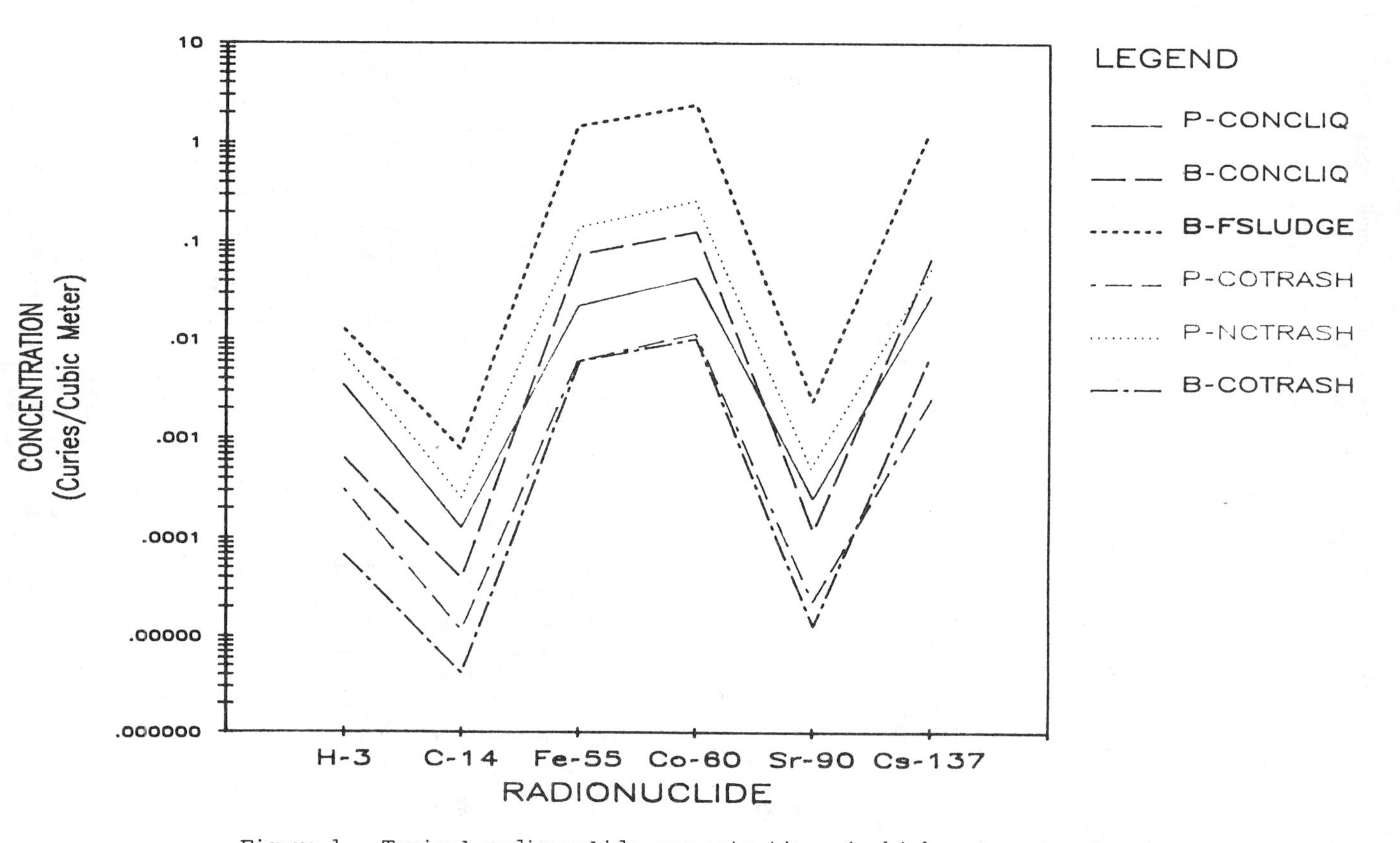

Figure 1. Typical radionuclide concentrations in high-volume low-level radioactive waste streams. P=Pressurized Water Reactor, B=Boiling Water Reactor, CONCLIQ=Concentrated Liquid Waste, FSLUDGE=Filter Sludge, COTRASH=Compressible Trash, NCTRASH=Non-compressible Trash (Based on Ref. 18).

Table I. Factors that may influence the release of radionuclides from LLW (Modified from Table 1.1, pg. 2, of Ref. 4).

System Factors	Radionuclide Factors
Temperature	Radionuclide
Pressure	Solubility
Radiation field	Solid speciation and form
Leachant resident time	Atomic radius
Leachant contact	Physical state
Ratio of waste form surface area to leachant volume	
Waste Form Factors	**Leachant Factors**
Composition	Leachant composition
Surface characteristics	Complexing agents
Effective diffusivity	pH and Eh
Corrosion rate	Percolation rate
Porosity	Biologic activity
Structural integrity	
Curing time	
Mixing ratio (waste/matrix and matrix/water)	
Biodegradation	

RELEASE

This paper focuses on a source term release approach and computer code entitled RELEASE. RELEASE is being developed and assessed by NRC staff along with other alternative approaches, including site-specific application of the source term algorithm of the Impacts Methodology [11] and a diffusion-limited approach developed by investigators at Brookhaven National Laboratory for estimating releases from LLW [12]. The objective of the NRC staff evaluation is to select a preferred approach or set of approaches for estimating radionuclide releases from LLW.

The RELEASE approach has been incorporated into a computer program of the same name, which was written by the author in advanced BASIC for execution on a personal computer. RELEASE interactively queries the user for input information and calculates incremental and cumulative fractional releases of individual radionuclides and annual releases in curies. The author is currently revising RELEASE to enhance its computational efficiency and user options. This section briefly reviews RELEASE's analytical approach, input requirements, and output characteristics, and compares the performance of RELEASE with other source term estimation approaches for a sample problem.

Analytical Approach

RELEASE is intended for site-specific application using the characteristics of the waste, disposal facility, and disposal environment. RELEASE provides limited flexibility to consider engineered barriers and other types of disposal facility enhancements intended to improve LLW isolation.

The analytical approach used in RELEASE assumes that radionuclide releases from LLW are dominated by rinsing of surface contamination from LLW trash and diffusion of radionuclides out of solidified waste forms. In addition, RELEASE approximates radionuclide releases from activated metals. RELEASE combines these processes to estimate total releases of a given radionuclide from individual LLW disposal units. The equations used in RELEASE to describe these processes were developed based on laboratory experiments and field observations at several LLW disposal facilities in the United States [4].

The analytical approach used in RELEASE is based on several fundamental assumptions. RELEASE assumes that (1) advective release of radionuclides from LLW trash can be represented by a rinsing model, (2) radionuclide releases from solidified wastes (e.g., cementitious or bituminous waste forms) are controlled by diffusion through the solid matrix, and (3) releases of radionuclides from activated metals can be approximated as a constant fraction of the diffusional releases. RELEASE also assumes that no liquid or gaseous wastes are included in the disposal units and that wastes do not degrade to form liquid or gaseous wastes other than leachate within the disposal unit. In addition, RELEASE does not explicitly simulate corrosion of waste packages or activated metal components. Further, RELEASE does not consider biodegradation of the LLW or secondary release mechansims, such as gaseous tritium transport with subsequent deposition in the unsaturated zone and transport into the saturated zone. The equations used in RELEASE were developed in References 4, 12, and 14.

The rinse model was developed in Reference 14 by Brookhaven National Laboratory to simulate releases of radionuclides from surface-contaminated trash in carbon steel drums at the Sheffield LLW disposal facility. The model is intended to simulate the mixing of leachant that occurs within a waste package after the package has been initially breached by corrosion or mechanical failure, but while the waste package still limits the exposed surface area of the waste. Thus, mixing occurs in a finite volume that is dependent upon the sum of the volumes of the individual waste packages. For each infiltration event during the year, RELEASE assumes that the wetting front propagates through the waste and displaces a finite volume of leachate from the waste packages. The rinse model is only appropriate if most of the radionuclide releases occur during the lifetime of the waste packages. Once the waste packages fail completely, advection of radionuclides through the bulk waste may dominate over the simple mixing process within individual waste packages. Therefore, the rinse model in RELEASE is best suited for estimating releases from unstabilized, low activity trash containing relatively short-lived radionuclides.

Based on the equations derived in References 4 and 14 for the rinse model, radionuclide releases from LLW trash may be approximated as follows:

$$An = XZ(1-Z)^{(n-1)}$$, where An = activity release per cycle,
n = number of rinse cycle,
Z = ratio between the rinse volume [L] and standing bath volume [L], and
X = source activity corrected for radioactive decay = $X(\exp(-Lt))$, where X = initial activity, L = decay constant [1/T], and t = age of waste [T].

RELEASE assumes that each infiltration event results in a constant displacement of mixed leachate from all waste packages. In addition, the model also assumes that the radionuclides are contained in the trash as surface contamination. Thus, each infiltration event releases an ever decreasing incremental release of the radionuclide; RELEASE sums these incremental releases to approximate the annual activity release rate.

RELEASE does not incorporate the "tilt" factor discussed in the original rinse model [14]. The tilt factor was originally used to decrease the annual releases proportional to the cosine of the angle of the long axis of the waste packages with respect to the vertical. Because of uncertainty about waste package configuration in disposal units, the reduction of releases by the tilt factor does not appear to provide conservative estimates of radionuclide releases and is, therefore, not included in RELEASE. RELEASE conservatively assumes that all waste packages are vertically oriented, which maximizes the rinse volume per infiltration cycle.

In addition to estimating radionuclide releases from unstabilized LLW trash, RELEASE can be used to estimate releases from stabilized LLW that has been solidified. As previously discussed, RELEASE assumes that releases from solidified wastes are dominated by diffusion of the radionuclides through the waste matrix. Once they reach the outer surface of the waste form, RELEASE assumes that they are instantaneously released into the leachant. RELEASE partially accounts for containment by waste packages by assuming that radionuclide releases are proportional to the exposed surface area of the waste, which increases as a function of time. Unlike the rinse model, however, the diffusion model in RELEASE does not account for alternating wetting-drying cycles. Such cycling has been shown to increase releases of certain radionuclides under specific conditions [12]. Future testing of the RELEASE approach will include evaluation of the need to consider wet-dry cycling in estimating radionuclide releases.

The equation used in RELEASE to estimate radionuclide releases via diffusion is developed in Reference 14 and may be summarized as:

$$CFR = (A/S)(C_1X + C_2X^2)$$, where CFR = cumulative fractional release,

$A = S + \sqrt{Kt}$, where
- A = surface area function,
- S = initially exposed surface area [L],
- K = container area constant, and
- t = leaching time [T],

S = surface area of waste form [L],

$X = (S/V)(\sqrt{Dt})$, where
- V = waste form volume [L],
- D = effective diffusivity [L/T], and

$C_1 = 1.3441$ and $C_2 = -0.4416$.

This equation is based on a quadratic power series fit for the solution to the infinite plane sheet diffusion equation as used in ANS Standard 16.1 [8, 14].

The incremental fractional release rates may be calculated as

IFR = (CFR(t) - CFR(t-1))(exp(-Lt')), where L = radioactive decay constant [1/T], and
t'= time since emplacement of waste.

The decay corrected CFR's are then calculated by summing the IFR's; incremental releases are calculated by multiplying the IFR's by the initial activity of the radionuclide in the waste. In the original reference [14], t' was assumed to be significantly greater than t because leaching only occurs during a small fraction of the total time (t') since emplacement of the waste. However, RELEASE invokes the conservative assumption that leaching time equals time since emplacement. Although it may be possible to estimate t on a site-specific basis using complicated moisture migration modeling within the waste, the uncertainties associated with predicting moisture migration in a highly heterogenous disposal unit may preclude such estimates. Therefore, RELEASE assumes that radionuclide diffusion through the waste matrix begins upon emplacement of the LLW in the disposal unit.

RELEASE also estimates releases from activated metals in the LLW by multiplying the IFR's for solidified waste by the initial activity contained in the activated metals and by a constant ratio of 0.01. Although the rates of radionuclide release from activated metals are not well understood, most approaches assume that these rates are controlled by the corrosion rate of the activated metal components. These rates are generally low because the components are constructed out of corrosion resistant metals. Many of the activated metal components are composed of high-alloy materials (materials with a high non-ferrous metallic component). The radionuclides contained within the activated metal components are gradually released as the metal matrix slowly corrodes.

Corrosion of these materials occurs at low rates such as 7.3E-6 g/cm/yr for high allow stainless steels [11, 16, and 18]. Although release rates based on metal corrosion are a function of geometry of the metallic component, these rates are generally lower than those estimated for solidified wastes using RELEASE. Thus, RELEASE estimates radionuclide releases from activated metals by multiplying the diffusion-dominated release rates by a constant factor of 0.01 to account for the slower rate of radionuclide release from activiated metals.

The author has incorporated several features in RELEASE to consider the increased isolation of LLW provided by engineered barriers. For example, covers that minimize infiltration into the waste in conventional trenches can be considered by reducing rinse volume. Special waste packages intended to contain the waste and minimize its contact with leachant can be considered by delaying initial release time and reducing or eliminating the initial surface area. RELEASE can also consider structural components such as concrete vaults and monoliths by delaying the initial release time and reducing rinse volumes. Further, these barriers may provide additional protection by containing leachate after radionuclides are released from the LLW and before the leachate is released to the general environment. Selection of these input parameters should be based on accelerated leach and durability testing in the laboratory; field observations of actual waste form, waste package, and disposal unit performance; and information on the reliability of engineered components such as waste packages.

Input Requirements and Output Characteristics

Users enter a variety of input data into RELEASE in response to a series of interactive queries. The input requirements are listed in Table II. The input requirements have been minimized so that RELEASE can be effectively implemented in assessing potential LLW disposal sites given the types and amounts of information available for the sites.

Based on input supplied by the user, RELEASE estimates radionuclide releases from the waste form into the disposal unit in terms of incremental and cumulative fractional releases and annual activity releases. RELEASE provides separate output for releases from LLW trash, solified waste, and activated metals. The releases are then summed to estimate the annual activity releases into individual disposal units. These estimated releases may then be used as the radionuclide source term for evaluations of transport out of the disposal units and into the surrounding environment.

Table II. Input Requirements for RELEASE

Input Type	Units
Radionuclide name	--
Decay constant	1/yr
Site name/unit number	--
Initial activity	Ci
Fraction of activity in trash	--
Fraction of activity in activated metals	--
Time of first pit/crack	yr
Total volume of waste packages	m
Rinse volume per cycle	m
Initial surface area	m
Total surface area of solidified waste	m
Effective diffusivity	cm/s

Comparison of Results

Figure 2 compares releases of cesium-137 from a hypothetical LLW disposal unit estimated using RELEASE and four other computer codes. Two of the codes, IMPACTS and DIFFUSE, are based on alternative approaches [11 and 12] for estimating source term releases and were written by the author as computer programs in BASIC to support the comparison of the approaches. The "IMPACTS" model is basically the submodel used in NRC's generic rational for 10 CFR Part 61 and is based on the algorithm used by the Impacts Analysis Methodology to calculate the waste form and package barrier factor [11]. The "DIFFUSE" model solves the infinite sheet diffusion-limited release model [12]. The last two models listed in Figure 2 are the "RINSE" and "PACK" (diffusion-limited) submodels that have been incorporated in RELEASE. Their results have been included in the figure to assess the relative release contribution from LLW trash and solidified waste for the sample problem. The input information for RINSE and PACK has been modified from that for the sample problem to represent all waste as LLW trash for the RINSE model and as solidified LLW for the PACK model. These modifications account for the large differences in the estimated releases between RINSE, PACK, and RELEASE.

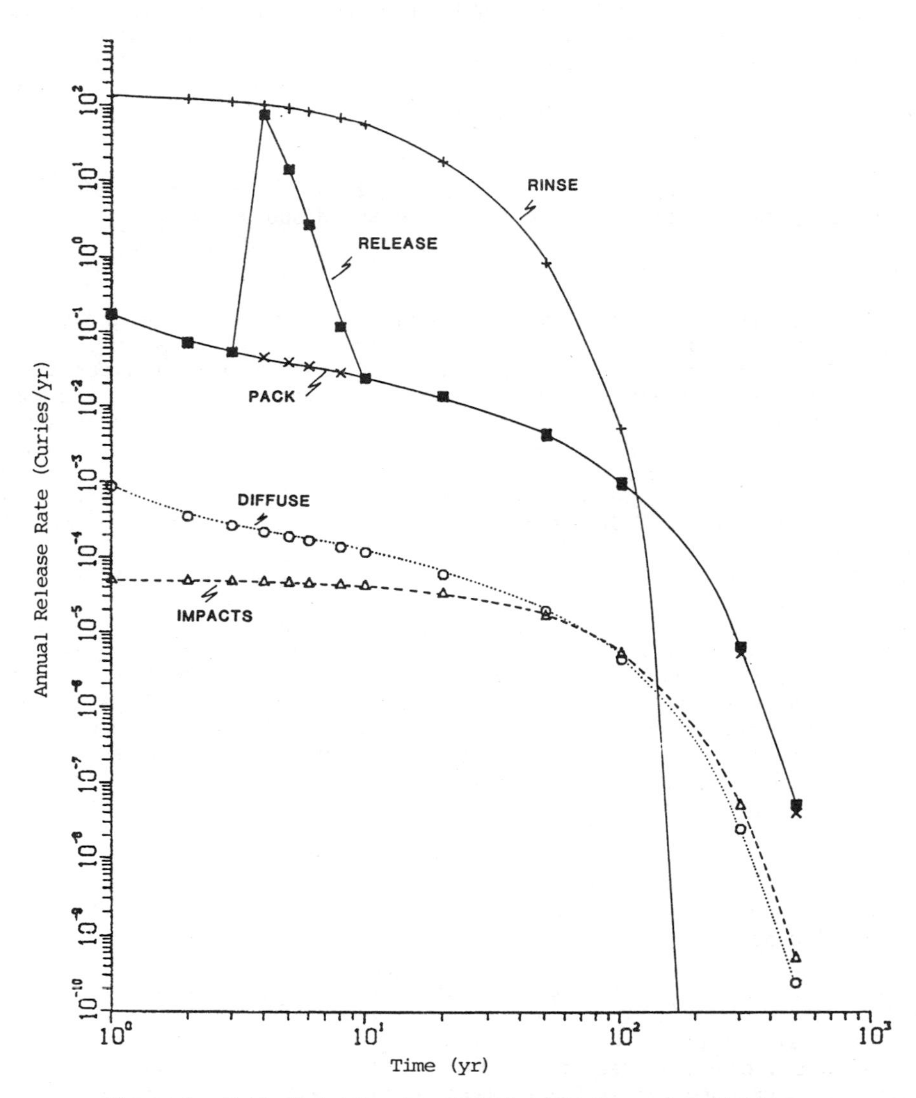

Figure 2. Estimated releases of Cesium-137 from a hypothetical LLW disposal unit. See text for discussion of models RELEASE, RINSE, PACK, DIFFUSE, and IMPACTS.

Figure 2 indicates that RELEASE provides relatively conservative estimates of radionuclide releases compared with the other models. The RINSE model provided the most conservative estimates, as expected, because it assumes that the entire radionuclide inventory is contained in the LLW as surface contamination. This approach is clearly conservative if much of the cesium-137 inventory is contained in solidified waste forms and is not readily accessible. Release estimates from RINSE decrease below the other estimates after 100 years because most of the waste activity has already been released. The IMPACTS model provides the lowest estimated source term releases for the first 100 years. It is important to recognize that the radionuclide releases estimated by IMPACTS for the sample problem are not necessarily representative of the results of generic applications of the Impacts Analysis Methodology. Although the source term releases estimated by IMPACTS are lower than the estimates from the other models, the relative conservatism of the Impacts Analysis Methodology is influenced by conservatisms introduced by such components as disposal unit design and performance, and environmental transport. The comparison of releases in Figure 2 does not account for these components because the models described only estimate releases from the waste form into the disposal unit.

FUTURE WORK

The RELEASE model described in this paper is being developed and tested by the NRC staff to provide necessary capabilities for estimating or bounding radionuclide releases from LLW. RELEASE represents one of several alternative approaches that are currently being evaluated by the NRC staff. Although preliminary comparisons of the performance of RELEASE with other approaches indicates that RELEASE may provide relatively conservative estimates, additional work is necessary to determine whether RELEASE provides radionuclide release estimates that are adequate and defensible in reviewing license applications for LLW disposal. Ongoing NRC studies of commercial LLW disposal facilities may provide information about actual releases of radionuclides that may be compared with releases estimated using RELEASE, IMPACTS, and other models. For example, NRC staff and contractors are developing transport models for the inactive, commercial LLW disposal facility at Sheffield, Illinois, which may be useful in estimating actual releases by inverse modeling. These estimates could then be used in calibrating and validating RELEASE and other approaches for estimating radionuclide releases from LLW. Further evaluation of the performance of RELEASE and other approaches is warranted to ensure that the approaches provide reasonably conservative or best estimates of radionuclide releases that are consistent with observations of radionuclide releases from LLW. The purpose of the staff's evaluation is to select a preferred approach that can be integrated into a comprehensive performance assessment methodology for LLW disposal facilities. In the future, the NRC staff may propose such an approach in regulatory guidance or as part of standard review procedures for LLW disposal license applications.

Prior to selection or rejection of RELEASE as an estimation approach, the NRC staff will further assess the performance of RELEASE to evaluate its capacity to provide reasonably conservative or best estimates of radionuclide releases from LLW based on reasonable input information. In particular, the staff intends to compare the estimates from RELEASE with laboratory and field measurements of releases from LLW. The staff also intends to evaluate alternative and more rigorous approaches to assess the release of radionuclides from activated metal components, which comprise most of the projected activity in LLW in the United States. Further, the staff intends to assess the importance of processes that RELEASE does not currently consider such as biodegradation and wet/dry cycling. The goal of these assessments is to enhance the capabilities of RELEASE or alternative approaches to provide reasonably conservative or best estimates of radionuclide releases from LLW.

SUMMARY

Estimation of the source term for radionuclide releases to the environment is required to assess long-term performance of disposal facilities for LLW. The processes and phenomena that control radionuclide releases from LLW are relatively poorly understood. The diversity of waste, facility, and environmental characteristics that influence radionuclide releases currently preclude use of complicated approaches to estimate releases. Simple approaches that can be demonstrated to provide reasonably conservative estimates are, therefore, preferable to more complicated approaches. Based on laboratory leach tests and actual performance of LLW disposal units, the NRC staff and its contractors have developed an approach for for estimating radionuclide releases. This approach has been developed into a computer code entitled RELEASE. The NRC staff is currently assessing the capability of RELEASE to provide reasonably conservative or best estimates of radionuclide releases from LLW. Preliminary comparison of results from RELEASE with two other approaches indicates that RELEASE can provide relatively conservative estimates of radionuclide releases. Prior to selection of a preferred approach for estimating radionuclide releases, the NRC staff intend to further evaluate the performance and capabilities of RELEASE and other approaches.

The author wishes to express his appreciation for NRC staff and contractor support of this effort. Because of the preliminary nature of the work described in this paper, the conclusions provided above are those of the author and do not necessarily represent those of the U.S. Nuclear Regulatory Commission or its staff.

REFERENCES

1. U. S. Nuclear Regulatory Commission: "Licensing Requirements for Land Disposal of Radioactive Waste", 10 CFR Part 61, 1982.

2. U. S. Nuclear Regulatory Commission: "Standard Review Plan for the Review of a License Application for a Low-Level Radioactive Waste Disposal Facility", NUREG-1200, 1987.

3. Siefkin, D., Pangburn, G., Pennifill, R., and R. J. Starmer: "Branch Technical Position on Site Suitability, Selection, and Characterization", U.S. Nuclear Regulatory Commission, NUREG-0902, 1982.

4. Sullivan, T., and C. R. Kempf: "Low-Level Waste Source Term Evaluation: Review of Published Modeling and Experimental Work, and Presentation of Low-Level Waste Source Term Modeling Framework and Preliminary Model Development", U.S. Nuclear Regulatory Commission, NUREG/CR-4897, 1987.

5. U. S. Nuclear Regulatory Commission: "Draft Environmental Impact Statement on 10 CFR Part 61 'Licensing Requirements for Land Disposal of Radioactive Waste'", NUREG-0782, 1981.

6. U. S. Nuclear Regulatory Commission: "Final Environmental Impact Statement on 10 CFR Part 61 'Licensing Requirements for Land Disposal of Radioactive Waste'", NUREG-0945, 1982.

7. International Atomic Energy Agency: "Conditioning of Low- and Intermediate-Level Radioactive Wastes", Technical Reports Series No. 222, 1983.

8. American Nuclear Society: "Measurement of Leachability of Solidified Low-Level Radioactive Wastes", American Nuclear Society Standards Committee, Final Draft of Standards ANS-16.1, 1984.

9. Dougherty, D. R., Fuhrman, M., and P. Columbo: "Accelerated Leach Test(s) Program Annual Report", Brookhaven National Laboratory, BNL-51955, 1985.

10. Hespe, E. D.: "Leach Testing of Immobilized Radioactive Waste Solids: A Proposal for a Standard Method", Atomic Energy Review, v. 9, pp. 195-207, 1971.

11. Oztunali, O. I., Re, G. C., Moskowitz, P. M., Picazo, E. D., and C. J. Pitt: "Data Base for Radioactive Waste Management: Impacts Methodology Report", U.S. Nuclear Regulatory Commission, NUREG/CR-1759, v.3, 1981.

12. Arora, H., and R. Dayal: "Leaching Studies of Cement-Based Low-Level Radioactive Waste Forms", U.S. Nuclear Regulatory Commission, NUREG/CR-4756, 1986.

13. Essington, E. H., Fuentes, H. R., Polzer, W. L., Lopez, E. A., and E. A. Stallings: "Leaching of Solutes from Ion-Exchange Resins Buried in Bandelier Tuff", U.S. Nuclear Regulatory Commission, NUREG/CR-4592, 1986.

14. MacKenzie, D. R., Smalley, J. F., Kempf, C. R., and R. E. Barletta: "Evaluation of the Radioactive Invetory in, and Estimation of Isotopic Releases from, the Waste in Eight Trenches at the Sheffield Low-Level Waste Burial Site", U.S. Nuclear Regulatory Commission, NUREG/CR-3865, 1985.

15. International Atomic Energy Agency: "Deep Underground Disposal of Radioactive Wastes: Near-Field Effects", Technical Report Series No. 251, 1985.

16. U. S. Nuclear Regulatory Commission: "Draft Environmental Impact Statement on 10 CFR Part 61 'Licensing Requirements for Land Disposal of Radioactive Waste': Appendices G-Q", NUREG-0782, v. 3, 1981.

17. Oztunali, O. I., and G. W. Roles: "Update of Part 61 Impacts Analysis Methodology", U.S. Nuclear Regulatory Commission, NUREG/CR-4370, v. 1, 1986.

18. Wild, R. E., Oztunali, O. I., Clancy, J. J., Pitt, C. J., and E. D. Picazo: "Data Base for Radioactive Waste Management: Waste Source Options Report", U.S. Nuclear Regulatory Commission, NUREG/CR-1759, v. 2, 1981.

MODELING OF RADIONUCLIDE RELEASE FROM THE SFR REPOSITORY FOR LOW- AND INTERMEDIATE-LEVEL WASTE

I. Neretnieks, L. Moreno, and S. Arve
Department of Chemical Engineering
Royal Institute of Technology
S 100 44 Stockholm, Sweden

ABSTRACT

The SFR repository of low- and intermediate-level waste is located in caverns and drifts underneath the Baltic Sea. The mechanisms and escape rate of the radionculides from the silo have been studied. The transport out of the silo is modeled as Fickian diffusion through the concrete and backfill walls to the groundwater passing in the surrounding rock. During the last few years several observations in the field and laboratory have been made which indicate that water flow is very unevenly distributed in fractured crystalline rocks. The transport in the rock from the repository to the bottom of the Baltic has been studied. A simple model has been made based on the assumption that the channels are independent over a certain distance. The transport properties in such a bundle of channels have been shown to be different in some important respects from the properties of a highly fractured medium.

MODELISATION DU REJET DE RADIONUCLEIDES A PARTIR DU DEPOT SFR DE DECHETS DE FAIBLE ET DE MOYENNE ACTIVITE

RESUME

Le dépôt SFR de déchets de faible et de moyenne activité est situé dans des excavations et des galeries pratiquées sous la Mer Baltique. Les mécanismes et les taux de fuite des radionucléides à partir du silo ont été étudiés. Le transport hors du silo est représenté par un modèle de diffusion de Fick à travers les parois de béton et de remblayage jusqu'à l'eau souterraine en passant par la roche réceptrice. Au cours des dernières années, plusieurs observations effectuées sur le terrain et en laboratoire ont montré que l'écoulement de l'eau est très inégalement réparti dans les roches cristallines fissurées. Le transport dans la roche à partir du dépôt en direction du fond de la Baltique a été étudié. On a établi un modèle simple fondé sur l'hypothèse selon laquelle les cheminements sont indépendants à une certaine distance. Il a été démontré que les propriétés de transport d'une telle série de cheminements sont différentes sur certains points importants des propriétés d'un milieu fortement fissuré.

1 INTRODUCTION

A low– and intermediate–level waste repository is under construction at Forsmark, Sweden. The repository is located in crystalline rock underneath the sea. The distance between the top of the repository caverns and the sea bottom is 50 m.

The waste that contains about 95% of the activity will be deposited in a silo. The remaining waste will be divided among four tunnels of three different types of construction: BMA (medium active waste), BTF (waste in concrete tanks), and BLA (low–level waste).

The silo is a vertical cylinder 53 m high and 27.6 m in diameter with 0.8 m thick walls. The interior of the silo is divided into vertical cells. The silo is surrounded by a backfill 1.2 m thick consisting of bentonite clay. The tunnels are 160 m long and have widths between 14.7 and 19.6 m. The heights of the tunnels vary between 9.5 and 16.5 m. No backfill surrounds any of the tunnels.

The silo has a rather complex geometry from the transport point of view. The water inside the silo is very mobile in the vertical direction because the cells which extend from the bottom to the top are filled with a porous concrete to allow the gas generated by corrosion to escape to the vents in the top. Previous calculations have shown that the release of nuclides from the boxes is "fast" with a time constant of less than 100 years for nonsorbing nuclides [1]. For the present purposes when the time considered is well in excess of 1000's of years, it is permissible to assume that all the nuclides from the boxes are quickly dissolved in all the water that is present in the silo. It is also assumed that all the nuclides are "immediately" soluble in the water in the silo since it cannot be guaranteed that the solubility limitations exist in the waters due to the potential presence of various complexing agents [2]. For release purposes the silo is thus modeled as a well mixed tank and the transport of the nuclides out of the silo as Fickian diffusion through the walls out to the passing water in the surrounding rock. The major limiting factor to the escape is the slow diffusion out to the water flowing in the rock. The diffusional resistance in the concrete walls and backfill is nearly negligible in comparison. The barriers' integrity is of importance to ensure that there is no flow through the silo.

Many of the short–lived nuclides decay to insignificant concentrations during the diffusion process out through the barriers. Other nuclides will eventually escape but at lower rates. The radionuclide escape is mainly governed by the low flowrate which may carry the nuclides away from the repository and for many of the species by the large amount of sorbing matter in the silo. The maximum escape rates of the nuclides to the passing water have been calculated.

A detailed mapping of the flow to the tunnels and drifts has shown that the water flows in sparse isolated spots from the walls of the drifts [3]. This has been taken as an indication that there may be flow in isolated narrow channels with a limited contact surface between the mobile water and the rock which may give only little surface for the sorbing nuclides to sorb on. A channeling model has been used to calculate the radionuclide migration in the rock. The results are compared with the conventional Advection–Dispersion–Matrix Diffusion model using channels of similar properties.

2 RADIONUCLIDE TRANSPORT IN THE BARRIERS SURROUNDING THE SILO

2.1 Transport through the outer wall and backfill

The hydraulic conductivity of the bentonite backfill outside the silo is so low that transport by diffusion is faster by one to two orders of magnitude than transport by flow for the hydraulic gradient expected to occur in the silo repository. Thus there is no need to calculate the transport by advective flow. It is sufficient to calculate the transport by diffusion only. In time the concrete will degrade from primarily sulphate attack, this and other degradation processes of the barriers are described by Neretnieks *et al.* [4]. It cannot be ruled out that cracks may form in the still intact concrete wall. The diffusion rate in such cracks is expected to be faster than in the pore waters in the concrete. However, the total increase in the rate of transport is expected to be practically negligible [4].

Since the transport is dominated by diffusion in the clay as well as in the concrete, then the transport calculations have been made using Fick's first and second laws. For the instationary diffusion in a porous medium Fick's second law is written

$$\frac{\partial c_p}{\partial t} = D_a \frac{\partial^2 c_p}{\partial z^2} - \lambda c_p \qquad (1)$$

where c_p is the concentration in the pore water (mol/m^3), λ is the decay constant (s^{-1}), t is time (s), z is distance (m), and D_a is the apparent diffusivity (m^2/s). The apparent diffusivity, D_a, includes the effects of retardation due to the sorption of the nuclides in the porous material. It is determined from the following expression:

$$D_a = D_e/(\epsilon + (1 - \epsilon)K_d\rho) \qquad (2)$$

where K_d is the sorption coefficient (m^3/kg), ϵ is porosity, and ρ is the density of the solid material (kg/m^3). Equation (2) applies to all three media: the "intact" concrete, the degraded concrete, and the clay in the buffer. The three media are coupled using the condition that what comes from one medium must go into the next.

The initial condition is that there are no nuclides in the barriers at time zero and that the concentration in the water in the silo is c_o. The boundary condition at the inside of the barrier is

$$W \frac{\partial c}{\partial t} = D_e A \frac{\partial c_p}{\partial z}\bigg|_{z=0} - W\lambda c \qquad (3)$$

The effective diffusivity, D_e, and the concentration in the pore water, c_p, apply to the concrete. The volume of water in the silo, W, and the nuclide concentration in that water, c, at z=0 indicate that the gradient is evaluated at the inner surface of the concrete.

The potential sorption on the concrete in the silo and of the other areas capable of sorption have not been explicitly included in Equation (3). This sorption capacity may be directly included by defining W as follows:

$$W = W_{water} + \sum(K_{d_i}\rho_i W_i)_{sorption} \qquad (4)$$

For the sorbing nuclides this effect may be considerable as will be shown later. The boundary condition at the outside of the bentonite barrier at $z=z_2$ is

$$N = -D_e A \frac{\partial c_p}{\partial z}\bigg|_{z=z_2} = Q_{eq}(c_w - 0) \qquad (5)$$

Equation (5) indicates that what diffuses out of the bentonite will be carried away by the flowing water. The water flowrate is Q_{eq} and has a concentration equal to zero as it approaches the bentonite and a concentration c_w as it leaves the bentonite. At low flowrates c_w will be near the concentration at the clay surface. At large flowrates the capacity of the water to carry away the nuclide is large and the concentration in the leaving water will be very low, bordering on zero.

Instationary period

Before the nuclides can reach the water outside of the backfill, the nuclides must penetrate the concrete and backfill. During this period some decay will take place. If a nuclide is retarded by sorption and has a short half-life it may decay to insignificant concentrations. The competition between penetration and decay has been analyzed by Neretnieks [5]. It was found that a very simple criterion could be derived which tells when decay is significant and when decay is insignificant. The analysis is applicable irrespective of what the conditions are outside the boundary. The criterion can be expressed that when $\Delta z^2/D_a t_{1/2} > 1000$ the nuclide decays to insignificance (10^{-9}), and when $\Delta z^2/D_a t_{1/2} < 1$ the nuclide escapes without significant decay to the next barrier (> 0.1).

Stationary period

For those nuclides which have not decayed an appreciable amount during the instationary period, two bounding mechanisms that will limit the radionuclide release are identified. The first is the diffusional resistance of the concrete and the bentonite barriers. The maximum release for nondecaying or slowly decaying species through the barriers will occur when there is enough flowing water, Q_{eq}, to carry away the escaping nuclides. When steady state has been reached, that is the concentration profiles are fully developed in the barriers, then the radionuclide release is directly obtained by integrating Equation (3). At steady state the concentration decay, c, in the waters in the silo is

$$c = c_o \exp(-t/\tau) \qquad (6)$$

where

$$\tau = 1/(Q_{eq}^*/W + \lambda) \quad \text{and} \quad Q_{eq}^* = \frac{1}{\frac{\Delta z_{conc}}{AD_{e,c}} + \frac{\Delta z_{back}}{AD_{e,b}} + \frac{1}{Q_{eq}}}$$

Q_{eq}^* may be thought of as the water flowrate which will have a concentration c_o as it leaves the repository.

At steady state the release from the silo to the passing water is

$$N = c_o \cdot Q_{eq}^* \cdot \exp(-t/\tau) \qquad (7)$$

The concentration c_o is obtained from

$$c_o = M_o/W \qquad (8)$$

where M_o is the original inventory of nuclides in the silo and W is defined as follows:

$$W = W_{water} + (K_d \rho W)_{concrete} + (K_d \rho W)_{backfill} \qquad (9)$$

When there is no decay, the decrease of concentration in the silo means that the nuclide has been released to the outside. The residence time in the concrete and bentonite is neglected. If there is abundant water flow outside the barriers to carry the escaping nuclides away, i.e. $Q_{eq} \to \infty$, the potential of the concrete and bentonite barriers to delay the escape can be assessed. With W_{water} equal to 7250 m^3, the time constant τ is on the order of 0.001 yr^{-1}. This means that it takes on the order of 1000 years to let out half of the content of a nonsorbing nuclide from the silo. If the half-life of a nuclide is considerably less than 1000 years, the nuclide will decay much faster than it is released, even considering that the instationary build-up of the concentration profile in the barriers has been neglected. On the other hand, nuclides with half-lives considerably longer than 1000 years will leave the silo at a rate which is on the order of half the momentary inventory every 1000 years. The delay in passing the barriers may, because of retardation effects due to sorption, considerably decrease the release of some nuclides because it will allow more time for decay. This is accounted for in full calculations by Neretnieks *et al.* [4].

For the other bounding mechanism, the impact of a limited water flowrate in the rock outside the repository is investigated. At steady state when the concentration inside the repository is c_o and that in the flowing water as it leaves the repository is c_w, then dc_p/dz becomes $(c_o - c_w)/\Delta z$ and Equation (5) can be rewritten as

$$\frac{D_e A}{\Delta z Q_{eq}} = \frac{c_w - 0}{c_o - c_w} = \frac{R_{water}}{R_{barrier}} \qquad (10)$$

The equation shows the relative importance of the flowrate of water and that of the sum of the barriers. R_{water} and $R_{barrier}$ indicate the "resistance" to transport in the water and barrier, respectively.

The flowrate Q_{eq} is the flowrate of water that passes the repository and becomes contaminated. This flowrate will transport away nuclides at a rate N equal to $c_w \cdot Q_{eq}$. The flowrate Q_{eq} is obtained from the data in the hydraulic calculations [6]. For the first 1000 – 2000 years the flow is expected to be upwards with a flux less than 1 $l/m^2 \cdot y$. After the land lift, the Baltic has receded and the flow direction will become downward and the flowrate will increase to no more than 10 $l/m^2 \cdot y$. For the calculations the values of the effective diffusivities, D_e, for the concrete and backfill are $3 \cdot 10^{-11}$ m^2/s and 10^{-10} m^2/s, respectively. The barrier thickness for concrete is 0.5 m and for the backfill is 1.2 m. The entity $D_e/\Delta z$ for the two barriers in series is $0.4 \cdot 10^{-10}$ m/s. The projected area of the silo is about 1000 m^2 so if it is assumed that all the water that were to flow through this area becomes contaminated then Q_{eq} would be 1000 l/yr.

The surface area for the diffusion of A is about 5000 m^2. With the data above it is found from Equation (10) that $R_{water}/R_{barrier}$ is about 6 which means that the limited water flowrate would decrease the nuclide release by a factor of about 6 compared to the case where an unlimited supply of water is available.

For low flowrates the limited water flow can thus considerably lower the release of the nuclides. To use this effect with confidence a more detailed study of the flowrate which passes the repository and which is actually contaminated must be made.

2.2 Diffusion to the slowly moving water in the rock

The water moving in the rock will not move through the backfill and concrete as previously shown. It will be diverted around the backfill. The water will only be in contact with the backfill over a very limited area which is mainly the area that makes up the fracture openings. The water contains no nuclides as it approaches the backfill. During its slow passage past the backfill, the nuclides will diffuse into the mobile water in the fractures. When the water leaves the vicinity of the backfill, the water will have picked up nuclides to a certain distance (the penetration depth by diffusion) out in the fracture. Therefore, the flowrate of nuclides from the backfill to the water is to a good approximation [7,8],

$$Q_{eq} = A_w \cdot \epsilon_{rock} \cdot (4 D_w / \pi t_{res})^{0.5} \qquad (11)$$

where A_w is the surface area between the rock and the backfill (about 5000 m^2), ϵ_{rock}

is the flow porosity, D_w is the diffusivity of the nuclides in water, and t_{res} is the residence time of the water in contact with the backfill.

Now it is assumed that the rock is described as a porous medium, i.e. the fractures are rather close compared to the thickness of the backfill barrier. The residence time, t_{res}, is directly obtained from the flowrate, porosity, and length of pathway. The flux u_o is on the order of 1 l/m^2y. With a porosity of 10^{-4} and a pathlength of 50 m, the residence time is 5 years. The diffusivity of water, D_w, is typically $2 \cdot 10^{-9}$ m^2/s. With these values Q_{eq} is equal to 63.6 l/yr which is the flowrate of water that will be contaminated.

Using Equation (10), the ratio of transport resistances at steady state is equal to 100, which means that the resistance to transport is totally dominated by the diffusion in the slowly moving water in the fractures with the rock. The diffusion in the concrete and backfill plays no role in practice. The slowly passing water picks up the nuclides so that an equivalent flowrate, Q_{eq}, is contaminated leaving the repository with a concentration c_w. In this case when the barriers have a small influence on the release, the concentration c_w is practically the same as the momentary concentration in the silo.

Because of the square root dependence in Equation (11) of the residence time and porosity, the results are not very sensitive to the data. If the data were in error by a factor of 10 for the flowrate as well as for the porosity at the same time, increasing both values, the equivalent flowrate Q_{eq} would increase by a factor of 10. Thus, the slow diffusion to the passing water would still limit the release of nuclides from the silo.

2.3 Summary of the nuclide release results from the repository

Table I summarizes the previously obtained results. The results are expressed as Q_{eq}^* and as the time constant τ. Moreno and Neretnieks [1] describe and summarize a series of calculations of radionuclide release performed with parameter values which differ from the case presented here in order to assess the sensitivity to the parameters. The calculations have been performed accounting for the retardation and decay in the concrete and backfill barriers, the limited transport capacity of the water in the rock, but not accounting for the sorption capacity of the concrete in the silo.

Table I Summary of the Nonsorbing Nuclides Release Calculations From the Silo to the Passing Water.

Barrier(s)	Q_{eq}^* (l/y)	τ (y)	Item
Concrete wall 0.5 m	9450	770	1
Backfill 1.2 m, porous medium case	13,140	550	2
Concrete and backfill	5500	1320	3
Flowrate in 1000 m^2 is contaminated	1000	7250	4
Diffusion to passing water if rock is porous medium	64	115,100	5
Combined effects of diffusion to passing water, concrete, and backfill (items 1,2, and 5)	64	115,100	6

3 DESCRIPTION OF THE FLOW SYSTEM AND CHANNEL OBSERVATIONS

3.1 Observations and interpretations of observations at SFR

Bolvede and Christiansson [3] have made a detailed survey of the water flow in the rock in the SFR drifts and tunnels. It was found that under the prevailing hydraulic gradient the ceilings of the four main repository tunnels have a water inflow of 30 l/min. This flowrate emerges from 164 different spots. 41 of the spots are at fracture intersections or small holes (points). Figure 1 shows the tunnel system in SFR and the spots where flow was observed. The total area of observation in these tunnels was 14,000 m^2.

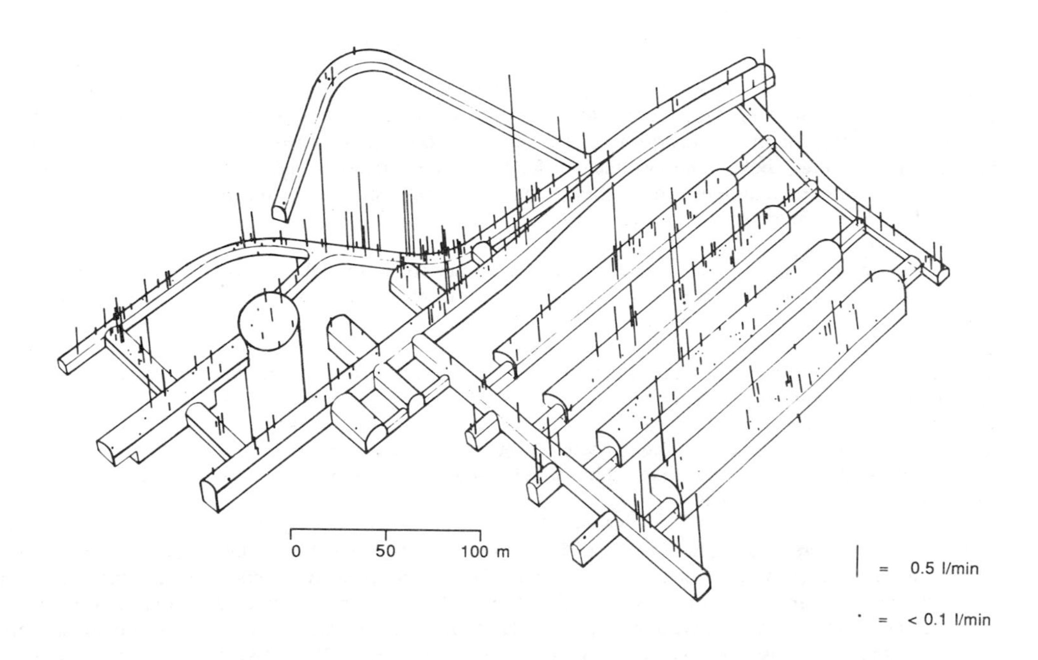

Figure 1 Overview over the SFR site (bar lengths are proportional to flowrate).

The flowrate in the different spots were measured and grouped in six different ranges. In Table II, the results show that nearly 50% of the flowrate can be accounted for in 18 spots over an area of 14,000 m^2.

Table II Summary of the flowrates in different flowrate ranges in the four main tunnels of SFR.

Flowrate range, l/min	Number of spots	Flowrate l/min	Fraction of spots	Fraction of flow
<0.1	67	2.53	0.41	0.08
0.1-<0.2	38	3.85	0.23	0.13
0.2-<0.4	41	9.30	0.25	0.31
0.4-<0.8	12	6.30	0.07	0.21
0.8-<1.6	4	4.50	0.02	0.15
>=1.6	2	4.00	0.01	0.13
Total	164	30.48	1.00	1.00

Interpretations of the observations at SFR show that on the average there is only one channel per 85 m^2 at SFR. Of these channels, a few carry most of the water. 50% of the flow takes place in less than 10% of the channels. It was noted that many of the channels with the highest flowrates were very narrow. The water emerged essentially in a small point in the rock. The wetted surface of the channels per volume of rock is very small if only those channels which carry most of the water are accounted for in this case. The hydraulic conductivity, K_p, was estimated to be $8 \cdot 10^{-9}$ m/s.

The observations above indicate that a conceptual model of flow in the channels is that most of the water flow takes place in a limited number of channels. The channels are seldom wider than a few meters and are often much narrower, not infrequently being small holes. If the observations reflect the flow paths in the bulk of the rock, there is considerably less sorption surface than if all the fractures are accessible to water flow. It is not known how far the channels extend before meeting other channels. It seems reasonable to assume that the channels will not meet for at least a distance equal to the average distance between channels. The distance between intersections is probably much larger because if the channels are like thin widely spaced ribbons they are unlikely to intersect even for very long distances.

If the channels do not intersect for a certain distance then they may be modeled as a bundle of independent channels each with its own flowrate and wetted surface. Those channels with a large flowrate and small sorption surface will carry the tracers rapidly and in large amounts, thus dominating the nuclide transport. If the water in the channels mixes regularly, then a mass of nuclides entering one channel will statistically move through all types of channels and the Advection–Dispersion case would apply.

4 CONCEPTUAL MODELS FOR FLOW AND TRANSPORT IN FRACTURED ROCK

4.1 Advection–Dispersion model

The concentration of a species in the water is described by

$$\frac{\partial c}{\partial t} + u \frac{\partial c}{\partial z} = D_L \frac{\partial^2 c}{\partial z^2} \qquad (12)$$

where c is the concentration in the liquid (mol/m^3), D_L is the dispersion coefficient (m^2/s), z is distance in the direction of flow (m), and t is time (s). Equation (12) can be used to calculate how a radionuclide which does not interact with the solid material will move with the water. Decay is not accounted for in Equation (12) but is easily added when needed. When the dissolved species sorb on the surfaces of the solid material it will be retarded in relation to the water velocity. For a linear instantaneous equilibrium Equation (12) is modified to become

$$R_a \frac{\partial c}{\partial t} + u \frac{\partial c}{\partial z} = D_L \frac{\partial^2 c}{\partial z^2} \tag{13}$$

where Ra is the retardation factor due to surface sorption ($R_a = 1 + K_a \cdot a$), K_a is the surface sorption coefficient, and $a = 2/\delta$ is the surface to volume ratio where δ is the fracture aperture. If the water velocity is u, the nuclide velocity will be

$$u_N = \frac{U_o S}{\delta(1+K_a \frac{2}{\delta})} \simeq \frac{U_o S}{2K_a} = \frac{U_o}{K_a a'} \qquad \text{for} \quad \frac{2K_a}{\delta} >> 1 \tag{14}$$

where S is the average fracture spacing, U_o is the Darcy velocity, and a' is the wetted fracture area per volume of rock (a'= 2/S). For the sorbing nuclides the term $2K_a/\delta$ usually is >> 1, therefore the nuclide velocity is independent of the water velocity in the rock for a given flux (Darcy velocity). The nuclide velocity is then primarily determined by the flux, surface sorption coefficient, and by the wetted fracture surface in the rock. It may be noted that the flow porosity of the rock does not influence the velocity of sorbing nuclides for a given Darcy velocity and fracture spacing under these circumstances.

4.2 "Bundle of channels" model

The "channeling model" states in essence that if the channel characteristics are known as well as the frequency of channels which carry a certain flowrate, the transport in every channel is calculated individually and the sum of the effluents is calculated by adding the effluent concentration of all channels. In this case where the channels have been grouped in flowrate ranges, the effluent from every group is calculated and multiplied by the number of channels in every group

$$C = \Sigma f_i \cdot c_i \tag{15}$$

where f_i is the fraction of flowrate in a group and c_i is the effluent concentration from a channel in group "i".

The concentration of the effluent from a channel will depend on the retardation and possible decay of the nuclides traveling through the channel. The main retardation effect is due to sorption within the rock matrix. Hydrodynamic dispersion is neglected in this case because "dispersion" due to the flowrate distribution will be considerably larger than that due to hydrodynamic dispersion.

The effluent concentration from a single channel has been modeled [9]. In this case there is, however, the added influence of the porous matrix. The nuclide will diffuse into the micropores of the matrix and sorb on the inner surfaces of the matrix. This is given by

$$\frac{\partial c_f}{\partial t} + u_f \frac{\partial c_f}{\partial z} = \frac{2D_e}{\delta} \cdot \left.\frac{\partial c_p}{\partial x}\right|_{x=0} - \lambda c_f \tag{16}$$

where c_f is the concentration in the liquid in the fracture (mol/m^3), c_p is the concentration in the liquid in the pores (mol/m^3), δ is the fracture aperture (m), u_f is the water velocity in a fracture (m^2/s), x is distance into the rock from a fracture (m), and λ is the decay constant (s^{-1}). The diffusion into (and out of) the rock matrix is given by

$$\frac{\partial c_p}{\partial t} = D_a \frac{\partial^2 c_p}{\partial x^2} - \lambda c_p \tag{17}$$

The initial and boundary conditions state that initially there is no nuclide in the water in the fracture and also no nuclide in the pore waters of the rock matrix. The boundary conditions state that at time zero the nuclide concentration is suddenly raised to c_o at the inlet of the fracture. If the inlet concentration decays by radioactive decay the solution to the above equations with the water residence time (t_w = channel volume/flowrate in channel) introduced, becomes

$$c = c_o \cdot e^{-\lambda t} \cdot \mathrm{erfc}\left[\frac{(D_e K_d \rho_p)^{1/2} LW_{fr}}{Q(t-t_w)^{1/2}}\right] \tag{18}$$

The wetted fracture surface is the important entity because it is through the surface in contact with the flowing water that the nuclides diffuse into and sorb in the rock matrix. For sorbing nuclides and for contact times longer than the water residence time, the water residence time has little influence. The most important entities to assess are the water flowrate in a channel, Q; the wetted surface of a channel, $2LW_{fr}$; the diffusivity in the rock matrix, D_e; and the sorption coefficient in the matrix, K_d.

4.3 Frequent mixing model

The channeling model assumes that water flows in channels which are widely separated with no connections in between and each channel transports radionuclides with a given velocity. This model predicts an equivalent dispersion coefficient which increases linearly with distance.

Rasmuson [10] studied the dispersion caused by channeling and determined the minimum number of mixing cells required to obtain an approximate Gaussian response. The minimum is a function of the channels aperture distribution (or of the flowrate). If the number of connections is large, then the channels loose their identity by mixing with other channels. The outlet concentration may then be calculated with the mean flowrate. Therefore, the entity that is important to determine is the ratio of wet surface to the water flow for all the channels.

For a short distance it may be expected that the number of mixing occasions is small, but for long distances a large number of mixing may be expected to occur. Then the outlet concentration will be determined by the existence of mixing between the channels.

5 APPLICATION TO SFR

Transport of nuclides from the SFR repository is calculated for different migration distances, flowrates, and sorption constants. Two models are used. The first considers the transport of nuclides through a bundle of channels with different properties and with no mixing between each other. In the second model, it is assumed that the transport occurs through the same channels, but with frequent mixing between the channels.

The calculation of nuclide migration by channeling requires that the geometrical characteristics of the channels and the flow distribution between the channels are known. It is assumed that the flowrate distribution observed on the ceiling of the four main tunnels at the SFR repository may be extended to the rock mass between the repository and the biosphere.

For calculation purposes the channels are divided into 6 groups with flowrate ranges increasing in a geometric progression. The number of channels and the fraction of the total flow in each channel group are needed for these calculations. The flow in each channel is calculated from the fraction of flow and the number of channels in each group and the total flux in the rock. Other additional data used in the calculations are the rock effective diffusivity, $5 \cdot 10^{-14}$ m^2/s; the channel width, 1 m; and the channel aperture, 1 mm.

For the larger times the channel aperture and water residence time have little or no impact, but for short times it has an impact. Calculations have been made for ^{240}Pu at a migration distance of 50 m for two different channel apertures, 0.01 and 1.0 mm [11]. The results are identical for times larger than 10 years. The channel aperture influences only the results for times less than a few years. Therefore in the calculations the channel widths are to be 1 m.

Table III summarizes the fraction of the flowrate which is carried in a certain fraction of the channels. The third column shows what the flowrate per channel would be when there is a hydraulic gradient of $4 \cdot 10^{-3}$ m/m which gives an average flux of 1 l/m^2y.

5.1 Calculated results for both channeling and mixing models

To illustrate the influence on the effluent concentration from the different channels, breakthrough curves for ^{240}Pu at a distance of 50 m and for a flux of 11al/m^2y are calculated. The outlet concentration for each group of channels and the effluent concentration for all the channels are shown in Figure 2. The results show that channels in group 1 are few but allow little decay. Channels in group 2 carry about half as much water per channel and allow some more decay. Channels in group 6 are the channels which carry the least water per channel and there the nuclides have considerable decay. The total effluent concentration from all channels is obtained from Equation (12).

The maximum concentration at a certain distance for the release of ^{240}Pu have been calculated [11]. The results show the impact of the water flux. The concentration is insignificantly increased with an increase of the water flow for short distances but has a considerable impact for long distances. The water flux of 10 l/m^2y is expected to occur at long times, longer than 1000 y, for a vertical downward flow near the repository. When the water flow is horizontal at large distances from the repository, the water flux for long times is expected to be somewhat smaller, about 5 l/m^2y.

Table III Fraction of the total flowrate which flows in the different groups of channels.

Fraction of flowrate	In fraction of channels	Would be flowrate per channel $m^3/s \cdot 10^{-8}$	Flowrate range
0.131	0.012	2.940	1= largest
0.148	0.024	1.653	2
0.207	0.073	0.772	3
0.305	0.250	0.333	4
0.126	0.232	0.149	5
0.083	0.409	0.056	6= smallest

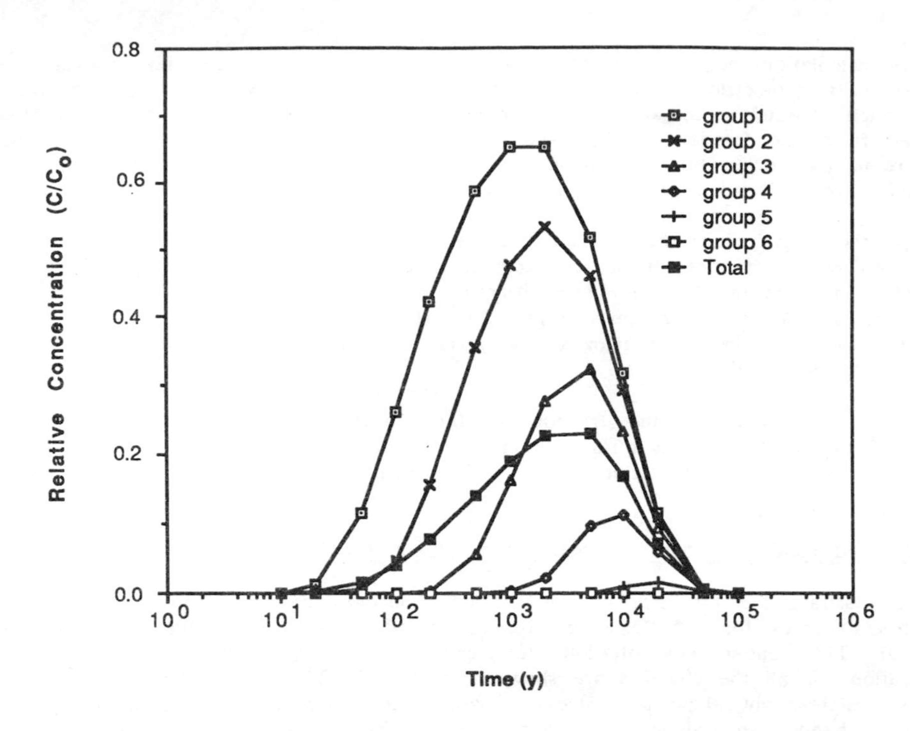

Figure 2 Effluent concentration in the different groups of channels for ^{240}Pu.

Calculations for channels which intersect many times in their path were performed by Moreno *et al.* [11]. The channeling model assumes that channels do not intersect each other from the repository to the bottom of the sea. As it is discussed above, for long distances it is expected that the channels intersect to some extent along their paths. If the channels intersect many times, the identity of the channels is lost. In this case the transport of radionuclides is determined by the ratio of the total wet surface to the water flux. The release then is much less than the outlet concentration when nonintersecting channels are used.

5.2 Calculations for a time limited nuclide inventory

In the calculations above, it is assumed that the concentration of the nuclide release from the silo is constant. This means that the release of contaminants from the silo is small if compared with the amount of nuclides in the repository. In these calculations a concentration decreasing only by decay during 100,000 years or larger is used. If the release of radionuclides is large compared with the initial amount of radionuclides in a part of the repository, the concentration in the silo will decrease with the time faster than just due to decay. After a long time this concentration may be low or very low. This situation was simulated assuming a pulse release of contaminent, therefore the concentration is zero before and after this time period.

Moreno *et al.* [11] calculated the maximum relative concentration of ^{240}Pu at the outlet for different pulse durations: 100, 400, and 1600 years. For long distances the effect of the release duration is more significant. For short distances the effect is important only for a release time of 100 years and water flow of 1 l/m^2y.

5.3 Impact of changes in K_d, W, and Dispersion

For sorbing radionuclides the concentration at the outlet at a given time is determined by the value of

$$\frac{LW(D_e K_d)^{1/2}}{Q} \qquad (19)$$

The effect of t_w is not accounted for. From Equation (19) it is seen that an increase in L by a factor 2 is equivalent to an increase in W by the same factor or an increase of K_d by a factor 4.

The sorption coefficient parameter requires less accuracy in its determination. This is due to that a change in a factor 10 in this parameter is equivalent to a change in a factor 3.3 in the other parameters. The same also applies for the diffusivity in the rock matrix.

In the calculations above, it is assumed that the dispersion in each channel is negligible if compared to the dispersion caused by channeling. Simulations were made using a value of 10 for the Peclet number. This value is based on compiled results from different laboratory tests and an in–situ tracer test [12].

The results show a slight increase of the concentration at the outlet when dispersion in the channel is considered. Since the calculations that include dispersion require more computing effort, the solutions without dispersion are sufficient for the accuracy required in these calculations.

6 DISCUSSION AND CONCLUSION

The silo and its surrounding backfill and rock cavern are a rather complex structure from the point of view of transport calculations. The escape through the walls of the silo and through the backfill takes place by diffusion because the bentonite backfill and the concrete walls of the silo ensure that flow is slow through the silo. All of the short–lived nuclides and some nuclides with intermediate half–lives but with some sorption should decay to insignificance during their transport. Other more long–lived nuclides will survive to a smaller or larger extent and reach the mobile water outside. Once the nuclides have escaped to the mobile water in the rock the nuclides may be transported to the biosphere.

It was shown that the most important limitation to nuclide release is the slow diffusion out to the passing water. Because the barriers are in series, resistances are added and the barriers which have large Q_{eq}'s will give a small contribution to the overall resistance. A very long–lived or nondecaying, nonsorbing nuclide will have a release rate equal to $8.8 \cdot 10^{-6}$ yr^{-1} of the momentary inventory, once steady state is reached. Sorbing nuclides will have considerably smaller releases because of the sorption in the concrete within the silo and the surrounding barriers of the silo.

Experimental observations indicate that water flows very unevenly distributed in crystalline rock. Water flows in channels of different widths and with different flowrates. For the transport of radionuclides from the silo to the biosphere two channeling concepts were used in these calculations: (a) separate channels transport the water and radionuclides without connections to each other, and (b) the water and radionuclides are transported by channels which mix their water many times along their paths.

There are four entities which are of importance in the determination of the contamination from the repository. The entities are the sorption capacity of the rock matrix, the width of the channels, the channel frequency, and the water flowrate or water flux.

The observations for the water flow in the rock presented are very recent an indicate that there may be less wetted surface in the rock than previously thought. Th models are also new and have not been extensively applied. Thus the presented results ar based on new observations, interpretations, and models, and should be used with caution We feel, however, that at least for the larger distances they should not give too lo release rates.

REFERENCES

[1] Moreno, L., and I. Neretnieks: "Gas, Water, and Contaminant Transport from a Fina Repository for Reactor Waste", SFR 85–09, Svensk Kärnbränslehantering AB, Stockholm 1985.

[2] Andersson, K., and B. Allard: "The Chemical Conditions Within a Cement Containin Radioactive Waste Repository", SFR 86–09, Svensk Kärnbränslehantering AB, Stockholm 1986.

[3] Bolvede P., and R. Christiansson: "SKB Forsmarksarbetena SFR. Vattenförande Spricko inom Lagerområdet", VIAK AB, Stockholm, 1987. (in Swedish)

[4] Neretnieks, I., S. Arve, L. Moreno, A. Rasmuson, and M. Zhu: "Degradation o Concrete and Transport of Radionuclides from SFR Repository for Low– an Intermediate–Level Nuclear Waste", SFR 87–11, Svensk Kärnbränslehantering AB Stockholm, 1987.

[5] Neretnieks, I.: "Diffusivities of Some Constituents in Compacted Wet Bentonite Cla and the Impact on Radionuclide Migration in the Buffer", Nuclear Technology, 71, (1985) 458–470.

[6] Carlsson, L., B. Grundfelt, and A. Winberg: "Hydraulic Modelling of the Fina Repository for Reactor Waste (SFR). Evaluation of the Groundwater Flow Situation at SFR" KEMAKTA AR 87–1, Swedish Geological Co., KEMAKTA Consultant Co., Stockholm 1987.

[7] Neretnieks, I,: "Diffusion in the Rock Matrix: An Important Factor in Radionuclid Retardation?", J. Geophys. Res., 85, (1980), 4379–4397.

[8] Bird, R. B., W. E. Stewart, and E. N. Lightfoot: Transport Phenomena, Wiley, Ne York (1960).

[9] Neretnieks, I., T. Eriksen, and P. Tähtinen: "Tracer Movement in a Single Fissure i Granitic Rock: Some Experimental Results and their Interpretation", Water Resources Res., 18, (1982), 849–858.

[10] Rasmuson, A.: "Analysis of Hydrodynamic Dispersion in Discrete Fracture Network Using the Method of Moments", Water Resources Res., 21, (1985), 1677–1683.

[11] Moreno, L., I. Neretnieks, and S. Arve: "A Note on the Possible Retardation o Radionuclides Escaping from the SFR Repository", Svensk Kärnbränslehantering AB Stockholm, 1987.

[12] Neretnieks, I.: "Transport in Fractured Rocks", Memoires of the 17th Internationa Congress of IAH, Tucson, AZ, vol. XVII, 1985, 301–318.

MODELLING OF RADIONUCLIDE TRANSPORT IN THE NEAR FIELD OF A SWISS L/ILW REPOSITORY

Wiborgh M., Pers K., Höglund L.O. / Kemakta, Sweden
van Dorp F. / Nagra, Switzerland

ABSTRACT

The major processes determining the radionuclide transport from a repository are considered to be diffusion and/or advection. Which of the transport mechanisms dominates depends on the properties of the host rock, the repository design, the properties of the engineered barriers and their degradation. Different processes may dominate for different time periods. The effects of the different properties of the host rock and the engineered barriers on the radionuclide release are presented. Comments are given on the effects of sorption and different boundary conditions on the radionuclide release for the separate transport mechanisms.

MODELISATION DU TRANSPORT DE RADIONUCLEIDES DANS LE CHAMP PROCHE D'UN DEPOT SUISSE DE DECHETS DE FAIBLE ET DE MOYENNE ACTIVITE

RESUME

On estime que les principaux processus régissant le transport de radionucléides à partir d'un dépôt sont la diffusion, l'advection + la diffusion, et l'advection. Parmi ces mécanismes, ceux qui influent le plus sur le transport dépendent des propriétés de la roche réceptrice, de la conception du dépôt, des propriétés des barrières ouvragées et de leur dégradation. Les processus prédominants peuvent varier dans le temps. Les effets des différentes propriétés de la roche réceptrice et des barrières ouvragées sur le rejet de radionucléides sont exposés. Des commentaires sont formulés à propos des effets que la sorption et des conditions différentes aux limites exercent sur le rejet de radionucléides pour chacun des mécanismes de transport considéré séparément.

INTRODUCTION

Performance assessment studies have been carried out for a Swiss L/ILW repository. In the present paper, some results of near-field radionuclide transport calculations are presented. The examples are based on a highly simplified geometry of a repository "sandwich" concept with conditioned waste in rock tunnels backfilled with concrete. Results are given as fractional release rates to enable direct comparison between different processes and assumptions.

PROCESSES CONSIDERED

The near-field of the Swiss repository for L/ILW waste consists of immobilised waste of different types [1,2,3] and concrete structures. Potential host rocks are a relatively soft marl and crystalline, both of which will restrict the groundwater flow rate through the repository due to low hydraulic conductivity. The hydrodynamic properties of intact concrete are likewise favourable for storage of radioactive waste due to very low hydraulic conductivity.

The major transport mechanism for radionuclides in a L/ILW repository depends on the relative importance of diffusive transport and the advective transport by flowing pore water. A measure on the relative importance of the different processes can be formulated by the Peclet number Pe;

$$Pe = \frac{U \cdot L}{De} \qquad \text{(eq. 1)}$$

U = waterflow rate (= large scale conductivity times gradient) $[m^3 \cdot m^{-2} \cdot s^{-1}]$
L = characteristic dispersion length, here: repository radius [m]
De = dispersion coefficient, here: effective diffusivity $[m^2 \cdot s^{-1}]$

The predominant transport mechanism for high Peclet numbers is advection and for low Peclet numbers diffusion will dominate. For Peclet numbers around unity (e.g. 0.01 - 100) both advection and diffusion must be considered.

The waterflow rate through the repository has been modelled taking into account the large scale hydraulic properties of the host rock as well as the hydraulic properties of the near-field barriers.

From potential theory it can be predicted that the maximum waterflow rate in a long tunnel is twice the flow in the surrounding rock if the flow is perpendicular to the tunnel axis [4]. More accurate estimations can be made by analytical and numerical models, but for this paper the range of possible waterflow rates in the repository has been assumed to be in the range from zero up to twice the flow rate in the host rock.

For waterflow rates exceeding $5 \cdot 10^{-9}$ $m^3 \cdot m^{-2} \cdot s^{-1}$, the transport by diffusion is negligible in comparison with advection for all possible values of the effective diffusivity.

Fresh concrete has, in general, a low hydraulic conductivity and low diffusion coefficients. This results in low radionuclide release rates from the near-field into the surrounding rock. Because of the large dimensions of the repository and the low diffusivities, steady-state develops only slowly. Thus, the release of radionuclides has to be modelled as a transient process. The chemical properties of concrete are also advantageous, e.g. high sorption capacity for many cations, high pH porewater giving low solubilities of some radionuclides and a reasonably high resistance to chemical attack from species in the ground water.

Due to ageing, the hydrodynamic properties of the concrete barriers could significantly change with time. Ageing can be caused by leaching, recrystallisation of the metastable calcium silicate gel phase (which gives the mechanical strength to the concrete) and the possible chemical attack of some species (e.g. carbonate, sulphate, magnesium) with subsequent swelling leading to fracturing and enhanced leaching. The actual time scale for deterioration is difficult to assess but can be estimated using pessimistic assumptions. It cannot be excluded that diffusivities and hydraulic conductivity will increase by one to three orders of magnitude with time [3,5,6,7]. Also, the sorption capacities can be expected to decrease with time [8].

Other processes can also produce gradients for radionuclide transport from the repository e.g.; gas formation in the repository giving rise to a pressure gradient [2,5,9], and heat evolution due to decaying radionuclides and heat released during cement hydration which will give rise to a temperature gradient (Soret effect). These effects will not be treated in this paper.

Solubility limitations can limit the release from the near-field but are not been treated in this paper.

METHODS

The methods used for the modelling of the radionuclide transport by different transport mechanisms are described below.

Conceptual model

The one-dimensional transport of solutes in direction x by water flowing with velocity v can be described by:

$$R \frac{\partial c}{\partial t} = Dp \frac{\partial^2 c}{\partial x^2} - v \frac{\partial c}{\partial x} \qquad \text{(eq. 2)}$$

$R = 1 + \frac{Kd \cdot \rho}{\epsilon}$, the chemical retention factor [-]
c = porewater concentration [$\text{amount} \cdot m^{-3}$]
t = time [a]
Dp = pore diffusivity or dispersion coefficient [$m^2 \cdot a^{-1}$]
Kd = distribution coefficient [$m^3 \cdot kg^{-1}$]
ρ = density of barrier material [$kg \cdot m^{-3}$]
ϵ = porosity [$m^3 \cdot m^{-3}$]

To solve (eq. 2) initial and boundary conditions are required. Initial conditions can be the radionuclide source term, e.g. given as initial concentrations or solubility limits. Boundary conditions are often chosen at the interface with the host rock, for example a zero concentration boundary condition. Radioactive decay is treated separately.

<u>Geometrical simplifications</u>

The effect of geometrical simplifications made can be demonstrated by performing the calculations for different model interpretations of the real system. An example from [3] is given in Figure 1.

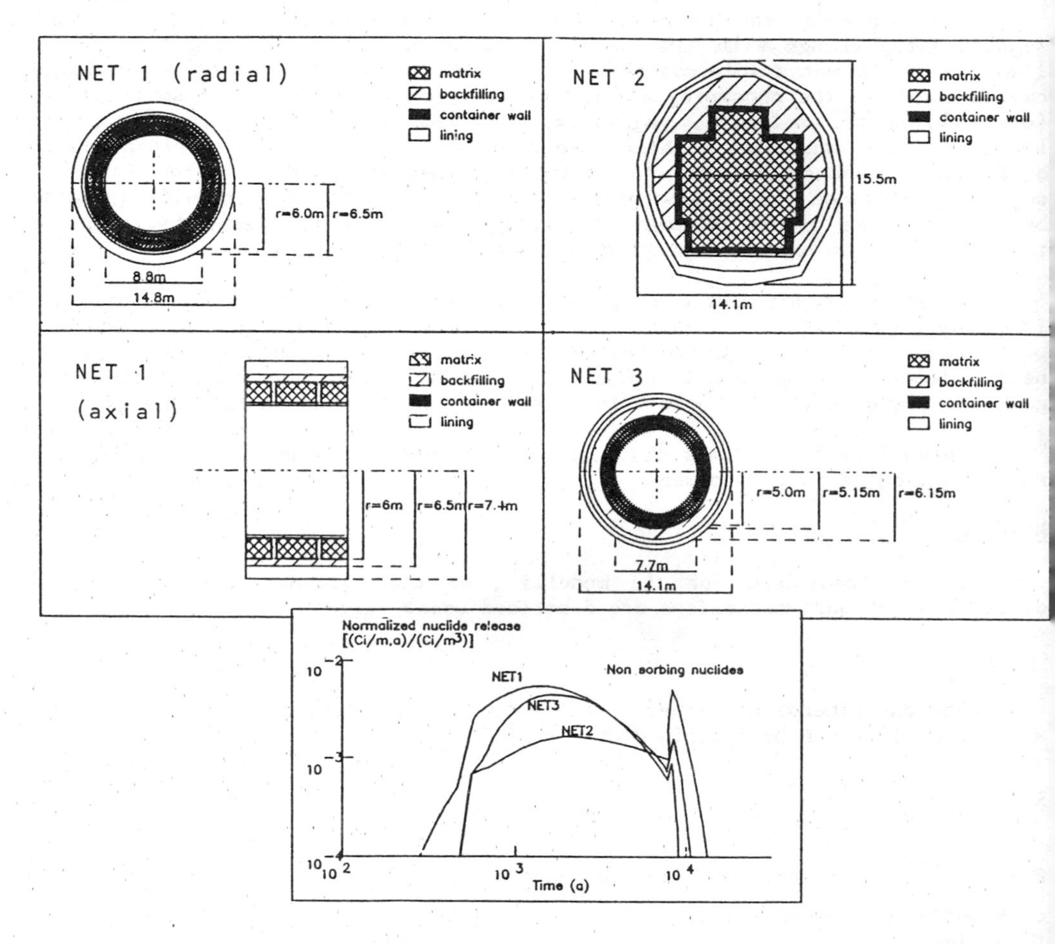

Figure 1 Three different model simplifications (NET1, NET2 and NET3) of the L/ILW near-field barriers.

The discretisation in a numerical model must be sufficiently detailed to guarantee required accuracy and to maintain numerical stability [10]. Normally, a number of simplifications are made to reduce the computational efforts. However, when simplifying a system, the following should be taken into account:

- the travel distance of radionuclides must not be overestimated in the model (this complicates radial simplifications, see Figure 1)

- surface/volume ratios of the different barriers and areas available for transport should resemble the reality if possible
- the volume of the immobilised waste should be identical to the real waste if solubility limitations are to be modelled
- the volume of the different barrier materials must be close to reality to describe the sorption capacity correctly
- all major transport paths must be included in the model.

Source terms for the model

The source term for the near-field model in a L/ILW repository is assumed to be the pore water concentration of the different radionuclides. For easily dissolved radionuclides in a porous material such as concrete, the pore water concentration is given by the amount of radionuclide per unit volume of immobilised waste times the porosity. For non-porous materials like bitumen, the source term will be determined by the rate of water penetration and the dissolution rate of the radionuclides in the penetrating water. At present these effects for bituminised waste are not accounted for but concrete properties are also assigned to bitumenised waste. Thus for all waste the radionuclides are assumed to be dissolved in the pore water. The concentrations of waste matrices are reduced by sorption and solubility limitations. Sorption has been modelled using distribution coefficients. Solubility limits for the different elements can be determined by chemical equilibrium calculations [3,11,12].

The release of radionuclides from the near-field has been modelled in four separate ways:

- release controlled by diffusion
- advective-diffusive release
- pure advective release, i.e. plug flow (not treated in this paper)
- no diffusion resistance, i.e. a mixing tank model.

Boundary conditions

For modelling diffusive transport the following boundary conditions for the release into the surrounding rock can be chosen:

- zero concentration at the near-field/rock interface
- a convective boundary condition, i.e. diffusion into the ground water flowing around the repository [13].

The zero concentration boundary condition is applicable for situations where the flow rate of ground water in the rock is high. For low and moderate flow rates in the host rock, the convective boundary condition may be more realistic.

Advective-diffusive transport in the near-field is modelled using one of the following boundary conditions:

- zero concentration at the near-field/rock interface, applicable when high water flowrates are expected in the rock

- no concentration gradient at the near-field/rock interface, i.e. discharge into a semi-infinite medium by a finite water flow rate, applicable for low and medium waterflow rates in the rock.

Instead of using these two boundary conditions, the transport in the rock can also be included in the model. However, this requires a large element mesh and an increased computational effort.

For pure advective transport, the only applicable boundary condition is the discharge into a semi-infinite medium with a zero concentration gradient at the near-field/rock interface.

Codes

Radionuclide releases from a L/ILW repository can be predicted by analytical and numerical models. The calculations presented in this paper have been carried out with the numerical model TRUMP [14]. This code is capable of solving one to three dimensional advective-diffusive transport problems and uses the integrated finite difference method. Figure 2 presents a comparison of the numerical results with an analytical solution to the advective-diffusive transport in a semi-infinite medium .

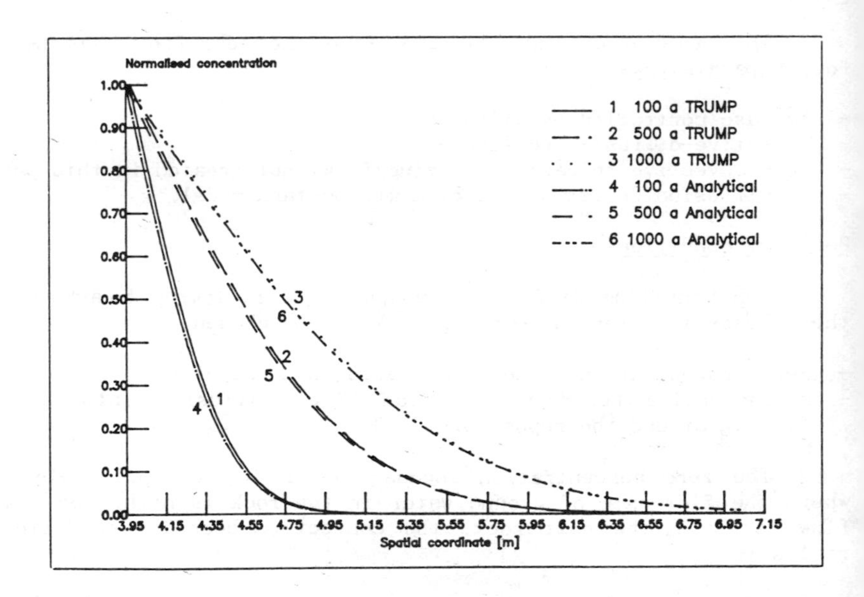

Figure 2 Comparison of the results of the numerical model TRUMP with an analytical solution. The different curves show the concentration profiles in a column at different times.

CASE DESCRIPTION

The model assumes the waste to be contained in a cementitious matrix material surrounded by a highly porous concrete. An optional high quality barrrier (HQ-barrier) surrounds the backfilling. Finally, a mechanical support lining is in direct contact with the rock walls. In Table I the data assumed for the calculations are presented.

Table I Data assumed and modelling approach in the calculations.

Case	Boundary condition	Host rock conductivity $[m \cdot s^{-1}] \cdot 10^{-10}$	Advective flow in barriers $[m^3 \cdot m^{-2} \cdot s^{-1}] \cdot 10^{-10}$	Effective diffusivity in Matrix $[m^2 \cdot s^{-1}] \cdot 10^{-11}$	Backfill	HQ	Lining
1	No gradient	10	10	3	60	0.3	4
2	Zero conc.	High	10	3	60	0.3	4
3	Zero conc.	High	Low[4]	3	60	0.3	4
4	No gradient	10	10	3	High[3]	0.3	4
5	No gradient	10	10	High[3]	High[3]	0.3	4
6	No gradient	10	10	High[3]	High[3]	High[3]	High[3]
7	No gradient	1	1	3	60	0.3	4
8	No gradient	0.1	0.1	3	60	0.3	4
9	Convective	10	Low[4]	3	60	0.3	4
10	Convective	1	Low[4]	3	60	0.3	4
11	No gradient	1	1	3	60	Absent	4
14[1]	No gradient	10	10	3	60	0.3	4
15[1]	No gradient	10	10	3	60	Absent	4
17[1]	No gradient	1	1	3	60	0.3	4
18	No gradient	10	0 - 10				
	Time<300 a	Flow in lining		3	60	0.3	4
	Time<1000 a	Flow in lining+HQ barrier		3	High[3]	0.3	4
	Time<10000 a	Flow in all materials		High[3]	High[3]	0.3	4
	Time>10000 a	Flow in all materials		High[3]	High[3]	High[3]	High[3]
19[2]	No gradient	10	0.1 - 10				
	Time<100 a	10	0.1	3	60	0.3	4
	Time<300 a	10	1	3→10	60	0.3→1.5	4→20
	Time<1000 a	10	1	10→30	60	1.5→3	20
	Time>1000 a	10	10	30	60	3	20

[1] Retention factor R=1000
[2] Continuously changing diffusivities are indicated by arrows
[3] High refers to the use of a mixing tank model for this material
[4] Low indicates transport by diffusion only.

In case 18 the transport process is assumed to gradually change with time from advective-diffusive transport to a transport with no diffusion resistance in the engineered barriers modelled as a mixing tank release, see Figure 3. In case 19 continuous degradation is assumed to change the transport properties of the engineered barriers with time and is modelled as an advective-diffusive transport.

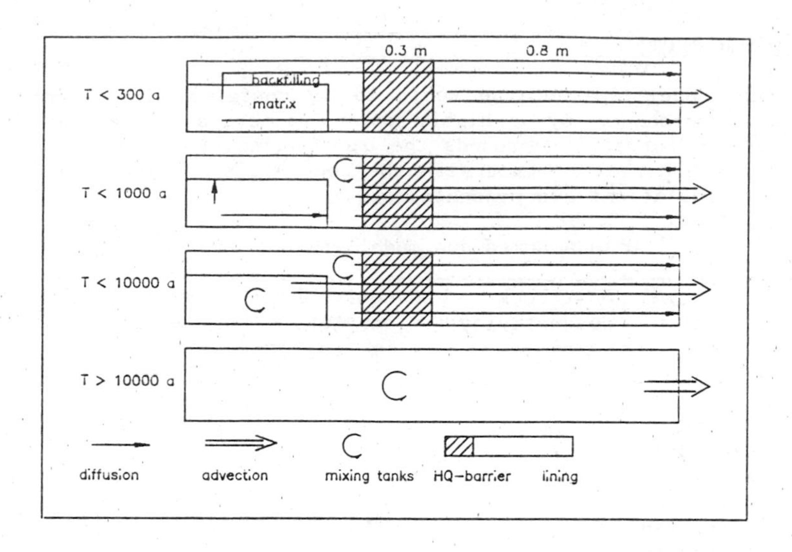

Figure 3 Schematic picture of the change of release mechanism with time as examplified by curve 7 in Figure 6.

RESULTS AND DISCUSSION

Diffusive release from the near-field

The effect of different boundary conditions on the rate of diffusive release from the repository is shown in Figure 4.

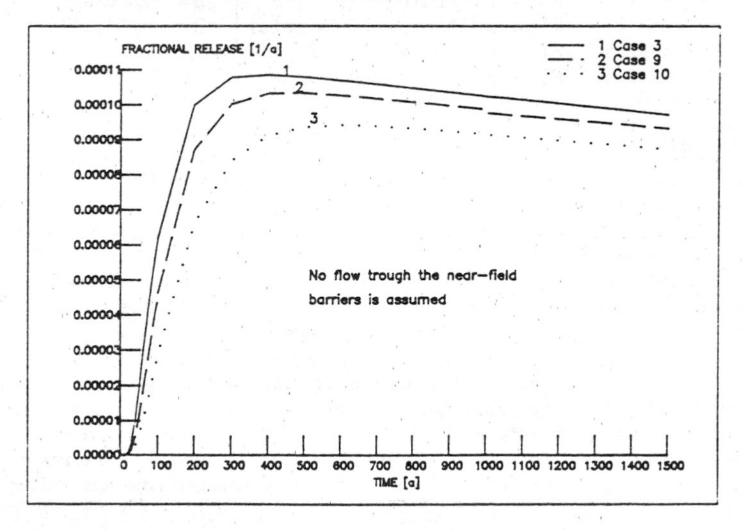

Figure 4 The effect of different ground water flow rates (requiring different boundary conditions) in the host rock on the release rates of radionuclides from the near-field by diffusion.

Advective - diffusive release from the near-field

For Peclet number rangning from 0.01 to 100 combined advective-diffusive release of radionuclides from the repository must be modelled.

The effect of different water flow rates in the engineered barriers as well as the effect of appropriate boundary conditions on the release into the host rock is demonstrated in Figure 5.

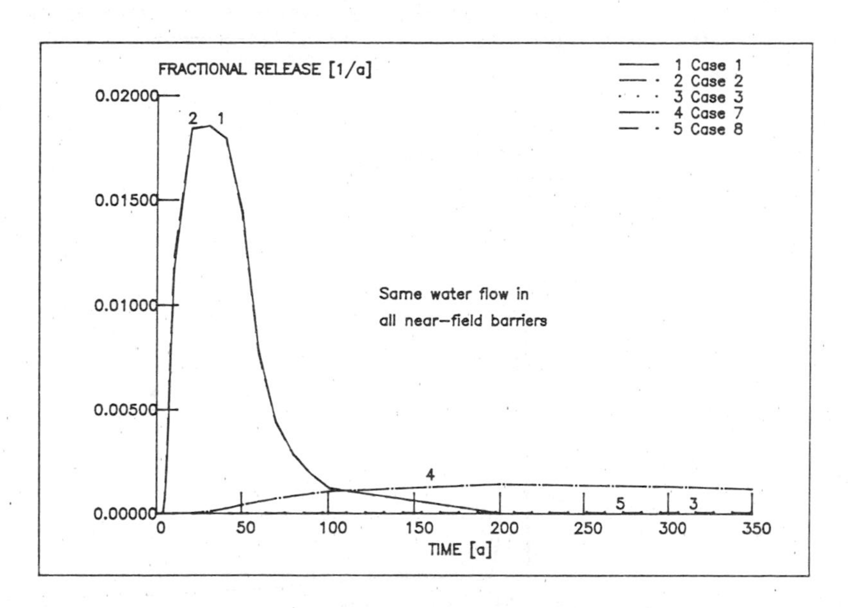

Figure 5 The effect of different water flow rates on the rate of release from the near-field. Also the effect of the boundary condition assumed for the interface between the near-field barriers and the host rock is shown.

It is clear from Figure 5 that, for the highest water flow rate (curve 1 and 2), the release is almost completely determined by advection. The effect of different assumptions for the boundary condition is small for the highest flow rate. For low and moderate flow rates, however, both diffusive and advective transport are important.

The simplification made in the conceptual model of the different barriers can also influence the calculated release from the near-field. Figure 6 (curves 2, 3 and 4) shows the effect of modelling one or more of the barriers as a mixing tank.

Surprisingly we find that, for the mixing tank, the release is retarded due to continued upstream mixing with incoming non-contaminated water when the water flow rate is moderate or high. This shows that the apparently pessimistic assumption of a barrier without any diffusion resistance can in fact become too optimistic if applied to the wrong situation.

The impact of a high quality barrier on the radionuclide release rate is displayed in Figure 6 by curves 7 and 8. For these curves the high quality barriers are assumed to be intact and restricts the water flow rate through the repository during the initial period, whereafter the barrier transport properties are assumed to gradually change with time. A schematic picture of the change of release mechanisms with time corresponding to curve 7 is given in Figure 3, the model used is a mixing tank. In addition, the effects of barrier properties changing with time are shown by curve 8 in Figure 6. Here, successive barrier degradation is assumed to result in continuously increasing diffusivities and stepwise increasing hydraulic conductivities visible as a peak at 1000 years (modelled for the total time period as advective-diffusive transport).

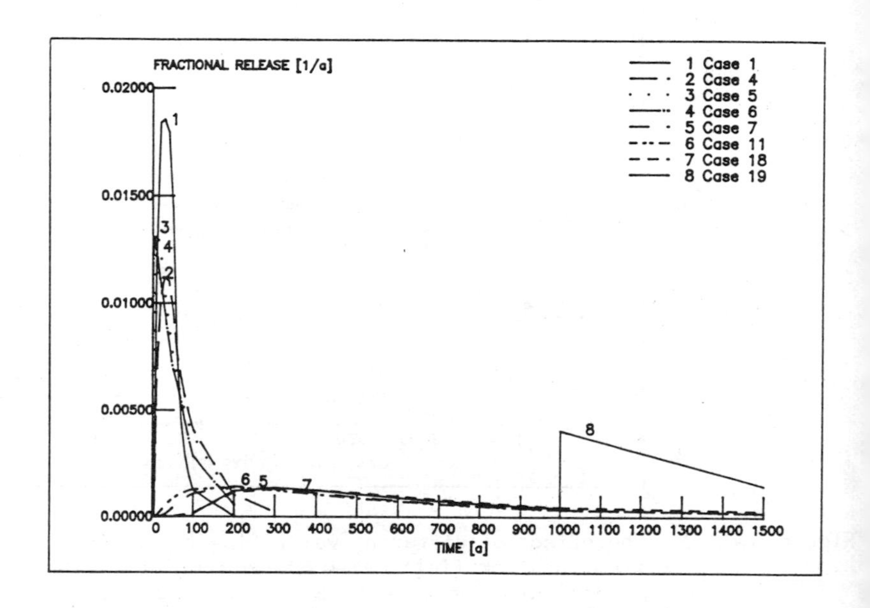

Figure 6 The effect of different assumptions regarding the transport mechanism on the release of radionuclides. The effect of transport mechanisms that change with time is also demonstrated in two different cases.

From the results it can be concluded that the high-quality barrier will have a considerable effect during the early period if the barrier remains intact and can contain many of the short-lived radionuclides long enough to allow them to decay within the barrier system. The ability of the near-field to contain and retard long-lived radionuclides, especially those that are strongly sorbed by cementitious materials, will not be significantly improved by the presence of high-quality barriers. The important factor for the retardation of such radionuclides is the sorption. In Figure 7 the effect of sorption on the cementitious barriers is shown.

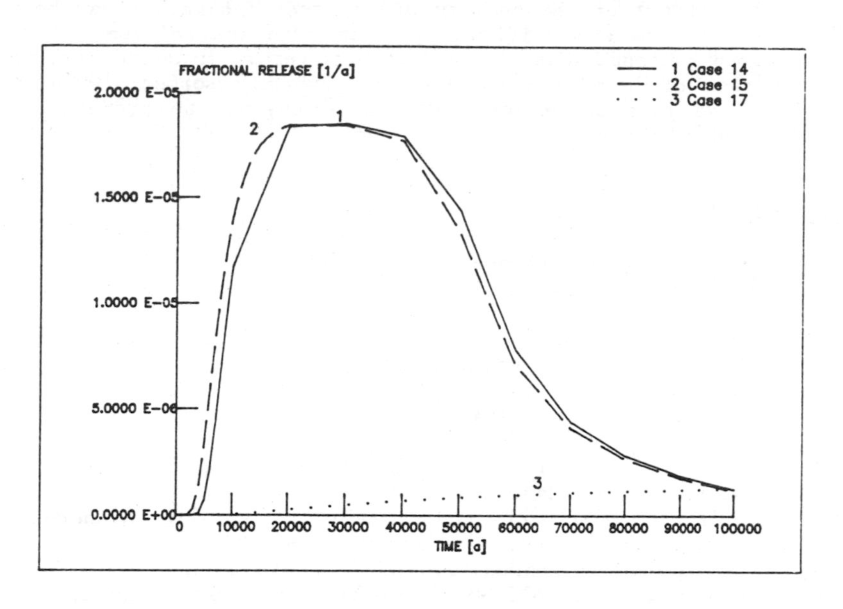

Figure 7 The influence of high-quality barriers on the release of sorbed radionuclides.

CONCLUSIONS

The effect of the quality of the near-field barriers is different depending on the quality of the host rock. For low permeable host rock, the near-field transport will be governed by diffusion whatever hydraulic properties the engineered barriers have. For a highly permeable host rock, the hydraulic properties of the engineered barriers determine the near-field release rates. For engineered barriers with low hydraulic conductivity, diffusive properties determine the release rate.

Diffusivity and hydraulic conductivity could increase with time by one to three orders of magnitude because of interaction between engineered barriers, waste and ground water. These changes will cause increased release by advective transport with time.

High porosity materials can be modelled as a mixing tank for low water flow rates. It was found that the apparently conservative mixing tank approach reduced the radionuclide release rate compared to the advective-diffusive release rate for high waterflow rates. This effect can be explained by the continuous upstream mixing with incoming non-contaminated ground water which results in dilution. This shows that the mixing tank model is not always a conservative approach.

The effect of the quality of engineered barriers has been demonstrated. The results show a significant influence on the release rate of short-lived, non-sorbed radionuclides if the barriers remain intact during an early period. The effect of barrier properties on strongly sorbed radionuclides is weak, because the time scales of release are beyond the provable lifetime of the engineered barriers.

REFERENCES

[1] Endlager für schwach- und mittelaktive Abfälle: Das System der Sicherheitsbarrieren, NGB 85-07, NAGRA, Baden, Switzerland, 1985.

[2] Wiborgh M., Höglund L.O., Pers K., "Gas Formation in a L/ILW Repository and Gas Transport in the Host Rock", Nagra Technical Report NTB 85-17, NAGRA, Baden, Switzerland, 1986.

[3] Karlsson L.G., Höglund L.O., Pers K., "Nuclide Release from the Near-Field of a L/ILW Repository", Nagra Technical Report NTB 85-33, NAGRA, Baden, Switzerland, 1986.

[4] Bird R.B., Stewart E.W., Lightfoot E.N., Transport Phenomena, Wiley International Edition, 1960.

[5] Endlager für schwach- und mittelaktive Abfälle: Sicher-heitsbericht, NGB 85-08, NAGRA, Baden, Switzerland, 1985.

[6] Höglund L.O., "Degradation of Concrete in a LLW/ILW Repository", Nagra Technical Report NTB 86-15, NAGRA, Baden, Switzerland, 1987.

[7] Berner U., Jacobsen J., McKinley I.G.,"The Near-Field Chemistry of a Swiss L/ILW Repository", Proceedings of the OECD/NEA Workshop on Near-Field Assessment of Repositories for Low and Medium Level Waste, Baden, Switzerland 23-25 November, 1987.

[8] Allard B., Anderson K.,"Chemical properties of radionuclides in a cementitious environment", SFR 86-09, SKB, Stockholm, 1987.

[9] Zuidema P., Höglund L.O.,"Impact of production and release of gas in a L/ILW repository - A summary of the work performed within the Nagra programme", Proceedings of the OECD/NEA Workshop on Near-Field Assessment of Repositories for Low and Medium Level Waste, Baden, Switzerland 23-25 November, 1987.

[10] Rasmuson A., Narasimhan T.N., Neretnieks I.,"Chemical Transport in Fissured Rock: Verification of a Numerical Model", Water Resources Research, Vol. 18, No. 5, 1982.

[11] Parkhurst D.L.,Thorstenson D.C., Plummer L.N., "PHREEQE - A computer program for geochemical calculations", U.S. Geological Survey, Water Resources Division, Rept. No. USGS/WRI-80-96, To be obtained at OECD/NEA, November 1980.

[12] Wolery T.J., "EQ3NR A Computer Program for Geochemical Aqueous Speciation-Solubility Calculations: Users Guide and Documentation", UCRL-53414, To be obtained at OECD/NEA, 1983.

[13] Neretnieks I.,"Transport of oxidants and radionuclides through a clay barrier" SKBF/KBS Technical Report 79, SKB, Stockholm, 1978.

[14] Edwards A.L.,"TRUMP: A Computer program for transient and steady state temperature distributions in multidimensional systems", National Technical Informational Service, National Bureau of Standards, Springfield, Va. 1969.

MIGRATION AND DIFFUSION OF RADIONUCLIDES IN ENGINEERED BARRIER SYSTEM

K. Shimooka and Y. Wadachi
Department of Environmental Safety Research,
Japan Atomic Energy Research Institute,
Tokai-mura, Ibaraki-ken, Japan

ABSTRACT

This paper describes the present status of the multibarrier system performance tests to provide a preliminary assessment of nuclide migration in the engineered barriers for shallow land burial of the low-level radioactive waste. Migration of radionuclides with seeped water through backfill and in subsequent diffusion in concrete pit are considered in this study. The results of laboratory investigations of unsaturated flow in backfill and radionuclides migration / diffusion in engineered barrier system are described and the calculated distribution of the radionuclides in backfill is presented.

MIGRATION ET DIFFUSION DES RADIONUCLEIDES DANS UN SYSTEME DE BARRIERES OUVRAGEES

RESUME

La présente communication décrit l'état actuel des essais de performances d'un système à barrières multiples, destinés à permettre une évaluation préliminaire de la migration des nucléides dans les barrières ouvragées utilisées pour l'enfouissement à faible profondeur dans le sol de déchets de faible activité. La migration des radionucléides entraînés par l'infiltration d'eau à travers les matériaux de remblayage et leur diffusion ultérieure dans le silo de béton sont examinées dans cette étude. Les auteurs décrivent les résultats des recherches en laboratoire portant sur l'écoulement non saturé dans le remblai et la migration/diffusion des radionucléides dans le système de barrières ouvragées et présentent la distribution calculée des radionucléides dans le remblais.

The work presented here was done under research contract with the Science and Technology Agency of Japan.

Les travaux exposés dans cette communication sont effectués dans le cadre d'un contrat de recherche passé avec l'Agence pour la science et la technologie du Japon.

Introduction

The objective of this experiment is to provide a preliminary assessment of the migration of radionuclides leached from the waste form in an engineered barrier for shallow land burial of the low-level radioactive wastes. The two main mechanisms by which radionuclides may be released from a shallow land burial facility are considered in the study. These are the migration of the radionuclides with seeped rain water through a backfill and the diffusion in a concrete pit. In order to make clear these mechanisms, two kinds of experimental conditions are set up in concrete pits and the experiment has been performed since 1985. One concrete pit with the backfill is for studying the migration behavior of the radionuclides under realistic conditions. The other pit in which solidified waste forms and radionuclide solution have been infilled is for diffusion experiment. Decomposed granite and a mixture of sand and loam are investigated as backfill materials. Radionuclides used for the study are ^{137}Cs, ^{90}Sr and ^{60}Co.

In order to calculate the concentration profiles of the radionuclides in the engineered barriers, mathematical models are used. The results of the calculations are compared with the actual concentration profiles of the radionuclides in the demonstration facility.

Experiments using the facility are planned to be accomplished in 1989. The final results will be used in the assessment of the performance of the engineered barrier system.

Multibarrier system performance test

The demonstration facility mainly consists of two kinds of concrete pits as illustrated in Fig. 1. The size of each pit is 5 m length x 2.5 m width x 2.3 m depth. One pit (pit-A)is for measuring the diffusion coefficients of the radionuclides in large amount of concrete matrix. For this purpose, 2 cement solidified waste forms of 200 l size were immersed in water in the pit and the leached radionuclides in the water were monitored for 200 days. Leveled off activity of the leachate was found not enough high as expected for the diffusion experiment, therefore, 1.5 mCi of ^{137}Cs , 1mCi of ^{90}Sr and 1mCi of ^{60}Co were added in the leachate solution (1,400 l) of pit-A. Concentration of each radionuclide in the solution after the addition of radionuclides has been measured. It keeps nearly constant values of $1.3x10^{-4}$, $7.1x10^{-5}$ and $3.4x10^{-4}$ μCi/ml for ^{137}Cs, ^{90}Sr and ^{60}Co, respectively.

The performance of the backfill has been tested in the other concrete pit (pit-B). Decomposed granite and the mixture of sand and loam were used as the backfill materials since these soils are common and exist widely. For the analysis of radionuclide migration in different conditions, this pit is divided into two sections and the migration tests have been performed. Half of the pit (pit B-1) has been backfilled with decomposed granite and the radionuclide solution was applied to the surface of the backfill to adapt one dimensional analysis. In the rest part of the pit (pit B-2), 4 cylindrical waste forms of 200 l size were emplaced and backfilled with the mixture of sand and loam. Water flow and the nuclide migration in this pit is analysed two - dimensionally. Water supply of 40 l (10 mm equivalent to the amount of rain fall) for pit B-1 and 100 l (24 mm) for pit B-2 is given by sprinkler at the surface of the backfill every once a week and the seeped solutions from the bottom of each pit is periodically sampled.

No activity of the radionuclides is detected in the seeped solutions sampled until now during the experimental term of about 2 years.

Laboratory experiment on backfill material

The migration and dispersion behavior of the radionuclides in the backfill was studied.

Characteristics of the backfill such as permeability, moisture potential, water content and distribution coefficient related to the water movement and the radionuclide migration were measured.

Dispersion as well as retardation was considered in a mass transport model to calculate the radionuclide migration in the equation [1].

$$C/C_0 = \frac{1}{2}\left\{\mathrm{erf}\,\frac{H/2-(x-Vt/Rf)}{2\sqrt{Dt/Rf}} + \mathrm{erf}\,\frac{H/2+(x-Vt/Rf)}{2\sqrt{Dt/Rf}}\right\} \quad (1)$$

where, H : initial length of diffusion substance, V : average pore water velocity, D : dispersion coefficient of non-absorbing material, R_f : retardation factor, t : time

In order to know the sorption property of the backfill under the condition of water flow, column experiments were carried out and the results were compared to those of batch experiments. Backfill materials used in the experiments were the decomposed granite and the mixture of sand and loam. Compositions of the both materials are the same as those of backfilled in the pits for the multibarrier system performance test. Radionuclides, ^{137}Cs, ^{90}Sr and ^{60}Co were spiked on the surface of the soil column and the deionized water was continuously supplied at a constant velocity. After certain time, the soil column was disconnected in every 10 to 20 mm length and the soil was sampled. The activities of the ^{137}Cs and ^{60}Co in the samples were measured by Ge(Li) semiconductor detector and those of ^{90}Sr were measured by proportional counter.

Profiles of the activities in the soil column were shown in Fig. 2 and Fig.3 for the decomposed granite and the mixture of sand and loam, respectively as a function of penetrated depth. In the mixture of sand and loam, migration of ^{137}Cs was not seen in deeper sections more than 3 cm from the surface. All amount of ^{137}Cs has been sorbed within this section. Whereas, in the decomposed granite, ^{137}Cs was found in the deeper sections where it was not found in the mixture soil.

Migration behavior of ^{60}Co in the mixture soil was similar to that of ^{137}Cs in the decomposed granite, namely, rapid migration is seen in the range of low concentration and deep penetration was observed.

On the other hand, migration behavior of ^{90}Sr was very similar to that of ^{60}Co in the both backfill materials.

Estimate of dispersion coefficient and retardation factor

From the column data, retardation factor was determined and the dispersion coefficient was calculated by one dimensional diffusion theory based upon the following assumption.

$$D = Dm \times V \qquad (Dm = const. = 0.7) \qquad (2)$$

where, Dm : dispersivity

The value of 0.7 for the Dm was determined from the results of the migration of ^{60}Co in the decomposed granite column. In the following calculations, the value of 0.7 was used as a constant.

For the estimation of the radionuclide migration in the range of low concentration where rapid movement of the radionuclides was found, a fraction of each radionuclide which had a small retardation factor was assumed. Modification was made to treat this fraction in the equation (1) and the modified equation (3) was used in the calculation.

$$C/C_0 = F \cdot \frac{1}{2}\left\{ \mathrm{erf}\,\frac{H/2-(x-Vt/R_{f1})}{2\sqrt{Dt/R_{f1}}} + \mathrm{erf}\,\frac{H/2+(x-Vt/R_{f1})}{2\sqrt{Dt/R_{f1}}} \right\}$$

$$+ (1-F) \cdot \frac{1}{2}\left\{ \mathrm{erf}\,\frac{H/2-(x-Vt/R_{f2})}{2\sqrt{Dt/R_{f2}}} + \mathrm{erf}\,\frac{H/2+(x-Vt/R_{f2})}{2\sqrt{Dt/R_{f2}}} \right\} \qquad (3)$$

where, F : fraction of radionuclide, R_{f1} and R_{f2} : retardation factor for the radionuclide of the fraction F and (1-F), respectively

Fitting curves calculated by equation (1) and equation (3) were lined in the Figs. 2 and 3.

Distribution coefficient Kd was calculated by the equation (4).

$$Kd = (R_f - 1)\, f\, S \,/(1-f)\, \rho \qquad (4)$$

where, R_f : retardation factor, f : porosity, S : degree of saturation, ρ : density of soil particle

Calculated distribution coefficients from the retardation factors obtained by column method are compared to measured Kd values by batch method in Table I.

From the results of the column experiment, the following can be said.

Migration of the radionuclides under steady flow through the backfill was estimated fairly well by the equation (1) and equation (3).

Distribution coefficient derived from the results of the column experiment showed nearly equal and/or smaller value than that from batch results. This might be caused by the the difference of chemical forms of the nuclide in the solution and the difference of contact times in the experiments.

Migration behavior of the radionuclides could be represented by two retardation factors. For the range of 1/1,000 - 1/10,000 of C/C_0, the radionuclide migration was represented by small apparent retardation factor which is about 1/10 order of magnitude to that for high concentration range.

Preliminary estimate for water movement and nuclide migration in the backfill material

One dimensional analysis for pit B-1

Concrete pit B-1 is used for one dimensional analysis of the nuclide migration in the backfill. Decomposed granite has been filled in this pit as a backfill material. Radionuclides spiked on the surface has been released by water supply being sprinkled once a week. The actual activity profile in the backfill after about 2 years of water supply is measured by taking core samples from the pit.

In order to estimate the migration of the radionuclides in the backfill, one dimensional analysis was carried out using the mathematical simulation model of mass transport of radionuclides in a porous media based on the equation of continuity and Darcy's Law.

Water flow in the backfill was estimated by the saturated - unsaturated seepage analysis.

$$\mathrm{div}\,(K[\theta(\psi)]\,\mathrm{grad}\,\phi) = C\,\frac{\partial \psi}{\partial t} - q \qquad (5)$$

where, K : permeability, θ : water content by volume, ψ : pressure head, ϕ : hydrostatic head, C : specific moisture capacity

Using the permeability of 1.05×10^{-5} cm/sec at the saturated condition and parameters of the unsaturated, flow velocity and water content by volume in the decomposed granite were calculated.

For the calculation of the nuclide migration with water flow, average value of the flow velocity and the water content were used in the mass transport equation (6).

$$R_f\,\frac{\partial c}{\partial t} + (u \cdot \nabla)\,C + R_f\,\lambda C = \nabla \cdot (D \cdot \nabla)\,C + \frac{F}{\theta} \qquad (6)$$

where, R_f : retardation factor, C : concentration, t : time, u : flow velocity, λ : decay constant, F : leaching rate, θ : water content by volume

Calculation was performed for three nuclides in six cases as listed in Table II. Calculated results are given in Figs. 4, 5 and 6.

Two dimensional analysis for water flow and nuclide migration around the waste form

Two dimensional analysis was adopted to estimate the behavior of the nuclide migration in the pit B-2 in which 4 cylindrical solidified wastes with cement (200 l, 60 cm in diameter and 90 cm in length) were buried in the mixture of sand

and loam. Source term in this pit is leached radionuclides from the solidified wastes by the water supply of every once a week.

The patterns of water flow in the backfill were calculated in the similar way for the case of the pit B-1, the saturated - unsaturated seepage analysis by FEM. For the calculation of the nuclide migration, the average velocity of the water flow was used. Calculated water flow around the waste form is shown in Fig. 7.

The parameters which have been considered in the calculation of the radionuclide migration are listed in Table III. An example of the calculated activity profiles for ^{137}Cs, ^{90}Sr and ^{60}Co after 2 years of migration is shown in Figs. 8,9 and 10. Deformation of the circle is seen on the contour line of the activity around the waste form which is caused by irregular stream of water flow. Calculated results show that almost all amount of radionuclides remain within the distance of 2-4 cm from the surface of the waste forms.

The results of the analysis are summarized as follows :

(1) Water flow in the backfill can be simulated by the mathematical model and the laboratory data.

(2) Migration of the radionuclides with water flow in the backfill depends mainly on the retardation factor of each radionuclide and the dispersivity of the leachate.

(3) Water flow in the backfill is affected by the shape and arrangement of emplaced waste forms.

Diffusion coefficient of ^{137}Cs, ^{90}Sr and ^{60}Co in concrete

The rate of release of radionuclides from the pit is expected to be controlled by diffusion process and sorption in the intact concrete matrix. Overall diffusivity of radionuclides in concrete is largely related to sorption property, porosity and tortuosity of the concrete. Therefore, diffusion coefficients of the radionuclides were measured on ordinary construction concrete and the various properties of the concrete were measured as well.

Concrete block sample (10 cm in diameter and 10 cm in height) has been saturated with water prior to the diffusion experiment and contacted to the radionuclide aqueous solution of a constant concentration for certain time. Composition of the solution used in the diffusion experiment is shown in Table IV. Radioactivities were measured on the 0.1 or 0.2 cm fraction ground off the concrete block. The activity profile of ^{137}Cs and ^{60}Co in the concrete block at the contact time of 119 days is shown in Fig. 11. The profile for ^{90}Sr at the contact time of 89 days is shown in Fig. 12.

The estimation of the diffusion coefficient was based on Fick's law, according to the one dimensional diffusion equation

$$\frac{\partial C}{\partial t} = D^{*}\frac{\partial^2 C}{\partial x^2} \qquad (7)$$

where, C is concentration of diffusing nuclide in the concrete and D^{*} is the diffusion coefficient. When the boundary is kept at constant concentration C_0 ($C=C_0$, $x=0$, $t>0$), and the initial concentration is zero ($C=0$, $x=0$, $t=0$) in a semi-finite medium, the solution for $x>0$ is [1]

$$C = C_0 \operatorname{erfc} \frac{x}{2\sqrt{(D^*t)}} \qquad (8)$$

The diffusion coefficient was obtained by fitting the measured activity profile to the theoretical profile calculated by equation (8). The fitting was accomplished by a least squares method.

The measured activity profiles for ^{137}Cs and ^{60}Co look very similar each other as shown in Fig. 11. They seem to follow the theoretical curves fairly well except the beginning of the profile curves. Higher activity than the initial solution is seen on the both profiles at the vicinity of concrete surface. This higher peak is probably due to adsorption and deposition of the radionuclide on the concrete surface. The diffusion coefficient for ^{137}Cs and ^{60}Co has been calculated excluding the peak. Whereas, on the profile of ^{90}Sr no higher peak is seen as shown in Fig. 12.

Calculated diffusion coefficient for ^{137}Cs, ^{90}Sr and ^{60}Co show the same order of magnitude, $10^{-13} m^2/sec$. No big difference was found in the overall diffusion coefficient of each radionuclide. The diffusivities of the nuclides in water do not differ appreciably from each other and are of the same order of magnitude, $10^{-9} m^2/sec$. Therefore, the difference of $10^{-4} m^2/sec$ in diffusion coefficient between concrete and water may be attributed to the difference of sorption properties of concrete and tortuosity.

These parameters of the concrete may, however, change with time during the long term storage mainly due to carbonatization of the concrete. Changes in the diffusivity during the storage duration will be examined with carbonated concrete samples which are being prepared now.

Migration behavior of ^{137}Cs, ^{90}Sr and ^{60}Co in a fracture of concrete

Water flow experiment in a concrete fracture was performed and the correlation between leakage, hydraulic gradient, fracture width and actual flow velocity was studied. Retardation of the radionuclides passed through a fracture in the 15 cm cubic concrete block was measured. Studied fracture widths were from 0.05 to 0.25 mm. Aqueous solution for this study was prepared to simulate the composition of the solution in which concrete was immersed for 5 months. The composition of the solution is shown in Table IV.

The effect of leakage rate on the concentration of radionuclides passed through the fracture was seen. Figure 13 for ^{60}Co and Fig.14 for ^{137}Cs show the dependence of C/C_0 on leakage rate of the solution.

Future work

The multibarrier system performance test described in this paper is still under way and will be continued approximately for 2 years. The distributions of the radionuclides under the realistic conditions of the disposal system will be measured by taking samples from the demonstration facility. On the basis of these results, assessment models for engineered barrier system will be made valid and the performance of the engineered barrier system will be clarified.

Acknowledgements

The authors wish to acknowledge the work of our colleagues, Messrs. A. Ito, Y. Horie, M. Hirai and S. Okagawa and also wish to thank Messrs. H. Ii, F, Hirosue, M. Abe, H. Miyahara and K. Yamada for their experimental helps.

Reference

[1] Crank, J. : "The Mathematics of Diffusion" Second Edition, Oxford University Press, New York, 1986.

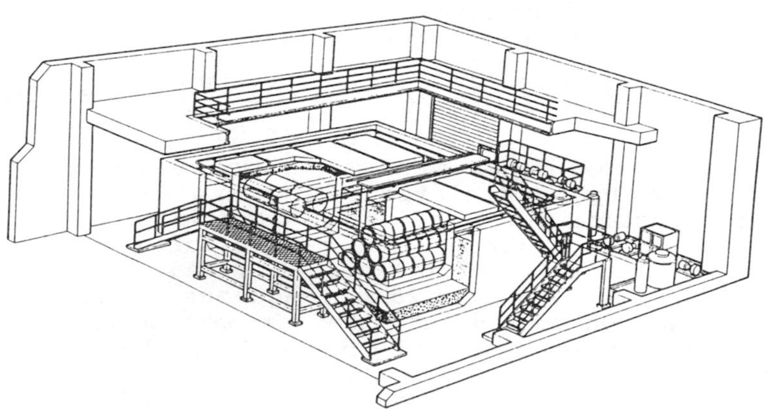

Fig. 1 The facility for the multibarrier system performance test

Table I Conditions of column experiment and the results

ackfill aterial	radio-nuclide	column length(cm)	porosity	degree of saturation	average pore water velocity (cm/sec)	period of water supply (day)	R_f	K_d(column method) (ml/g)	K_d(batch method) (ml/g)
composed anite	^{60}Co	10.2	0.497	0.92	3.3×10^{-4}	7.76	150	50	3000
	^{137}Cs						1400 - 80*	460	3000
	^{90}Sr	16.2	0.501	0.92	2.6×10^{-4}	16.8	240	80	400
xture of and and oam	^{60}Co	20.0	0.553	0.92	3.0×10^{-5}	64.1	600 40*	250	70~500**
	^{137}Cs						8C0	330	50~700**
	^{90}Sr	10.2	0.554	0.96	7.7×10^{-6}	15.0	120 20*	50	30~60 **

* low concentration range
** loam~sand

Table II Analysis case (pit B-1)

Case	Nuclide	R_f	D (cm^2/day)
1	^{137}Cs	6301	3.71
2	^{137}Cs	6301	0.371
3	^{90}Sr	841.2	3.71
4	^{90}Sr	841.2	0.371
5	^{60}Co	6301	3.71
6	^{60}Co	6301	0.371

Table III Analysis case (pit B-2)

Case	Nuclide	R_f	D (cm^2/day)
7	^{137}Cs	879	10.0
8	^{137}Cs	879	1.0
9	^{137}Cs	80*	10.0
10	^{90}Sr	74	10.0
11	^{90}Sr	74	1.0
12	^{90}Sr	10*	10.0
13	^{60}Co	586	10.0
14	^{60}Co	586	1.0

* Variation case

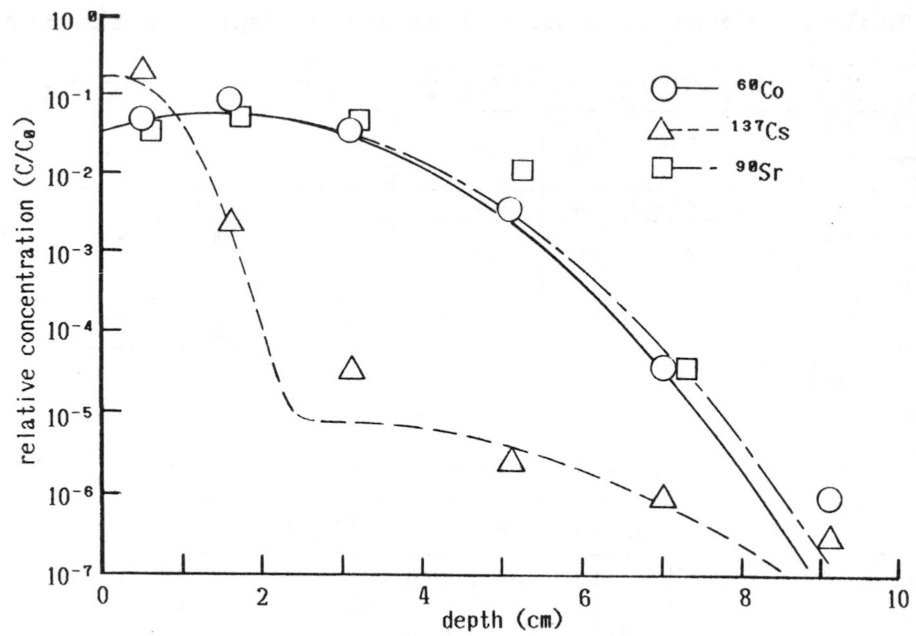

Fig. 2 Activity profiles for ^{60}Co, ^{137}Cs and ^{90}Sr in the decomposed granite

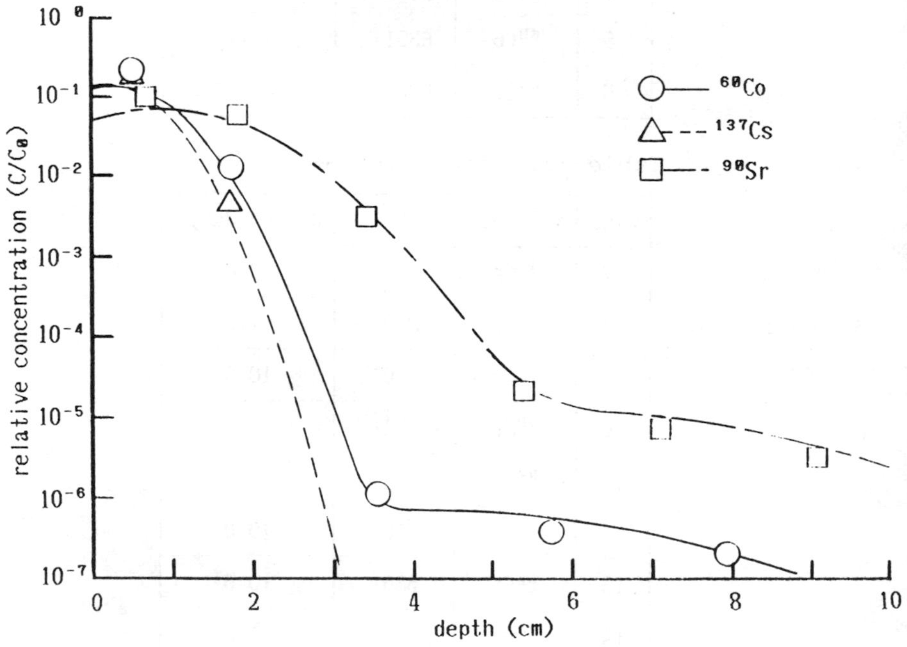

Fig. 3 Activity profiles for ^{60}Co, ^{137}Cs and ^{90}Sr in the mixture of sand and loam

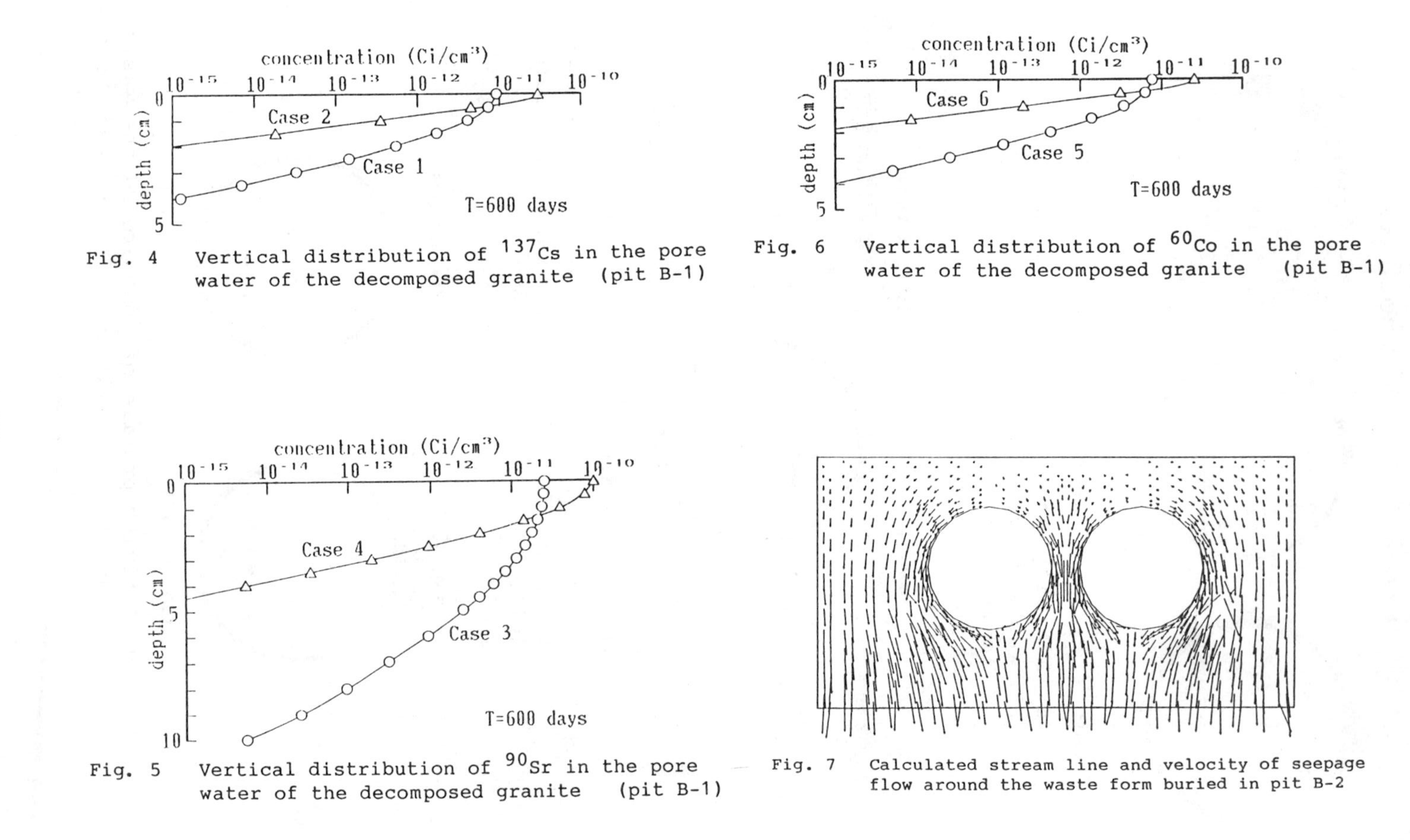

Fig. 4 Vertical distribution of ^{137}Cs in the pore water of the decomposed granite (pit B-1)

Fig. 6 Vertical distribution of ^{60}Co in the pore water of the decomposed granite (pit B-1)

Fig. 5 Vertical distribution of ^{90}Sr in the pore water of the decomposed granite (pit B-1)

Fig. 7 Calculated stream line and velocity of seepage flow around the waste form buried in pit B-2

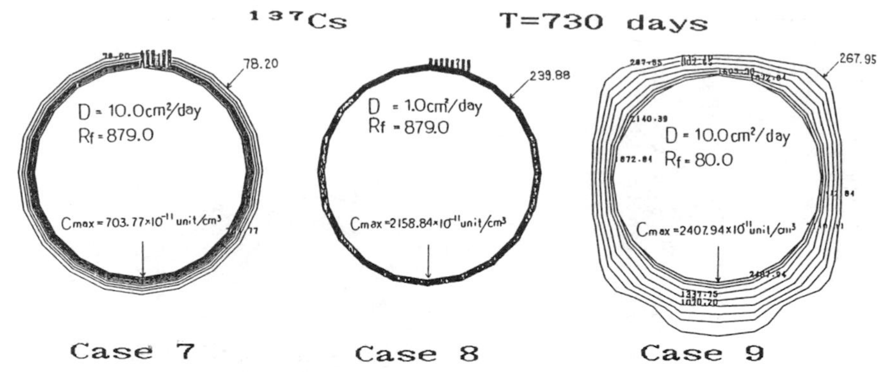

Fig. 8 Activity profile of ^{137}Cs around the waste form in the mixture of sand and loam (pit B-2)

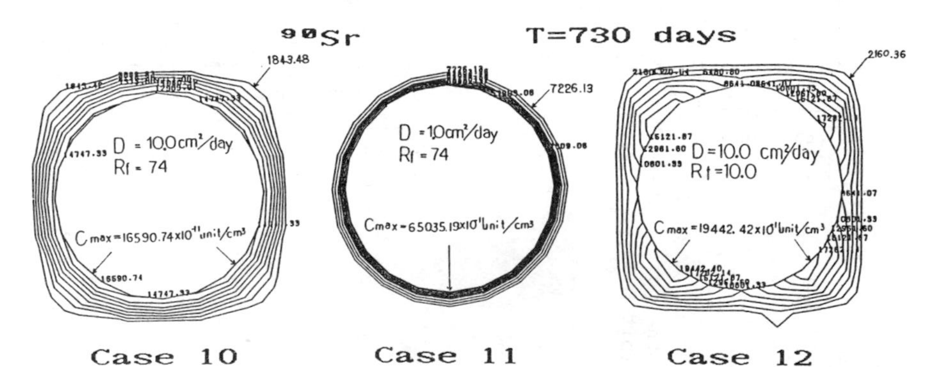

Fig. 9 Activity profile of ^{90}Sr around the waste form in the mixture of sand and loam (pit B-2)

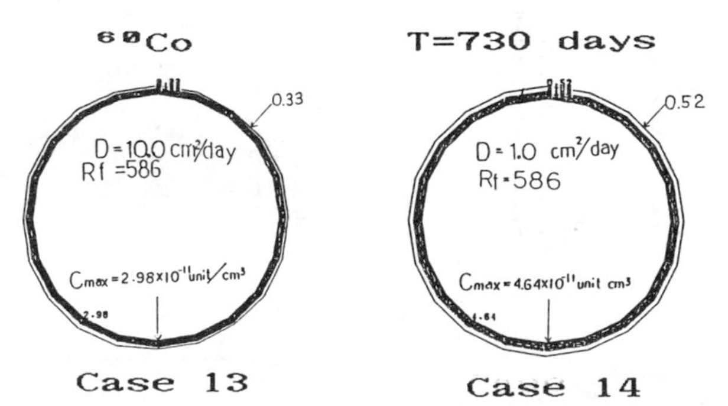

Fig. 10 Activity profile of ^{60}Co around the waste form in the mixture of sand and loam (pit B-2)

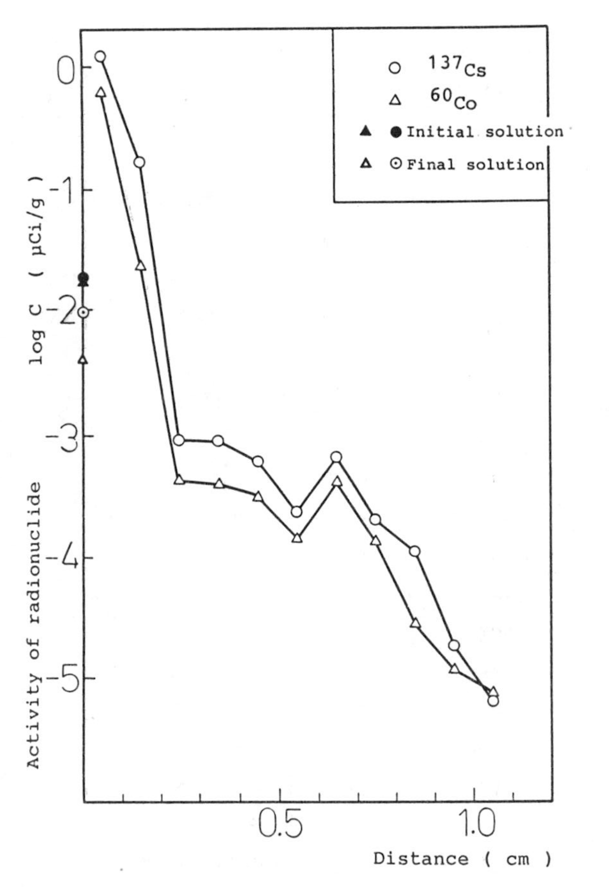

Fig. 11 Activity profile for ^{137}Cs and ^{60}Co in concrete at diffusion time of 119 days

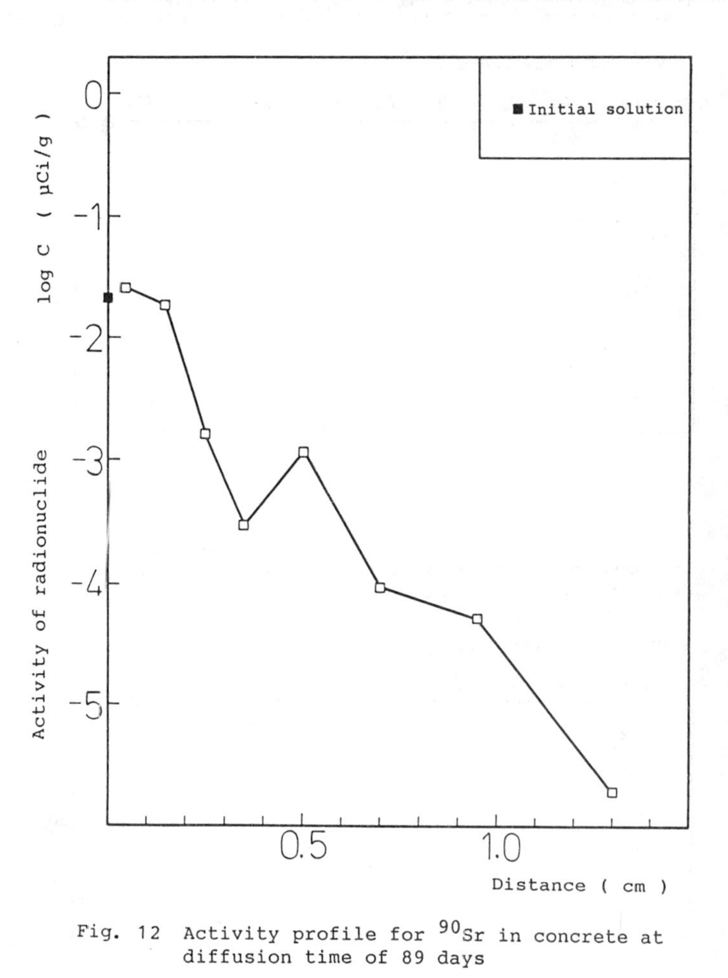

Fig. 12 Activity profile for ^{90}Sr in concrete at diffusion time of 89 days

Table IV Composition of saturated solution with concrete

composition	unit	value
Na^+	mg/l	14.6
K^+	mg/l	8.0
Ca^{2+}	mg/l	30.7
Mg^{2+}	mg/l	0.05
Cl^-	mg/l	0.5
$SO_4{}^{2-}$	mg/l	4.8
$CO_3{}^{2-}$	mg/l	22.9
OH^-	mg/l	64.4
SiO_2	mg/l	9.4
PH	21℃	11.6

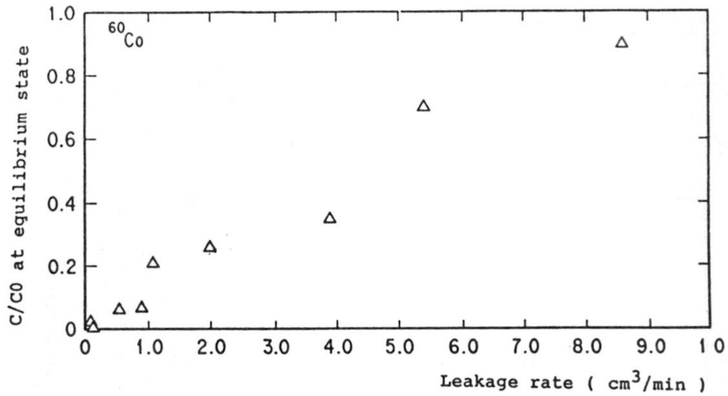

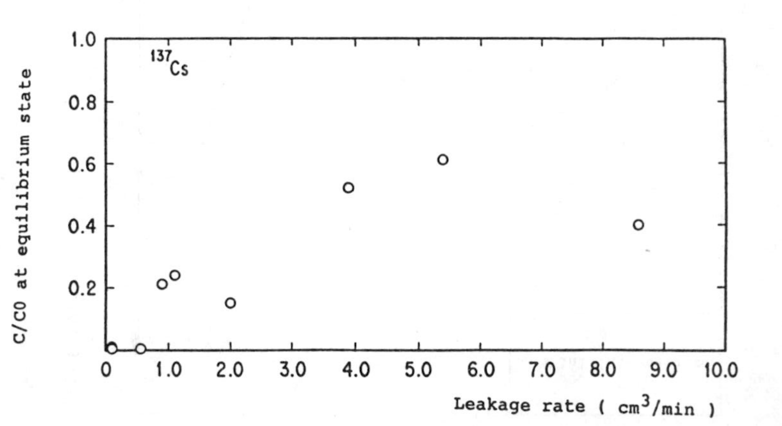

Fig. 13 Relationship between leakage rate and relative concentration of ^{60}Co in the solution passed through a fracture

Fig. 14 Relationship between leakage rate and relative concentration of ^{137}Cs in the solution passed through

MODELISATION DU TERME SOURCE DES CENTRES DE STOCKAGE DE DECHETS RADIOACTIFS DE FAIBLE ET MOYENNE ACTIVITE

Th. Foult
CEA-ANDRA
Paris (France)

ABSTRACT

After a short description of the french concept for shallow ground disposal of radioactive waste, main parameters involved in the "source term" modelling, are listed : radioactive inventory, engineered barriers (structures for waste disposal, waste package, cover, water collecting system). Performance of these barriers is time-dependant and the forseen scenarios are different according to the period of the disposal system lifetime : in the surveillance period the worst case is connected with waste leaching whereas in the unrestricted use period worst cases are connected with human intrusion.

RESUME

Après une brève description du concept français de centre de stockage de surface, on fait l'inventaire des paramètres nécessaires à la modélisation du terme source : inventaire radioactif, barrières ouvragées (couverture, colis, ouvrages, réseaux de collecte des eaux). Les performances de ces barrières évoluent au cours du temps et les scénarios envisagés ne sont pas les mêmes en phase de surveillance et en phase de banalisation. Il en est tenu compte dans la modélisation du terme source : en phase de surveillance le scénario enveloppe est un scénario de lixiviation, tandis qu'en phase de banalisation les scénarios dimensionnants sont les scénarios d'intrusion humaine.

1. LE CONCEPT FRANCAIS DE CENTRE DE STOCKAGE DE SURFACE.

1.1 Options de Sûreté.

Le respect des objectifs fondamentaux (protection des personnes et de l'environnement, banalisation du site au plus tard 300 ans après le début de la phase de surveillance) conduit aux bases de conception suivantes :

. Isoler la radioactivité pendant les phases d'exploitation et de surveillance.

. Limiter l'activité initiale stockée pour qu'à la fin de la période de surveillance l'activité résiduelle soit compatible avec la banalisation du site.

La première option conduit à concevoir un dispositif d'isolement capable d'assurer le confinement des déchets pendant au moins 300 ans.

La deuxième option conduit à limiter l'inventaire radioactif de façon telle que pour tous les scénarios envisagés dans les trois phases de la vie du centre, l'impact radiologique du stockage reste acceptable.

Quand il y a risque de lixiviation par l'eau des déchets, on cherche à évaluer la quantité d'activité pouvant s'échapper annuellement du stockage. Pour cela il faut connaître l'inventaire radioactif et les propriétés du dispositif d'isolement vis-à-vis du phénomène de lixiviation.

1.2 Le dispositif d'isolement.

L'isolement est assuré par un ensemble de dispositions qui limite la quantité de substances radioactives entrainées par l'eau en cas d'infiltration, à un niveau suffisamment faible pour que les conséquences radiologiques soient négligeables.

Les barrières qui composent ce dispositif sont :

- les colis et les ouvrages de stockage,
- la couverture,
- le système de collecte des eaux.

1.2.1 Les modules de stockage et les colis.

L'ensemble doit présenter une sûreté intrinsèque suffisante pendant au moins 300 ans et pour cela remplir les conditions suivantes :

. présenter une résistance mécanique suffisante pour assurer pendant 300 ans une bonne tenue des ouvrages et la stabilité du support sur lequel sera réalisée la couverture.

. ne pas présenter une quantité d'activité lixiviable annuellement (Q.A.L.) supérieure à une limite fixée, pour chaque radionucléide susceptible d'y être présent. Ces valeurs limites de Q.A.L. résultent des évaluations de l'impact radiologique du stockage, celui-ci devant rester acceptable dans toutes les phases de la vie du centre et dans toutes les situations raisonnablement envisageables.

Si le colis offre par lui-même cette sûreté intrinsèque suffisante, il peut être dirigé vers une structure d'accueil en plateforme et le module de stockage contenant les colis de cette nature est appelé tumulus.

Si le colis n'apporte pas à lui seul cette sûreté intrinsèque suffisante, il sera dirigé vers une structure d'accueil en alvéole ;le module de stockage contenant les colis de cette nature est appelé monolithe ; il apporte les compléments nécessaires pour assurer cette sûreté intrinsèque.

En pratique, les blocs béton contenant des déchets enrobés, de même que les fûts et les caissons contenant des déchets bloqués de très faible activité, sont évacués en tumulus ; les autres types de colis (enrobés en enveloppe périssable ou déchets non enrobés) sont généralement évacués en monolithes.

1.2.2 La couverture.

Les modules de stockage dont l'exploitation est terminée sont protégés de l'eau de pluie par une couverture qui doit être stable et suffisamment imperméable pendant 300 ans au moins, l'entretien de cette couverture devant être assuré jusqu'à la fin de la phase de surveillance.

La condition d'imperméabilité se traduit par la limitation de la quantité moyenne d'eau annuellement infiltrée à travers la couverture. L'objectif étant un taux d'infiltration moyen à travers la couverture compris entre 1,5 l/m^2/an et 15 l/m^2/an.

1.2.3 Le système de collecte des eaux.

Le réseau de collecte des eaux doit être fiable, pendant 300 ans ; on a choisi un système passif, conçu et maillé de façon à permettre de localiser facilement une éventuelle anomalie.

A cet effet, l'objectif fixé est de collecter à la base des ouvrages les eaux éventuellement infiltrées à travers la couverture puis de les diriger par gravité vers un bassin de contrôle ; le réseau ainsi constitué, indépendant du réseau pluvial, est dénommé réseau séparatif gravitaire enterré (R.S.G.E.).

2. EFFICACITE DU SYSTEME D'ISOLEMENT.

Par mesure de prudence, on ne tient pas compte de la collecte des eaux infiltrées dans le R.S.G.E. et on fait l'hypothèse que les eaux infiltrées percolent à travers le radier des ouvrages de stockage.

. En phase d'exploitation et de surveillance, l'efficacité du dispositif d'isolement dépend des propriétés des colis, des modules de stockage et de la couverture. On suppose que la lixiviation n'affecte que la fraction de colis dont l'enveloppe est dégradée.

. A la banalisation on suppose que toutes les barrières ouvragées ont perdu leur efficacité : les déchets sont mélangés au béton pulvérulent et aux matériaux de la couverture.

2.1 Vieillissement des colis.

Pour caractériser les enveloppes des colis vis-à-vis du vieillissement, on les range en deux catégories : les enveloppes en béton (coques) et les enveloppes en métal (caissons ou fûts).

2.1.1 Coques en béton :

La loi de dégradation des colis en béton est définie par une densité de probabilité

$$N(\tau) = \frac{1}{\sigma\sqrt{2}} \exp - \frac{(\tau - \tilde{\tau})^2}{2\sigma^2}$$

dans laquelle τ est le temps courant, $\tilde{\tau}$ la durée de vie moyenne des colis et σ l'écart type ou la dispersion autour de la valeur moyenne $\tilde{\tau}$

La fraction cumulée des colis dégradés à l'instant t s'obtient alors par intégration de la densité de probabilité $N(\tau)$,

$$F(t) = \int_{-\infty}^{t} N(\tau)\, d\tau$$

On caractérise la densité de probabilité $N(\tau)$ par un temps de latence moyen de 500 ans et un écart type de 170 ans ce qui conduit à un taux de dégradation initial $F(0) = 1{,}6.10^{-3}$.

Les résultats sont présentés sur la figure 1.

Figure 1 LOI DE DEGRADATION DES CONTENEURS EN BETON

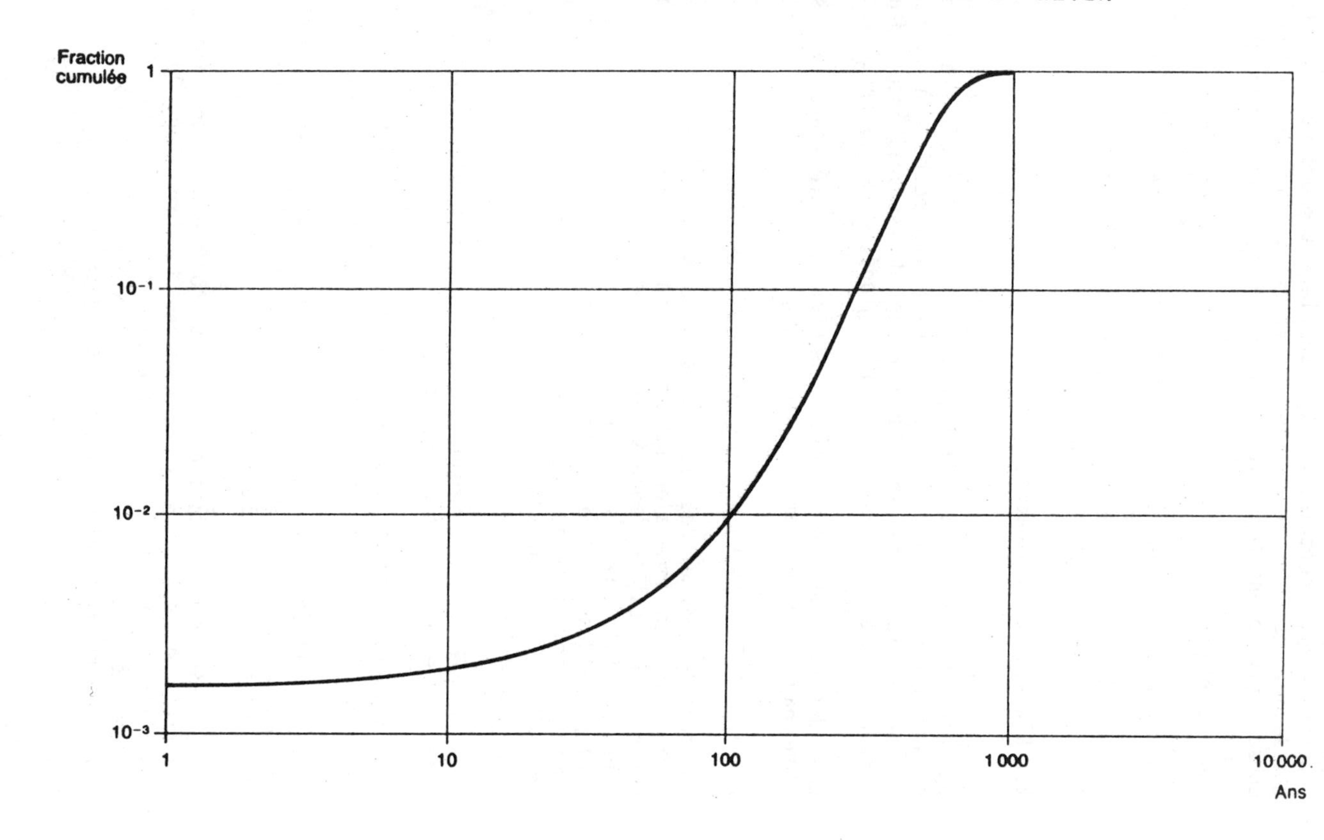

2.1.2 Fûts métalliques et caissons :

Même si l'on peut considérer que l'enveloppe métallique peut constituer dans certains cas une barrière temporaire à l'atteinte du déchet par les eaux de lixiviation, on fait l'hypothèse conservative d'une dégradation instantanée de cette enveloppe.

Dans certains cas les déchets conditionnés dans des fûts métalliques sont stockés dans des alvéoles. Ils sont protégés des eaux infiltrées par les matériaux de remplissage et les matériaux de structures des alvéoles ; on leur applique donc la loi de dégradation des installations en béton armé qui est la même que celle des conteneurs en béton.

2.2 Lixiviation des déchets.

La qualité des colis évolue dans le temps. Pour en tenir compte du point de vue de la lixiviation par l'eau des déchets enrobés, on est amené à considérer d'une part les déchets à liant peu ou pas dégradé, pendant la période de surveillance, d'autre part les déchets à liant fortement dégradé, après la banalisation.

Les déchets de très faible activité, inférieure au seuil d'enrobage, sont également pris en compte.

2.2.1 Déchets enrobés à liant peu ou pas dégradé.

En tenant compte des résultats des diverses expériences de lixiviation en immersion permanente faites, soit sur des blocs enrobés de taille industrielle, soit sur des échantillons représentatifs traités en laboratoire et portant sur différents radionucléides et différents matériaux d'enrobage (liants hydrauliques, bitume, polymères), on caractérise des valeurs de F.A.L. (fraction d'activité lixiviée annuellement) expérimentales.

Si les valeurs expérimentales ne sont pas disponibles, elles font l'objet d'évaluation spécifique.

Par ailleurs, on conçoit que les valeurs des F.A.L. des déchets en situation de stockage sont susceptibles de varier au cours des différentes phases de la vie du stockage, selon les quantités d'eau infiltrées susceptibles d'atteindre les colis de déchets ; ces quantités varieront en fonction du temps ; nous avons retenu pour les calculs le schéma suivant :

- en phase d'exploitation, si les conteneurs sont totalement abrités de l'eau de pluie par une toiture, leur lixiviation par l'eau est nulle. Par contre si les ouvrages sont recouverts d'une couverture provisoire les F.A.L. retenues correspondent à l'immersion totale (F.A.L. expérimentales),

- en phase de surveillance, l'ouvrage une fois constitué est protégé des eaux d'infiltration par réalisation d'une couverture globalement très peu perméable dont l'intégrité est contrôlée tout au long de la phase de surveillance. On suppose que pendant toute cette période le débit des eaux d'infiltration est de 1,5 l/m^2/an et on a de ce fait choisi alors comme F.A.L. le 1/100 des F.A.L. expérimentales. Pour le tritium, on a retenu le 1/10 de la F.A.L. de base pour tenir compte du fait que ce radionucléide peut s'échapper non seulement par lixiviation mais aussi par dégazage,

2.2.2 Déchets non enrobés.

On entend par déchets non enrobés les déchets conditionnés sans matrice qualifiée au regard du confinement des radionucléides. Ils sont par principe des déchets de très faible activité (activité massique inférieure au seuil d'enrobage) et de ce fait leur stockage en plateforme est justifié, à partir du moment où un traitement approprié de blocage assure la tenue mécanique du colis.

Pour la conduite des calculs d'activité lixiviée à partir des colis, on fait également intervenir la notion de F.A.L. ; il est bon de rappeler ici que le facteur n'est qu'un élément de calcul servant à déterminer la quantité d'activité lixiviée (Q.A.L.) qui est la grandeur effectivement représentative pour les évaluations d'impact radiologique du stockage. Par définition, la Q.A.L. est le produit de la F.A.L. par l'activité totale contenue dans les colis, ceci pour un radionucléide donné.

La diversité de nature des déchets non enrobés dans lesquels les radionucléides sont sous des formes chimiques plus ou moins solubles dans l'eau, rend difficile de fixer avec précision des valeurs hypothèses pour les F.A.L. de ces colis, comme pour les déchets enrobés, déterminer des F.A.L. de base puis des F.A.L. en situation de stockage jusqu'à la fin de la période de surveillance, lorsque les colis de déchets sont protégés par la couverture.

Si l'ensemble des valeurs de la F.A.L. de base ainsi définie pour les déchets enrobés et non enrobés conduit à un impact radiologique très faible en phase de surveillance, elles deviennent des spécifications pour les producteurs.

2.2.3 Lixiviation des déchets à la banalisation du site.

Le raisonnement suivant s'applique lorsque les barrières du dispositif d'isolement (colis, ouvrages, couverture) sont complètement dégradées.

Le centre n'est plus surveillé. La couverture redevient semblable au terrain d'origine ; le débit des eaux d'infiltration est de 150 $l/m^2/an$ et le liant des déchets enrobés est supposé fortement dégradé.

Le phénomène de lixiviation est gouverné par les processus de percolation et d'absorption qui s'établissent dans le milieu constitué par les matériaux de dégradation finement divisés. Dans cette situation, les phénomènes diffusifs n'étant plus limitant, on considère que le taux de lixiviation atteint une valeur maximale qui dépend du débit de l'eau qui percole à travers les matériaux lixiviés résultant de la dégradation et du pouvoir d'absorption de ces matériaux.

Dans une masse M de déchets dégradés contenant une activité A, arrosée périodiquement par les eaux d'infiltration avec un débit moyen F, l'activité lixiviée par unité de temps est égale au produit du débit de l'eau par son activité volumique. L'activité lixiviée est maximale lorsque l'équilibre est établi entre les phases solide et liquide.

Elle s'exprime alors par la formule :

$$a = FC = F \frac{A}{Mk_d}$$

dans laquelle :

F est exprimée en m^3/an,

M en tonne,

k_d est le coefficient de partage du radionucléide entre l'eau et le matériau de dégradation. La valeur de ce coefficient (en m^3/t) représente le rapport de l'activité massique du ciment (en Ci/t) à l'activité volumique de l'eau (en Ci/m^3) à l'équilibre.

La fraction d'activité lixiviée annuellement s'écrit alors :

$$\frac{a}{A} = \frac{F}{Mk_d} \qquad (1)$$

On a retenu les bases de calcul suivantes :

- la totalité des ouvrages et des colis est dégradée et réduite à l'état pulvérulent,

- la couverture a les caractéristiques de perméabilité du terrain naturel ce qui correspond à une infiltration de 150 $l/m^2/an$.

3. CALCUL DU TERME SOURCE.

3.1 Modélisation du stockage.

Pour tenir compte de la distribution surfacique des activités on partitionne le stockage. Chaque partie est affectée en son centre de gravité par la somme des activités qu'elle contient. On a ainsi pour l'évaluation de Sûreté une distribution multisource ponctuelle considérée comme équivalente à la distribution réelle.

Cet artifice permet ultérieurement d'utiliser la résolution analytique simple de l'équation du transfert par dispersion et convection dans le terrain.

3.2 Calcul du terme source.

3.2.1 Scénarios de lixiviation.

Le calcul de l'activité lixiviée chaque année et le calcul de l'activité résiduelle passe (pour chaque source ponctuelle) par la résolution de l'équation différentielle.

$$\frac{da_i^{jk}}{dt} = -(\lambda i + F_k \tau ij)\, a_i^{jk}$$

où : - a_i^{jk} (t) représente l'activité résiduelle stockée pour le radionucléide (i) sous la forme enrobée ou non enrobée (j = 1 ou 2).

- F_k (t) représente la fraction cumulée de colis à enveloppe dégradée, celle-ci pouvant être en métal ou en béton (k = 1 ou 2).

- τ ij (t) représente la F.A.L. après t années de stockage ; elle dépend de la nature et de la forme chimique du radionucléide (i), de la forme du déchet qui peut être non enrobé (j = 1) ou enrobé (j = 2) et des quantités d'eau susceptibles d'atteindre les colis de déchets ; ce dernier paramètre varie en fonction des caractéristiques d'imperméabilité de la couverture au cours des différentes phases de la vie du Centre ce qui explique la dépendance du coefficient τ par rapport au temps.

- λi est le facteur de décroissance radioactive pour le radionucléide (i).

3.2.2 Scénarios d'intrusion à la banalisation du site.

3.2.2.1 Transfert par inhalation.

Le risque de mise en suspension de poussières radioactives à partir du stockage peut intervenir après la période de surveillance du site. Il concerne des travailleurs effectuant des travaux de terrassement, de génie civil, des personnes du public vivant sur le site ou à proximité et susceptibles d'inhaler des poussières mises en suspension par le vent, des personnes se livrant à des activités diverses telles que le sport, jardinage, etc...

La masse de poussières radioactives inhalées par un individu est donnée par la relation :

$$m = p.\ f.\ v.\ t.$$

dans laquelle :

- m est la masse de poussières radioactives inhalées (exprimées en mg),
- p est le taux d'empoussièrement moyen de l'air en poussières inhalables, pour les conditions d'exposition envisagées et les conditions climatiques du site (exprimé en mg/m^3),
- f est la fraction de ces poussières provenant des déchets. Ce coefficient, inférieur à 1, dépend de la nature des déchets, de leur conditionnement et de leur mode de stockage,
- v est le débit respiratoire (exprimé en m^3/heure),
- t est la durée de l'exposition (en heures).

Compte tenu des scénarios envisagés, on peut estimer que la composition des poussières de déchets inhalées est semblable à la composition moyenne des déchets. La dose effective, exprimée en Sv, est alors égale à :

$$m \sum_i Q(i)\ A(i)$$

où :

- m est la masse de poussières inhalée exprimée en mg,
- A (i) est l'activité massique moyenne du radionucléide i dans les déchets concernés par le scénario (en Bq/mg),
- Q (i) est la dose par becquerel inhalé (en Sv/Bq).

La date à laquelle les activités A(i) doivent être calculées dépend du scénario envisagé. Par exemple, un scénario qui suppose que l'ensemble des structures du stockage est complètement réduit à l'état de graviers ne peut intervenir que longtemps après la fin de la phase de surveillance du site et le calcul des activités doit tenir compte non seulement de la décroissance radioactive mais également des pertes par lixiviation.

Cependant, puisque la phase de banalisation correspond à un usage sans restriction du site, une intrusion est possible dès le début de cette phase. C'est donc à 300 ans que l'on a situé les scénarios en supposant que les radionucléides présents à l'origine dans les déchets n'auront pas encore migré dans le sol. On a donc tenu compte, pour calculer l'activité présente dans 300 ans, de la décroissance et des filiations radioactives.

3.2.2.2 Exposition externe.

En phase de banalisation, le site est censé ne plus offrir de résistance à une intrusion humaine. On suppose que des travaux importants modifient profondément les lieux. Les déchets se retrouvent alors mêlés aux matériaux de structure du stockage dégradés et au sol d'origine. Au niveau de la surface, on admet que les déchets sont répartis de façon homogène dans le sol.

L'exposition externe résulte alors essentiellement des rayonnements gamma émis par le sol contaminé, qui compte tenu de son épaisseur peut être assimilé à un milieu semi infini.

4. CONCLUSION.

La modélisation du terme source est ainsi très différente en phase de surveillance (déchets à liant peu dégradé) et à la banalisation du site (déchets à liant très dégradé). Dans les deux cas une modélisation simple permet une quantification du terme source.

A moyen terme les efforts porteront sur une modélisation globale du terme source (couverture, ouvrages, colis, radiers) qui tient compte des conditions éventuelles de non saturation dans et sous le stockage.

THE DEMONSTRATION OF A PROPOSED METHODOLOGY FOR THE VERIFICATION AND VALIDATION OF NEAR FIELD MODELS

J.M. Laurens and M.C. Thorne

Electrowatt Engineering Services (UK) Ltd
Grandford House, 16 Carfax
Horsham, Sussex RH12 1UP, UK

Work conducted under contract to the
UK Department of the Environment and the
Commission of the European Communities

ABSTRACT

In order to assure confidence in the estimation of long term radiological performance of underground waste disposal, near-field models must be verified and validated. An international programme similar to "HYDROCOIN" or "BIOMOVS" is required to evaluate these models.

Recently, as part of the CEC PACOMA project, a methodology for verification and validation of near-field models has been proposed and successfully demonstrated on individual physical and chemical processes modelled in the UK Department of the Environment code "VERMIN". However, there appears to be no satisfactory experimental study in the literature that can be used to validate the entire near-field system model. An experimental programme is, therefore, proposed for this purpose.

DEMONSTRATION D'UNE METHODOLOGIE PROPOSEE POUR LA VERIFICATION ET LA VALIDATION DES MODELES DE CHAMP PROCHE

Travaux exécutés sous contrat
passé avec le Ministère de l'Environnement du Royaume-Uni
et la Commission des Communautés Européennes

RESUME

Pour que l'on puisse se fier aux estimations relatives aux performances radiologiques à long terme d'un dépôt souterrain de déchets, des modèles du champ proche doivent être vérifiés et validés. Un programme international analogue à HYDROCOIN ou à BIOMOVS est nécessaire pour évaluer ces modèles.

Récemment, dans le cadre du projet PACOMA de la CCE, on a proposé une méthodologie permettant de vérifier et de valider des modèles du champ proche et on a réussi à en faire la démonstration sur les divers processus physiques et chimiques modélisés dans le programme de calcul VERMIN du Ministère de l'Environnement du Royaume-Uni. Toutefois, il n'existe pas, semble-t-il, d'étude expérimentale satisfaisante dans les publications, qui puisse être utilisée pour valider l'ensemble du modèle de champ proche. Un programme expérimental est par conséquent proposé à cet effet.

INTRODUCTION

In order to assure confidence in assessments of the long-term radiological performance of underground waste disposal facilities, the mathematical models and computer codes of the different components of the containment system must be verified and validated.

Verification ensures that the models are mathematically correct and that the computer code is correctly solving the mathematical models implemented to simulate a process or combination of processes. Validation confirms that the model and the derived computer code provide a good representation of the actual processes occurring in the real system (1).

The fastest and most efficient way to verify a model and its implementation is to compare the results obtained with an already verified code. At present, there are a number of international programmes designed to justify models for water flow in the geosphere (HYDROCOIN [2]) and for radionuclide transport in the biosphere (BIOMOVS [3]) by intercomparison. A similar programme of international intercomparison for near-field models is the next logical step.

A general methodology for computer code validation has been developed and applied to the UK Department of the Environment (DoE) near-field model "VERMIN". This methodology was developed for use with already existing experimental data. However, in the process of application, it was recognised that the existing data are not sufficient for validating the combination of processes represented in VERMIN. New experimental protocols have been developed to provide the relevant data and a need has been identified to extend the validation methodology to include experimental programmes designed specifically for the purpose of validation. This revised methodology could usefully be employed in any international verification and validation exercise for near-field codes.

A more detailed report on these studies is available [4].

METHODOLOGY

Figure 1 illustrates the basic steps of the methodology for code verification and validation. The initial requirement is to identify the physical and chemical processes modelled. For verification and validation of individual processes or specific combinations of processes, it must be possible to eliminate, or at least minimise, the effects of other processes, by a suitable selection of input data. Modifications to the code for the purposes of verification and validation are not acceptable, since identical performance to the unmodified version cannot be guaranteed.

All processes represented in the code require verification, but only those which correspond closely to well-defined physical and chemical concepts can be subject to validation. If a process, or combination of processes, is to be validated, the data required for the simulation must be identified and acquired.

Following an attempt at verification against theory and/or other computer codes, revision of the code may be required, in order to provide a better representation of the process or processes under consideration. If such a revision is undertaken, previous verification and validation exercises should be repeated for the new version of the code.

Procedures for validation are very similar to those for verification, except that they are preceded by a gathering and sorting of available experimental results.

Validation using available experimental data is often difficult and of limited applicability. The experimental protocol may not be reproduced exactly in the simulation and the input data required may be outside the ranges normally used with the code. This is particularly the case in models relating to assessments of radioactive waste disposal, since timescales of interest are generally decades to millenia and spatial scales are typically metres to kilometres. In contrast, typical experimental timescales are weeks to months and spatial scales may be as little as a few millimetres.

VERMIN

The schematic vault geometry used in VERMIN is presented in Figure 2. The canisters containing the waste are embedded in buffer material surrounded by a liner. A clay layer surrounds the liner for shallow land burial options. Finally, a damaged zone is included between the clay layer and the host rock.

The various aspects of vault behaviour modelled are: degradation of barriers and water ingress, corrosion of canisters, leaching of radionuclides from the waste matrix, transport through the buffer and transport through the liner, clay layer and damaged zone. The output consists of the radionuclide fluxes entering the geosphere.

The water flow through the different parts of the vault is calculated from Darcy's law, using an equivalent resistance network approach.

The first stage of nuclide transport is the leaching from the waste matrix, which is represented by a semi-empirical model based on a leach-rate, the driving force being the difference in concentration between the waste and the buffer. A two-dimensional transport calculation is then used to model transport through the buffer. For speed, only five elements are considered. This transport involves diffusion and advective/dispersive processes for sorbed and non-sorbed species. The code can represent decay chains and variations in solubility limits with pH (see Ref. [5]). The pH variation with time is assumed to be determined by $Ca(OH)_2$ depletion with ground water flowing through the vault. Transport through the liner and clay layer (when present) is only considered in respect of the time delay introduced. Transport through the damaged zone is calculated with a one-dimensional model equivalent to the buffer transport calculation (on user request only).

VERMIN is the near-field module of UK DoE pra codes SYVAC A/C,D. Its verification and validation were, therefore, conducted in this context.

EXAMPLE OF A VERIFICATION TEST

For the purpose of clarity, a very simple example is chosen to illustrate the verification exercise. A simple simulation was run of suitable duration in order to see the total amount of nuclides leaving the vault. The half-life and solubility limit were set at high values. The same case was then run with a low, time-independent solubility limit. The expected effect of the low solubility limit is to maintain the release constant when the concentration of nuclides in the water reaches the solubility limit. Since the leaching from the waste matrix is also considered to be solubility limit controlled, the effect should be observable as soon as the release starts.

Figure 3 shows the rate of release for both the solubility unlimited and the solubility limited cases. This confirms that the rate rises in the solubility limited case until the concentration within the repository reaches the solubility limit. After this time, the rate remains constant. Figure 4, which shows the same results expressed as a cumulative fraction of the initial inventory released, demonstrates that, in the solubility limited case, only 0.55 of an initial inventory is lost from the repository over the study period.

EXAMPLE OF A VALIDATION TEST

As discussed in Section 2, validation of model representations of a process, or combination of processes, is usually more difficult than verification. The experimental conditions have to be simulated as accurately as possible, without modifying the code under test. Some of the difficulties are illustrated by the validation test of advective transport of sorbed and non-sorbed species described below.

In the experiment [6], slugs of finite volume saturated either with chloride (non-sorbed) or lithium (sorbed) passed down a column of sand/clay (Figure 5a). To simulate this experiment it was necessary to specify conditions which would result in a constant flux reaching a layer representing the sand/clay column. The damaged zone was taken to simulate the column and a constant flux was achieved using the solubility limit approach, as discussed in Section 4. Results of the simulation are presented in Figure 5b.

In both cases, the simulated results show an earlier front than the experimental. This difference is due to the difficulty of simulating a constant release at the edge of the damaged zone. Also, the simulation used a constant advective velocity, as required by the code. In contrast, the experiment was carried out under gravity, implying a changing head and a changing advective velocity in consequence.

The difficulties encountered in this example arise mainly because the experiment was designed to study the advective process, rather than to simulate conditions appropriate to a radioactive waste repository. Such difficulties become more acute when validation of combinations of processes is required. In these circumstances, the use of experiments designed specifically for the validation exercise is essential.

PROPOSED EXPERIMENTAL STUDIES

In order to simulate the interactions between various processes of interest in a geometry more relevant to a radioactive waste repository, the experimental configuration shown in Figure 6 is proposed. In this arrangement, the waste form is ^{22}Na adsorbed onto bentonite and high resolution localisation (~0.4 mm) of the activity is achieved by coincidence counting of the two annihilation photons.

In the experiment, the bentonite would, for convenience, be packed into a gelatin capsule. This capsule would be embedded in sand or silicaceous silt as a backfill material. This would not be completely analegous to the backfill materials which are likely to be used in practice, since it would be essentially non-sorbing for ^{22}Na. Above and below this material would be layers of sand or silt homogeneously mixed with kaolin to give a material of the required hydraulic conductivity, but exhibiting retardation by ion exchange. Cement and bentonite are unsuitable backfill materials for these experimental studies because their permeabilities are too low to permit rapid advective flow, unless very high hydraulic pressures are applied.

A hydraulic head would be maintained across the system and the volumetric water flow monitored using calibrated burettes. Initially, the system would be flushed with distilled water, which would dissolve the gelatin capsule, but not displace ^{22}Na from its binding sites on the bentonite. At a fixed time, this would be replaced by an aqueous solution containing K^+ or Ca^{2+} to induce ^{22}Na displacement. Displaced ^{22}Na would diffuse from the bentonite region into the sand/silt backfill, where it would be transported by advection/dispersion. Advective/dispersive transport would also occur in the kaolin/sand layers, but in combination with significant retardation. Water flow in the system would be monitored by introducing a dye into the source burette at the time that the ion-containing solution replaced the distilled water.

Because of the small dimensions and simple construction of the experimental system, 10 to 50 such systems could be readily accomodated in a single radiochemical laboratory, allowing adequate replication and studies at several different hydraulic heads covering the range from diffusive to advective dominated conditions. Aspects to be investigated would include:

- depletion of the source region;
- longitudinal distribution of activity within the backfill;
- interface effects at the backfill/kaolinite boundary;
- concentration changes near the outer kaolinite boundary;
- losses of activity in leachate.

Because of the high resolution obtainable, the system would be very sensitive to modelling assumptions. Also, because the experiment is designed specifically for validation purposes, it could be represented very accurately with various near-field models, which might usefully be compared in an international exercise.

CONCLUSIONS

Verification is an integral part of code development and should be undertaken for all the processes modelled. Validation, as normally practiced, involves identification of relevant existing experimental studies, many of which will have been undertaken for purposes unconnected with code validation. Experience in attempting to validate VERMIN has demonstrated that such experiments may be inadequate for several reasons:

- data may not be available for all the processes and combinations of processes modelled;
- the experiments may not relate closely to the geometrical, physical and chemical characteristics which can be modelled;
- the spatial and temporal scales of the experiments may require the code to operate on data which are outside its normal range of application.
- the need for an international intercomparison of near-field codes.

For these reasons, it is suggested that the development of appropriate new experimental protocols should be considered as part of any code validation exercise.

ACKNOWLEDGEMENTS

This work was conducted under contract to the UK Department of the Environment (Contract No. PECD 7/9/377) and to the Commission of the European Communities (Contract No. FI1W-0047-UK(H1)).

The results of this work will be used in the formulation of Government Policy, but at this stage they do not necessarily represent Government Policy.

Dr. Z.A. Gralewski was a major contributor to the work presented here. We would also like to thank Drs P. Kane and M.J. Plews for help in formulating new experimental protocols.

REFERENCES

1. IAEA, Radioactive Waste Management Glossary, IAEA-TECDOC-264. IAEA, Vienna, 1982.
2. HYDROCOIN, Progress Report No. 2, January-June 1985. Available from NEA/OECD, 38 Boulevard Suchet, Paris.
3. BIOMOVS Newsletters 1 to 6. Available from the National Institute of Radiation Protection, Box 602 04, S-104 01, Stockholm, Sweden.
4. Electrowatt Engineering Services (UK) Ltd. Validation of the Vault Computer Model 'VERMIN' for Post-closure Behaviour of Repositories in Geological Formations. DoE Report No. DoE/RW/87-079, October 1987.
5. Prike D.C. and Rees J.M. Understanding the behaviour of Actinides under Disposat Conditions: A comparison between calculated and experimental solubilities. AERE-R.12245, September 1986.
6. Kipp, K.L. Mathematical Modelling of Chemical Transport in Soil Columns. Harwell Report No. AERE-R9012, February 1978.

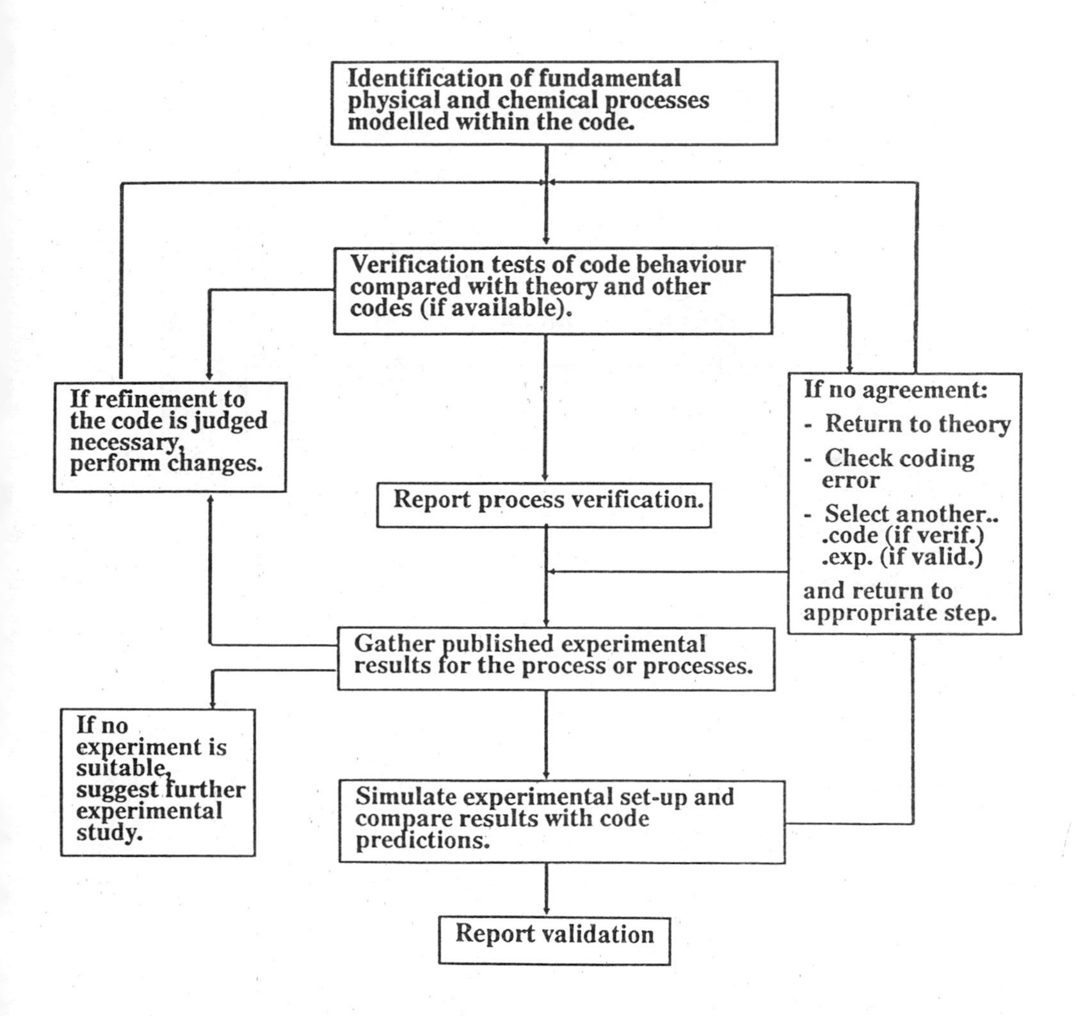

Figure 1: Basic steps of the verification/validation methodology.

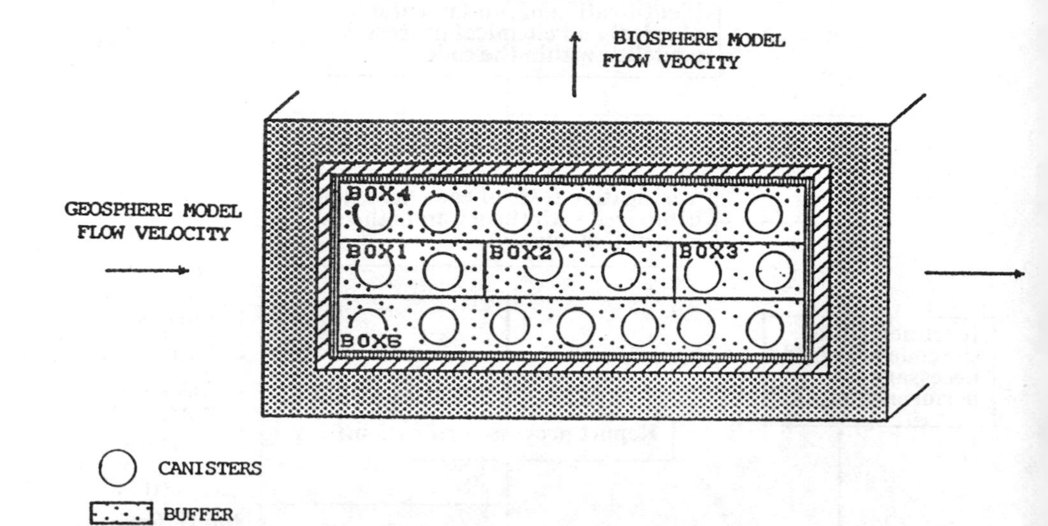

Figure 2. Vault Geometry used in VERMIN

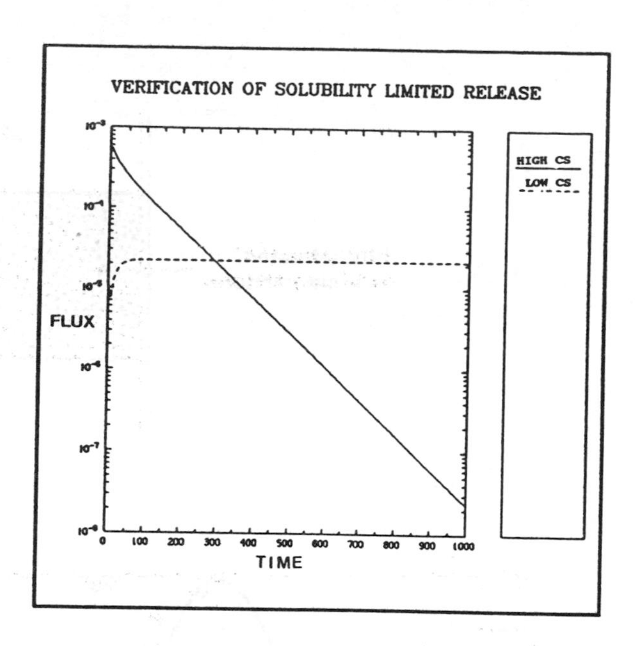

Figure 3.
Fluxes for solubility limited (dash) and non-limited (continuous) release from vault.

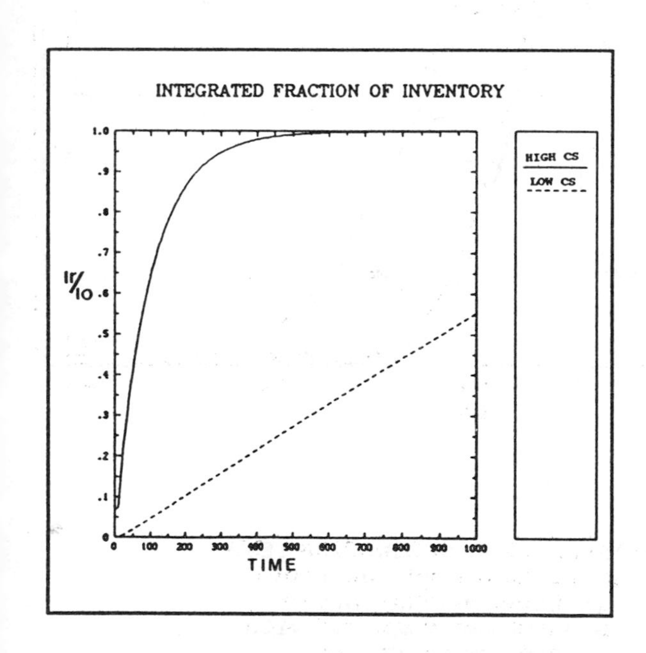

Figure 4.
Fraction of inventory released (Ir/Io) in solubility limited (dash) and non-limited (continuous) cases.

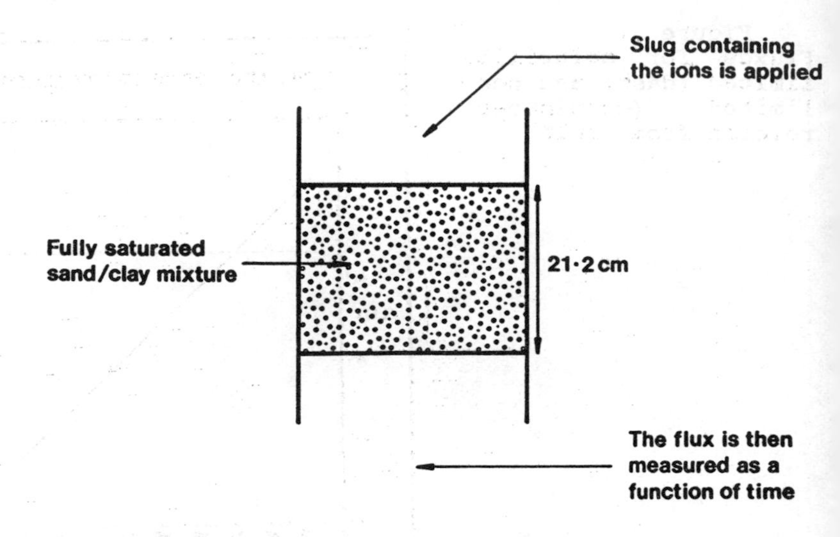

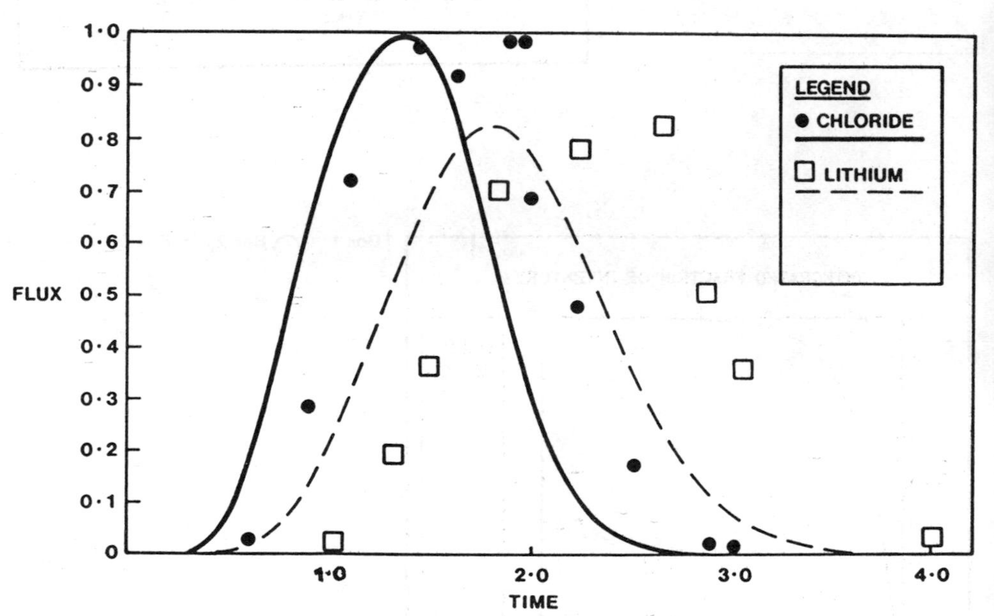

Figure 5. EXAMPLE OF A VALIDATION TEST
Advection/sorption model
a) Experimental set-up
b) Results compared with code predictions.

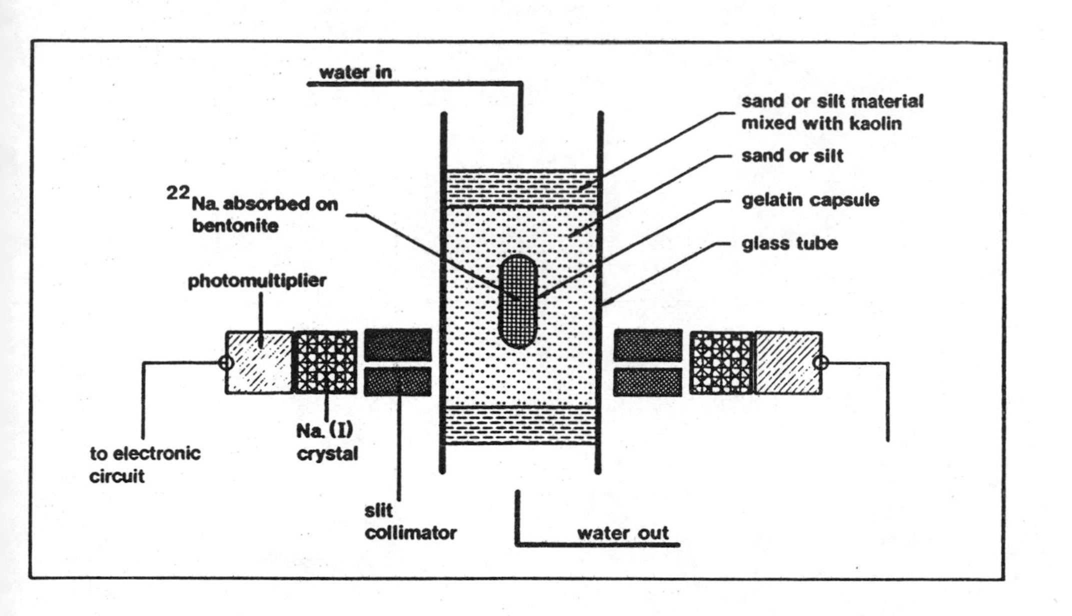

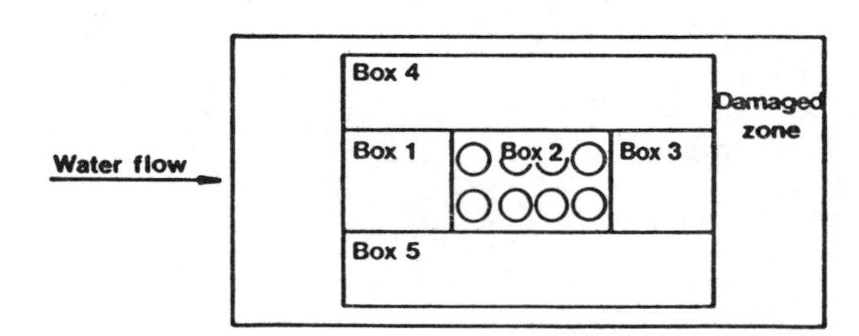

Equivalent representation of the experimental set-up for VERMIN

Figure 6. Proposed experimental set-up for vault simulation.

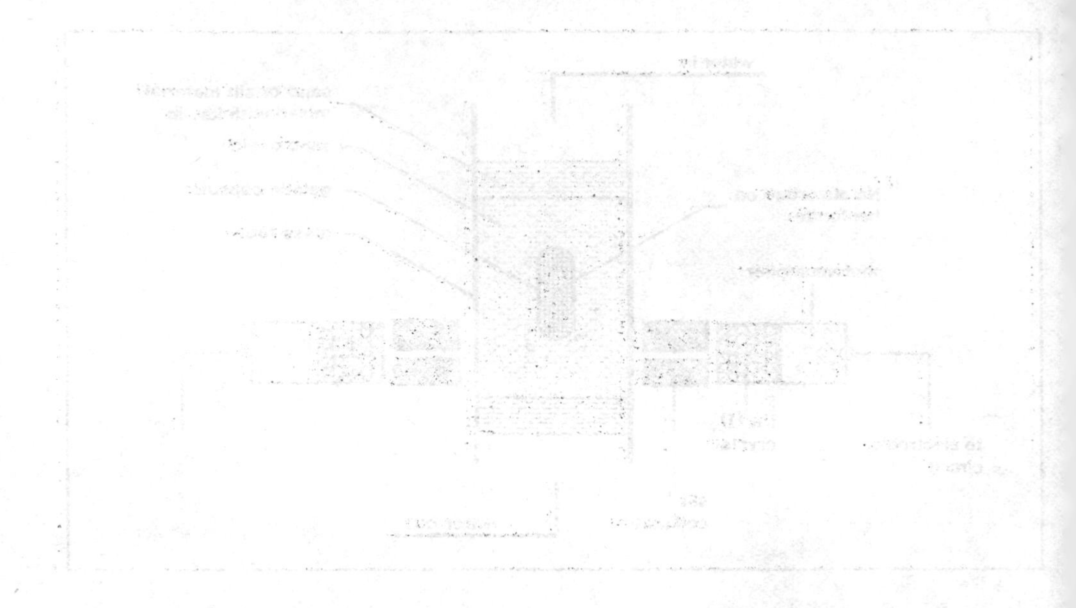

SESSION V

DISCUSSION, CONCLUSION

SEANCE V

DISCUSSIONS, CONCLUSIONS

Chairman - Président

C. McCOMBIE

(Switzerland)

Based upon the discussions, a summary and conclusions were prepared by the NEA Secretariat and the session chairmen and it is reported on page 10 of these proceedings.

*

* *

Sur la base des discussions de cette séance, le secrétariat et les présidents de chaque séance ont établi un résumé et des conclusions que vous trouverez en page 14 de ce compte rendu.

IST OF PARTICIPANTS LISTE DES PARTICIPANTS

BELGIUM - BELGIQUE

AN DE PUT, K., ONDRAF, Organisme National des Déchets Radioactifs et des Matières Fissiles, Boulevard du Régent 54, Boîte 5, B-1000 Bruxelles

CANADA

OROK J., AECL, Chalk River Nuclear Laboratories, Chalk River, Ontario KOJ 1J0

FEDERAL REPUBLIC OF GERMANY - REPUBLIQUE FEDERALE D'ALLEMAGNE

IRSEKORN, R.P., Gesellschaft für Strahlen- und Umweltforschung mbH München, Institut für Tieflagerung, Theodor-Heuss-Strasse 4, D-3300 Braunschweig

ORENTZ, J., Niedersächsisches Umweltministerium, Postfach 41 07, D-Hannover 1

ESTERLE, R., TÜV Hannover e. V., Postfach 81 07 40, D-3000 Hannover 81

ÖSEL, B., Niedersächsisches Landesamt für Bodenforschung, Stilleweg 2, D-3000 Hannover 51

TORCK, R., Gesellschaft für Strahlen- und Umweltforschung mbH München, Institut für Tieflagerung, Theodor-Heuss-Strasse 4, D-3300 Braunschweig

FINLAND - FINLANDE

ÄRKÖNEN, H.T., Imatran Voima Oy, P.O. Box 138, SF-00101 Helsinki

NELLMAN, M.V., Technical Research Centre of Finland (VTT), Reactor Laboratory, Otakaari 3A, SF-02150 Espoo

IENO, T.K., Technical Research Centre of Finland (VTT), Nuclear Engineering Laboratory, P.O. Box 169, SF-00181 Helsinki

FRANCE

CERNES, A.C., Commissariat à l'Energie Atomique, DAS/SAED, B.P. n° 6, 92265 Fontenay-aux-Roses

LOVERA, P., Commissariat à l'Energie Atomique, CEN/FAR, B.P. n° 6, 92265 Fontenay-aux-Roses

MOULIN, V., Commissariat à l'Energie Atomique, DRDD/SESD, B.P. n° 6, 92265 Fontenay-aux-Roses

JAPAN - JAPON

ISHIGURO, K., PNC, Tokai, Department of Chemical Engineering, The Royal Institute of Technology, S-100 44 Stockholm 70 (Sweden)

YUSA, Y., Waste Management Technology Division, Tokai Works, PNC, Tokai, Ibaraki 319-11

PORTUGAL

DA CONCEICAO SEVERO, A.G., LNETI, Departamento de Protecçâo e Segurança Radiologica, Estrada Nacional 10, 2685 Sacavem

SPAIN - ESPAGNE

ALONSO DIAZ-TERAN, J., ENRESA, Castellana 135, 28046 Madrid

ZULOAGA LALANA, P., ENRESA, Castellana 135, 28046 Madrid

SWEDEN - SUEDE

CARLSSON, J.S., SKB, Box 5864, S-102 48 Stockholm

HAEGG, C., National Institute of Radiation Protection, Box 60 204, S-104 01 Stockholm

NERETNIEKS, I., Department of Chemical Engineering, Royal Institute of Technology, S-100 44 Stockholm

RASMUSON, A., Department of Chemical Engineering, Royal Institute of Technology, S-100 44 Stockholm

IGORGH, M., Kemakta Consultants Co., Pipersgatan 27, S-112 28 Stockholm

INGEFORS, S., Swedish Nuclear Power Inspectorate, Box 27 106,
S-102 52 Stockholm

SWITZERLAND - SUISSE

ERNER, U., Eidgenössisches Institut für Reaktorforschung (EIR),
CH-5303 Würenlingen

cCOMBIE, C., Société Coopérative Nationale pour l'Entreposage de Déchets
Radioactifs (NAGRA), Parkstrasse 23, CH-5401 Baden

cKINLEY, I., Société Coopérative Nationale pour l'Entreposage de Déchets
Radioactifs (NAGRA), Parkstrasse 23, CH-5401 Baden

AN DORP, F., Société Coopérative Nationale pour l'Entreposage de Déchets
Radioactifs (NAGRA), Parkstrasse 23, CH-5401 Baden

IGFUSSON, J.O., Swiss Safety Inspectorate, CH-5303 Würenlingen

UIDEMA, P., Société Coopérative Nationale pour l'Entreposage de Déchets
Radioactifs (NAGRA), Parkstrasse 23, CH-5401 Baden

Observers - Observateurs

KSOYOGLU, S., Eidgenössisches Institut für Reaktorforschung (EIR),
CH-5303 Würenlingen

AEYENS, B., Eidgenössisches Institut für Reaktorforschung (EIR),
CH-5303 Würenlingen

RADBURY, M., Eidgenössisches Institut für Reaktorforschung (EIR),
CH-5303 Würenlingen

EGUELDRE, C., Eidgenössisches Institut für Reaktorforschung (EIR),
CH-5303 Würenlingen

ELLA CASA, A., Société Coopérative Nationale pour l'Entreposage de Déchets
Radioactifs (NAGRA), Parkstrasse 23, CH-5401 Baden

ADERMANN, J., Eidgenössisches Institut für Reaktorforschung (EIR),
CH-5303 Würenlingen

AWAMURA, H., Société Coopérative Nationale pour l'Entreposage de Déchets
Radioactifs (NAGRA), Parkstrasse 23, CH-5401 Baden

NECHT, B., Société Coopérative Nationale pour l'Entreposage de Déchets
Radioactifs (NAGRA), Parkstrasse 23, CH-5401 Baden

NOLD, A., Société Coopérative Nationale pour l'Entreposage de Déchets Radioactifs (NAGRA), Parkstrasse 23, CH-5401 Baden

WERNLI, B., Eidgenössisches Institut für Reaktorforschung (EIR), CH-5303 Würenlingen

WITTWER, C., Société Coopérative Nationale pour l'Entreposage de Déchets Radioactifs (NAGRA), Parkstrasse 23, CH-5401 Baden

UNITED KINGDOM - ROYAUME-UNI

EWART, F.T., Atomic Energy Research Establishment, Harwell, Oxon. OX11 ORA

HODGKINSON, D., INTERA-Exploration Consultants Ltd., Highlands Farm, Greys Road, Henley-on-Thames, Oxon. RG9 4PS

LAURENS, J-M., Electrowatt Engineering Services (UK) Ltd, Grandford House, 16 Carfax, Horsham, West Sussex RH12 1UP

LEVER, D.A., Atomic Energy Research Establishment, Harwell, Oxon. OX11 ORA

MOBBS, S.F., National Radiological Protection Board, Chilton, Didcot, Oxfordshire OX11 ORQ

TASKER, P.W., Atomic Energy Research Establishment, Harwell, Oxon. OX11 ORA

UNITED STATES - ETATS-UNIS

OTIS, M.D., EG+G Idaho, P.O. Box 1625, Idaho Falls, Idaho

WEBER, M.F., US NRC, Mail Stop 623-SS, Washington, D.C. 20555

COMMISSION OF THE EUROPEAN COMMUNITIES
COMMISSION DES COMMUNAUTES EUROPEENNES

SIMON, R., Commission des Communautés Européennes, 200 rue de la Loi (ARTS 2/14), B-1049 Bruxelles (Belgique)

NUCLEAR ENERGY AGENCY
AGENCE POUR L'ENERGIE NUCLEAIRE

THEGERSTRÖM, C., Division of Radiation Protection and Waste Management, 38 boulevard Suchet, 75016 Paris (France)

WHERE TO OBTAIN OECD PUBLICATIONS
OÙ OBTENIR LES PUBLICATIONS DE L'OCDE

ARGENTINA - ARGENTINE
Carlos Hirsch S.R.L.,
Florida 165, 4° Piso,
(Galeria Guemes) 1333 Buenos Aires
Tel. 33.1787.2391 y 30.7122

AUSTRALIA - AUSTRALIE
D.A. Book (Aust.) Pty. Ltd.
11-13 Station Street (P.O. Box 163)
Mitcham, Vic. 3132 Tel. (03) 873 4411

AUSTRIA - AUTRICHE
OECD Publications and Information Centre,
4 Simrockstrasse,
5300 Bonn (Germany) Tel. (0228) 21.60.45
Gerold & Co., Graben 31, Wien 1 Tel. 52.22.35

BELGIUM - BELGIQUE
Jean de Lannoy,
avenue du Roi 202
B-1060 Bruxelles Tel. (02) 538.51.69

CANADA
Renouf Publishing Company Ltd/
Éditions Renouf Ltée,
1294 Algoma Road, Ottawa, Ont. K1B 3W8
Tel: (613) 741-4333
Toll Free/Sans Frais:
Ontario, Quebec, Maritimes:
1-800-267-1805
Western Canada, Newfoundland:
1-800-267-1826
Stores/Magasins:
61 rue Sparks St., Ottawa, Ont. K1P 5A6
Tel: (613) 238-8985
211 rue Yonge St., Toronto, Ont. M5B 1M4
Tel: (416) 363-3171

DENMARK - DANEMARK
Munksgaard Export and Subscription Service
35, Nørre Søgade, DK-1370 København K
Tel. +45.1.12.85.70

FINLAND - FINLANDE
Akateeminen Kirjakauppa,
Keskuskatu 1, 00100 Helsinki 10 Tel. 0.12141

FRANCE
OCDE/OECD
Mail Orders/Commandes par correspondance :
2, rue André-Pascal,
75775 Paris Cedex 16
Tel. (1) 45.24.82.00
Bookshop/Librairie : 33, rue Octave-Feuillet
75016 Paris
Tel. (1) 45.24.81.67 or/ou (1) 45.24.81.81
Librairie de l'Université,
12*a*, rue Nazareth,
13602 Aix-en-Provence Tel. 42.26.18.08

GERMANY - ALLEMAGNE
OECD Publications and Information Centre,
4 Simrockstrasse,
5300 Bonn Tel. (0228) 21.60.45

GREECE - GRÈCE
Librairie Kauffmann,
28, rue du Stade, 105 64 Athens Tel. 322.21.60

HONG KONG
Government Information Services,
Publications (Sales) Office,
Information Services Department
No. 1, Battery Path, Central

ICELAND - ISLANDE
Snæbjörn Jónsson & Co., h.f.,
Hafnarstræti 4 & 9,
P.O.B. 1131 – Reykjavik
Tel. 13133/14281/11936

INDIA - INDE
Oxford Book and Stationery Co.,
Scindia House, New Delhi 110001
Tel. 331.5896/5308
17 Park St., Calcutta 700016 Tel. 240832

INDONESIA - INDONÉSIE
Pdii-Lipi, P.O. Box 3065/JKT.Jakarta
Tel. 583467

IRELAND - IRLANDE
TDC Publishers - Library Suppliers,
12 North Frederick Street, Dublin 1
Tel. 744835-749677

ITALY - ITALIE
Libreria Commissionaria Sansoni,
Via Lamarmora 45, 50121 Firenze
Tel. 579751/584468
Via Bartolini 29, 20155 Milano Tel. 365083
Editrice e Libreria Herder,
Piazza Montecitorio 120, 00186 Roma
Tel. 6794628
Libreria Hœpli,
Via Hœpli 5, 20121 Milano Tel. 865446
Libreria Scientifica
Dott. Lucio de Biasio "Aeiou"
Via Meravigli 16, 20123 Milano Tel. 807679
Libreria Lattes,
Via Garibaldi 3, 10122 Torino Tel. 519274
La diffusione delle edizioni OCSE è inoltre assicurata dalle migliori librerie nelle città più importanti.

JAPAN - JAPON
OECD Publications and Information Centre,
Landic Akasaka Bldg., 2-3-4 Akasaka,
Minato-ku, Tokyo 107 Tel. 586.2016

KOREA - CORÉE
Kyobo Book Centre Co. Ltd.
P.O.Box: Kwang Hwa Moon 1658,
Seoul Tel. (REP) 730.78.91

LEBANON - LIBAN
Documenta Scientifica/Redico,
Edison Building, Bliss St.,
P.O.B. 5641, Beirut Tel. 354429-344425

MALAYSIA/SINGAPORE - MALAISIE/SINGAPOUR
University of Malaya Co-operative Bookshop Ltd.,
7 Lrg 51A/227A, Petaling Jaya
Malaysia Tel. 7565000/7565425
Information Publications Pte Ltd
Pei-Fu Industrial Building,
24 New Industrial Road No. 02-06
Singapore 1953 Tel. 2831786, 2831798

NETHERLANDS - PAYS-BAS
Staatsuitgeverij
Chr. Plantijnstraat, 2 Postbus 20014
2500 EA S-Gravenhage Tel. 070-789911
Voor bestellingen: Tel. 070-789880

NEW ZEALAND - NOUVELLE-ZÉLANDE
Government Printing Office Bookshops:
Auckland: Retail Bookshop, 25 Rutland Stseet,
Mail Orders, 85 Beach Road
Private Bag C.P.O.
Hamilton: Retail: Ward Street,
Mail Orders, P.O. Box 857
Wellington: Retail, Mulgrave Street, (Head Office)
Cubacade World Trade Centre,
Mail Orders, Private Bag
Christchurch: Retail, 159 Hereford Street,
Mail Orders, Private Bag
Dunedin: Retail, Princes Street,
Mail Orders, P.O. Box 1104

NORWAY - NORVÈGE
Tanum-Karl Johan
Karl Johans gate 43, Oslo 1
PB 1177 Sentrum, 0107 Oslo 1Tel. (02) 42.93.10

PAKISTAN
Mirza Book Agency
65 Shahrah Quaid-E-Azam, Lahore 3 Tel. 66839

PHILIPPINES
I.J. Sagun Enterprises, Inc.
P.O. Box 4322 CPO Manila
Tel. 695-1946, 922-9495

PORTUGAL
Livraria Portugal,
Rua do Carmo 70-74, 1117 Lisboa Codex
Tel. 360582/3

SINGAPORE/MALAYSIA - SINGAPOUR/MALAISIE
See "Malaysia/Singapor". Voir «Malaisie/Singapour»

SPAIN - ESPAGNE
Mundi-Prensa Libros, S.A.,
Castelló 37, Apartado 1223, Madrid-28001
Tel. 431.33.99
Libreria Bosch, Ronda Universidad 11,
Barcelona 7 Tel. 317.53.08/317.53.58

SWEDEN - SUÈDE
AB CE Fritzes Kungl. Hovbokhandel,
Box 16356, S 103 27 STH,
Regeringsgatan 12,
DS Stockholm Tel. (08) 23.89.00
Subscription Agency/Abonnements:
Wennergren-Williams AB,
Box 30004, S104 25 Stockholm Tel. (08)54.12.00

SWITZERLAND - SUISSE
OECD Publications and Information Centre,
4 Simrockstrasse,
5300 Bonn (Germany) Tel. (0228) 21.60.45
Librairie Payot,
6 rue Grenus, 1211 Genève 11
Tel. (022) 31.89.50
United Nations Bookshop/
Librairie des Nations-Unies
Palais des Nations,
1211 – Geneva 10
Tel. 022-34-60-11 (ext. 48 72)

TAIWAN - FORMOSE
Good Faith Worldwide Int'l Co., Ltd.
9th floor, No. 118, Sec.2
Chung Hsiao E. Road
Taipei Tel. 391.7396/391.7397

THAILAND - THAILANDE
Suksit Siam Co., Ltd.,
1715 Rama IV Rd.,
Samyam Bangkok 5 Tel. 2511630
INDEX Book Promotion & Service Ltd.
59/6 Soi Lang Suan, Ploenchit Road
Patjumamwan, Bangkok 10500
Tel. 250-1919, 252-1066

TURKEY - TURQUIE
Kültur Yayinlari Is-Türk Ltd. Sti.
Atatürk Bulvari No: 191/Kat. 21
Kavaklidere/Ankara Tel. 25.07.60
Dolmabahce Cad. No: 29
Besiktas/Istanbul Tel. 160.71.88

UNITED KINGDOM - ROYAUME-UNI
H.M. Stationery Office,
Postal orders only: (01)211-5656
P.O.B. 276, London SW8 5DT
Telephone orders: (01) 622.3316, or
Personal callers:
49 High Holborn, London WC1V 6HB
Branches at: Belfast, Birmingham,
Bristol, Edinburgh, Manchester

UNITED STATES - ÉTATS-UNIS
OECD Publications and Information Centre,
2001 L Street, N.W., Suite 700,
Washington, D.C. 20036 - 4095
Tel. (202) 785.6323

VENEZUELA
Libreria del Este,
Avda F. Miranda 52, Aptdo. 60337,
Edificio Galipan, Caracas 106
Tel. 32.23.01/33.26.04/31.58.38

YUGOSLAVIA - YOUGOSLAVIE
Jugoslovenska Knjiga, Knez Mihajlova 2,
P.O.B. 36, Beograd Tel. 621.992

Orders and inquiries from countries where Distributors have not yet been appointed should be sent to:
OECD, Publications Service, 2, rue André-Pascal, 75775 PARIS CEDEX 16.

Les commandes provenant de pays où l'OCDE n'a pas encore désigné de distributeur peuvent être adressées à :
OCDE, Service des Publications. 2. rue André-Pascal, 75775 PARIS CEDEX 16.

71056-02-1988

PUBLICATIONS DE L'OCDE, 2, rue André-Pascal, 75775 PARIS CEDEX 16 - Nº 44302 1988
IMPRIMÉ EN FRANCE
(66 88 02 3) ISBN 92-64-03060-3

PROCEEDINGS OF AN NEA WORKSHOP ON NEAR-FIELD ASSESSMENT OF REPOSITORIES FOR LOW AND MEDIUM LEVEL RADIOACTIVE WASTE

COMPTE RENDU D'UNE RÉUNION DE TRAVAIL DE L'AEN SUR L'ÉVALUATION DU CHAMP PROCHE DES DÉPÔTS DE DÉCHETS RADIOACTIFS DE FAIBLE ET MOYENNE ACTIVITÉ

BADEN
SWITZERLAND/SUISSE
23-25 Nov. 1987

Organised by the
OECD NUCLEAR ENERGY AGENCY
in co-operation with the
National Cooperative for the Storage
of Radioactive Waste, NAGRA, Switzerland

Organisée par
L'AGENCE DE L'OCDE POUR L'ÉNERGIE NUCLÉAIRE
en coopération avec
La Société Coopérative Nationale pour
l'Entreposage de Déchets Radioactifs, CEDRA, Suisse

Pursuant to article 1 of the Convention signed in Paris on 14th December, 1960, and which came into force on 30th September, 1961, the Organisation for Economic Co-operation and Development (OECD) shall promote policies designed:

- to achieve the highest sustainable economic growth and employment and a rising standard of living in Member countries, while maintaining financial stability, and thus to contribute to the development of the world economy;
- to contribute to sound economic expansion in Member as well as non-member countries in the process of economic development; and
- to contribute to the expansion of world trade on a multilateral, non-discriminatory basis in accordance with international obligations.

The original Member countries of the OECD are Austria, Belgium, Canada, Denmark, France, the Federal Republic of Germany, Greece, Iceland, Ireland, Italy, Luxembourg, the Netherlands, Norway, Portugal, Spain, Sweden, Switzerland, Turkey, the United Kingdom and the United States. The following countries became Members subsequently through accession at the dates indicated hereafter: Japan (28th April, 1964), Finland (28th January, 1969), Australia (7th June, 1971) and New Zealand (29th May, 1973).

The Socialist Federal Republic of Yugoslavia takes part in some of the work of the OECD (agreement of 28th October, 1961).

The OECD Nuclear Energy Agency (NEA) was established in 1957 under the name of the OEEC European Nuclear Energy Agency. It received its present designation on 20th April, 1972, when Japan became its first non-European full Member. NEA membership today consists of all European Member countries of OECD as well as Australia, Canada, Japan and the United States. The commission of the European Communities takes part in the work of the Agency.

The primary objective of NEA is to promote co-operation between the governments of its participating countries in furthering the development of nuclear power as a safe, environmentally acceptable and economic energy source.

This is achieved by:

- *encouraging harmonisation of national, regulatory policies and practices, with particular reference to the safety of nuclear installations, protection of man against ionising radiation and preservation of the environment, radioactive waste management, and nuclear third party liability and insurance;*
- *assessing the contribution of nuclear power to the overall energy supply by keeping under review the technical and economic aspects of nuclear power growth and forecasting demand and supply for the different phases of the nuclear fuel cycle;*
- *developing exchanges of scientific and technical information particularly through participation in common services;*
- *setting up international research and development programmes and joint undertakings.*

In these and related tasks, NEA works in close collaboration with the International Atomic Energy Agency in Vienna, with which it has concluded a Co-operation Agreement, as well as with other international organisations in the nuclear field.

Pursuant to article 1 of the Convention signed in Paris on 14th December, 1960, and which came into force on 30th September, 1961, the Organisation for Economic Co-operation and Development (OECD) shall promote policies designed:

- to achieve the highest sustainable economic growth and employment and a rising standard of living in Member countries, while maintaining financial stability, and thus to contribute to the development of the world economy;
- to contribute to sound economic expansion in Member as well as non-member countries in the process of economic development; and
- to contribute to the expansion of world trade on a multilateral, non-discriminatory basis in accordance with international obligations.

The original Member countries of the OECD are Austria, Belgium, Canada, Denmark, France, the Federal Republic of Germany, Greece, Iceland, Ireland, Italy, Luxembourg, the Netherlands, Norway, Portugal, Spain, Sweden, Switzerland, Turkey, the United Kingdom and the United States. The following countries became Members subsequently through accession at the dates indicated hereafter: Japan (28th April, 1964), Finland (28th January, 1969), Australia (7th June, 1971) and New Zealand (29th May, 1973).

The Socialist Federal Republic of Yugoslavia takes part in some of the work of the OECD (agreement of 28th October, 1961).

The OECD Nuclear Energy Agency (NEA) was established in 1957 under the name of the OEEC European Nuclear Energy Agency. It received its present designation on 20th April, 1972, when Japan became its first non-European full Member. NEA membership today consists of all European Member countries of OECD as well as Australia, Canada, Japan and the United States. The commission of the European Communities takes part in the work of the Agency.

The primary objective of NEA is to promote co-operation between the governments of its participating countries in furthering the development of nuclear power as a safe, environmentally acceptable and economic energy source.

This is achieved by:

- *encouraging harmonisation of national, regulatory policies and practices, with particular reference to the safety of nuclear installations, protection of man against ionising radiation and preservation of the environment, radioactive waste management, and nuclear third party liability and insurance;*
- *assessing the contribution of nuclear power to the overall energy supply by keeping under review the technical and economic aspects of nuclear power growth and forecasting demand and supply for the different phases of the nuclear fuel cycle;*
- *developing exchanges of scientific and technical information particularly through participation in common services;*
- *setting up international research and development programmes and joint undertakings.*

In these and related tasks, NEA works in close collaboration with the International Atomic Energy Agency in Vienna, with which it has concluded a Co-operation Agreement, as well as with other international organisations in the nuclear field.

FOREWORD

Currently one of the most important areas of radioactive waste management is research and development directed towards asssessments of the long-term performance of radioactive waste disposal systems. In the present programme of the NEA Radioactive Waste Management Committee (RWMC), performance assessment-related activities play a major role. It has been recognised as a priority area under rapid development with a strong need for international co-operation and co-ordination. In order to focus work on the most appropriate topics in this field, a "Performance Assessment Advisory Group" (PAAG) has been set up to give guidance and advice to the RWMC within this area. PAAG consists of Member countries' representatives with key positions in the national performance assessment programmes, including the governmental and industry organisations responsible for the programmes and regulatory bodies. The ultimate aim is to promote the quality and credibility of safety assessment techniques for radioactive waste disposal.

Current NEA activities in this area include the setting up of geochemical databases, the promotion of models/codes development and verification/validation exercises, the sponsorship of international R&D projects and the organisation of workshops, working groups and courses on selected topical issues.

The PAAG recommended that NEA arrange a workshop on studies and modelling of the behaviour phenomena in the immediate vicinity (near-field) of repositories for low and medium-level wastes likely to influence the release of radionuclides into the environment. Many countries are presently in the process of planning or constructing repositories for low and medium-level wastes and the assessment of near-field performance is attracting much attention.

These proceedings reproduce the papers submitted for presentation at the workshop together with a summary and conclusions prepared by the NEA Secretariat and the session chairmen, based upon the discussions held at the workshop. The opinions expressed are those of the authors and in no way commit the Member countries of OECD.